Horace F. Parshall

Armature Windings of Electric Machines

Horace F. Parshall

Armature Windings of Electric Machines

ISBN/EAN: 9783337164072

Printed in Europe, USA, Canada, Australia, Japan

Cover: Foto ©berggeist007 / pixelio.de

More available books at **www.hansebooks.com**

ARMATURE WINDINGS

OF

ELECTRIC MACHINES

BY

H. F. PARSHALL

MEMBER AMERICAN INSTITUTE ELECTRICAL ENGINEERS, MEMBER INSTITUTION ELECTRICAL ENGINEERS
GREAT BRITAIN, MEMBER AMERICAN SOCIETY OF MECHANICAL ENGINEERS, ETC.

AND

H. M. HOBART, S.B.

NEW YORK
D. VAN NOSTRAND COMPANY
LONDON
ROBERT W. BLACKWELL
39 VICTORIA STREET, WESTMINSTER
1895

Typography by J. S. Cushing & Co., Norwood, Mass., U.S.A.

TABLE OF CONTENTS.

——∘◦⟡◦∘——

Multipolar commutating dynamos — Limits of bipolar dynamos — Considerations governing choice of windings — Cases in which two-circuit windings may be employed — Importance of symmetry — Extent to which symmetry may be departed from in certain cases — Gramme windings — Lack of symmetry introduced by spider arms — Utility of two-circuit, multiple windings — Conditions affecting voltage between adjacent commutator segments — Slotted armatures — Interdependence of re-entrancy, conductors per slot, number of slots, and number of poles — Interpretation of formulæ in case of coils consisting of several conductors bound together — Alternate-current armature windings.

PART I.

CONTINUOUS-CURRENT ARMATURE WINDINGS.

Characteristics — Methods of cross-connecting — Use of only two sets of brushes with multipolar dynamos — Methods of reducing the number of commutator segments relatively to the number of winding sections — Windings suitable for poorly balanced magnetic circuits — Diminution of sparking by use of resistances.

Multiple windings — Their advantages — Limiting conditions — Importance of symmetry with small numbers of conductors — Singly and multiply re-entrant windings — Importance of avoiding the use of interpolations and cross-connections.

Cases permitting the employment of two-circuit windings — Characteristics — Lack of symmetry of the armature coils — Short-connection and long-connection types — Effect of unequal air gaps — Use of long-connection type advisable for high potential armatures — Formulæ and tables for use with the long-connection gramme winding — Definition of "pitch," y — Table for use in determining permissible angular distance between brushes with different numbers of poles — Examples of two-circuit gramme windings — Chief objection to the short-connection type is the great difference of potential existing between adjacent sections of the winding — Modified types.

PART II.

WINDINGS FOR ALTERNATE-CURRENT DYNAMOS AND MOTORS.

PART III.

WINDING FORMULÆ AND TABLES.

LIST OF DIAGRAMS.

———•⚬❋⚬•———

PART I.

CHAPTER I. — SINGLE-WOUND GRAMME RINGS.

vii

CHAPTER XV. — THREE-PHASE WINDINGS.

INTRODUCTORY.

THE present treatise is the outcome of an investigation made a number of years ago, before the principles of the armature winding of multipolar commutating dynamos were generally understood by electricians. At that time it appeared that the demand for dynamos of greater current output could only be met satisfactorily by dynamos of the multipolar type, since with bipolars beyond a certain output the number of commutator segments compatible with freedom from sparking was found to be incompatible with the maximum armature reaction which experience has shown to be permissible. After some study it was concluded the only feature of the multipolar dynamo requiring special study was that of the armature windings.

A considerable number of diagrams were prepared and classified; the advantages and disadvantages of each, and the comparative fitness of these windings for different purposes, noted. Inasmuch as it was found convenient to refer to this data frequently, and on account of the comparative inaccessibility of such information when in the form of notes, we decided that it would be a great convenience to electricians generally if our notes were published in book form. We therefore proceeded to do this; but owing to the intervention of certain circumstances contingent to our position in an industrial concern, it became necessary to lay aside this work until those competent to judge of its nature should feel able to permit us to proceed as we had wished. The delay has not been disadvantageous, since in the meantime we have not laid the work aside; on the contrary, we have made a study of the properties of a number of the more important windings, so that the original manuscript has been largely added to.

In the section on continuous-current armature windings our endeavor has been to include only those windings that possess some practical merit, and we have frequently pointed out the advantages and disadvantages peculiar to certain classes of windings. The thought will probably occur to the reader, which one of these windings should be selected for a given voltage after the number of poles and the magnitude of the magnetic flux at the poles have been assigned a proper value. We cannot point out the fitness of each winding for a given purpose, since this is more or less dependent upon the magnetic characteristics peculiar to any particular design. Thus in some machines of particularly good characteristics two-circuit windings have been used in the generation of comparatively large currents with some success, when had the magnetic characteristics of the dynamos been ordinarily good, the use of the two-circuit winding would have been attended with results entirely unsatisfactory.

xiii

In general, we may state, the type of winding should be determined with reference to the magnitude of the current to be generated. Any deviation from a perfectly symmetrical arrangement of the armature conductors should be inversely proportional to the magnitude of the currents to be generated. When the currents to be generated are large, the coils should be similarly situated with respect to each other, and should all have the same resistance and inductance. It has been frequently found that when the conductors are dissimilarly situated with respect to each other or to any other body that can affect the armature conductors inductively, the wearing away of the commutator is uneven, the trouble increasing more and more as the currents in the conductors are increased, or the resistance of the collecting brushes diminished. Especially in armatures in which there are more than two coils in a slot this uneven wearing away of the commutator has been noticed. In this case the coils are of slightly unequal area, due to the progression of the winding from slot to slot.

In gramme windings the lack of symmetry may be due to some of the coils being longer than the others, or carried near the spider arms.

It may, therefore, be stated generally that when a given result has to be obtained without experimenting, such windings as these are to be avoided when the currents in the conductors have to be of any considerable magnitude.

The utility of the double, triple, and quadruple windings shown and described depends very largely upon the maximum arc upon the commutator over which uniform contact resistance can be obtained. With the thickness of segments now common in practice, only double and triple windings appear to be of practical value, since, in general, brushes cannot be relied upon to maintain a uniform contact over an arc of much more than three-quarters of an inch in width. When the width of the brush has to exceed this amount, it is found that it bridges imperfectly from commutator bar to commutator bar in the same winding, thereby causing sparking.

A feature peculiar to these windings, as well as to some of the two-circuit single windings, is that the voltage between adjacent commutator sections is affected by the angular distance between the different sets of collecting brushes. With some of these windings the voltage between adjacent commutator sections varies simply according to the field strength when the angle between the different sets of brushes corresponds to the angle between the centers of the poles. In other windings the voltage between adjacent commutator sections varies by jumps, but may be made to vary according to the field strength by slightly varying the position of some one set of brushes with respect to the other sets. This feature of the different windings is a subject for special investigation, and is of more or less importance, according to the nature of the winding and the average voltage between commutator bars.

We have frequently made mention of the number of slots. With respect to slotted armatures in general, it is to be remembered that an additional condition to that for smooth-core armatures has to be fulfilled; i.e. the total number of the conductors to suit the equations for re-entrancy has to be divisible by the number of conductors possible to place in a slot, this number being dependent upon the number of poles. The number of conductors permissible per slot for two-circuit windings for different numbers of poles is shown in a table.

We have omitted any reference to mechanical details of construction of armature windings, since these permit of great variety, without in any way modifying the results. Further, they are a part of the stock in trade of the electrical manufacturer.

The drum windings considered are principally those in which the end connections are interchangeable, and are in the form of evolutes, as in the Eickemeyer and Hopkinson windings, description of which will be found in Weymouth's "Drum Armatures and Commutators" ("The Electrician" Printing and Publishing Company, London, 1893). In general, such windings possess the advantages that all coils are of equal inductance and resistance, are equally accessible, have equal radiating surfaces, and are most easily repaired. When a coil consists of a number of conductors, bound together so as to be considered a single unit mechanically, it is so considered in the text, and in the formulæ for the arrangement of conductors.

These windings appear to have been invented by Bollmann, Desroziers, Fritsche, Pischon, Eickemeyer, and others; but inasmuch as it is a disputed question as to which of these inventors has the right to claim priority, and as there may be more or less litigation before the question is settled, we have considered it best to omit all discussion as to who may have invented any of the windings. Where with a winding is given the name of a supposed inventor, it is simply because that winding has been known under that name, and not because the writers possess any special evidence to show by whom the winding was invented. After the possibility of litigation has ceased we hope to do justice to all inventors concerned, giving to each his proper proportion of credit for the work he has done.

We believe that the tables on drum windings are a feature that should meet with especial favor, since after the number of conductors required for a given type of winding has been determined, the proper pitches for any style of winding can be found in the tables. Further, by referring opposite to this number of conductors in the different tables it may be ascertained at a glance whether, by slightly changing the end connections, the winding may be adapted to some other voltage. Such features, peculiar to certain numbers of conductors, are frequently in practice of the greatest importance. As a practical example take the following case: In a six-pole machine with 104 armature conductors, the winding may be connected for a two-circuit single winding by making the pitch 17 on each end, or for a two-circuit, doubly re-entrant double winding, by making the pitch 17 on one end and 19 on the other; this second arrangement being suitable for the same watt output as the first, at one-half the voltage.

In the section on alternate-current armature windings are included a number of windings that have now only a limited application in practice, as it is thought that, on account of the very limited literature on this subject, a description of all windings of any practical use will be appreciated.

With respect to the work in general, we should be glad to receive the suggestions and criticisms of all who are interested in this subject.

The following articles on armature windings have been consulted in the preparation of this book, and are mentioned here for reference: —

ARNOLD — Die Ankerwicklungen der Gleichstrom-Dynamomaschinen. Berlin, 1891.
FRITSCHE — Die Gleichstrom-Dynamomaschine. Berlin, 1880.
KAPP — Practical Electrical Engineering, Vol. II., p. 43. London, 1893.
KITTLER — Handbuch der Elektrotechnik, Vol. I. Stuttgart, 1892.
RECHNIEWSKI — L'Electricien, Vol. V. Jan. 14, 1893 et seq.
THOMPSON — Dynamo-Electric Machinery. London, 1892.
WEYMOUTH — The Electrician, Vol. XXV. Nov. 7 to Dec. 19, 1890.

Part I.

CONTINUOUS–CURRENT ARMATURE WINDINGS.

CHAPTER I.

SINGLE-WOUND GRAMME RINGS.

THESE are the simplest windings in use, and will require only a very few diagrams and explanations. Many complex connections have been proposed, but only such forms will be discussed as are of general practical use.

The plain gramme ring, with a single winding, is shown in Figs. 1 and 2, from which it may be seen that the construction, as far as concerns location of coils, connectors, and commutator segments, is independent of the number of poles. The number of coils should be a multiple of the number of poles in order to maintain

S

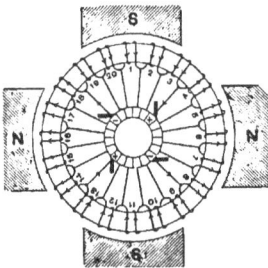

Fig. 1

FOUR-CIRCUIT, SINGLE-WINDING.

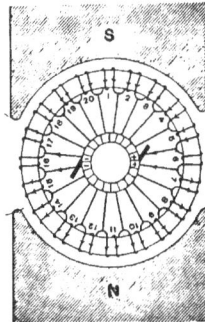

Fig. 2

TWO-CIRCUIT, SINGLE-WINDING.

symmetry among all the branches from brush to brush. The number of commutator segments is equal to the number of coils. It is desirable to minimize the turns per coil, and consequently the inductance of the short-circuited elements, by as large a number of segments as practicable.

A further discussion of these two diagrams would be superfluous, beyond calling attention to the progressive nature of the rise of potential around the ring, whereby the contiguous wires have only the small difference of potential of one turn, making the question of insulation very simple.

3

In cases where it is desirable to use but two brushes in multipolar rings with more than two circuits, the method of cross-connecting, shown in Fig. 3, may be used. The number of commutator segments remains equal to the number of coils. An inspection of the diagram will show that it really consists in connecting in parallel those coils occupying corresponding positions in the various fields.

It would seldom be desirable to utilize this method of connection, except in very small machines, as the use of only one pair of sets of brushes would necessitate lengthening the commutator in order to retain the proper extent of brush contact surfaces.

Fig. 3

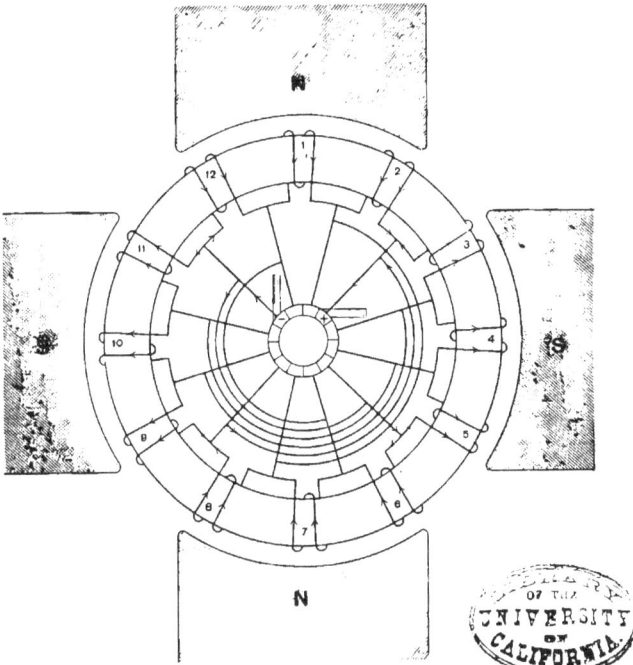

FOUR CIRCUIT, SINGLE WINDING.

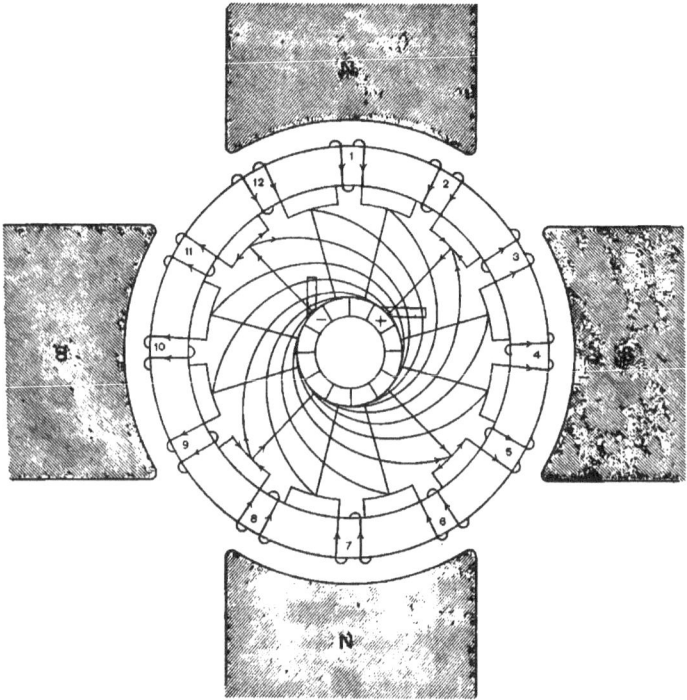

Fig. 4

FOUR CIRCUIT SINGLE WINDING

Figure 4 differs from Fig. 3 only in the use of two cross-connecting leads instead of one. This diagram would sometimes be of advantage, inasmuch as it utilizes the available space more completely and symmetrically. Each cross-connecting conductor could be of smaller cross-section than if only one were used.

Both this and the preceding method have the disadvantage that the two parallel sections have unequal resistance, due to one section having the long cross-connecting leads in series with it, and the other merely the regular short leads to the commutator.

Failure to give due attention to this point often causes serious trouble.

Figure 5 gives a winding which is *wrong*, but which has been given in the treatises of many of the specialists on windings, none of whom, except Herr Arnold, criticise it.

The fault is that the positions of the coils bear such a relation to the positions of their respective commutator segments, that during each revolution of the armature the position given in the figure is the only one in which the brushes are properly placed with regard to the diameter of commutation. In order that the brushes should always be in a position to properly perform their commutative function, they would have to be revolved in a direction opposite to that of the armature, and with a velocity equal to it.

The characteristic of the winding is that it brings together into one segment each pair of cross-connected segments of the previous diagram. As above stated, however, this diagram is worthless, except to call attention to its character, so that the text-books in which it is described shall not be misleading.

See ARNOLD — Die Ankerwicklungen der Gleichstrom-Dynamomaschinen, Fig. 42.

KITTLER — Handbuch der Elektrotechnik, 1892, Fig. 401 C.

FRITSCHE — Die Gleichstrom-Dynamomaschinen, Fig 64.

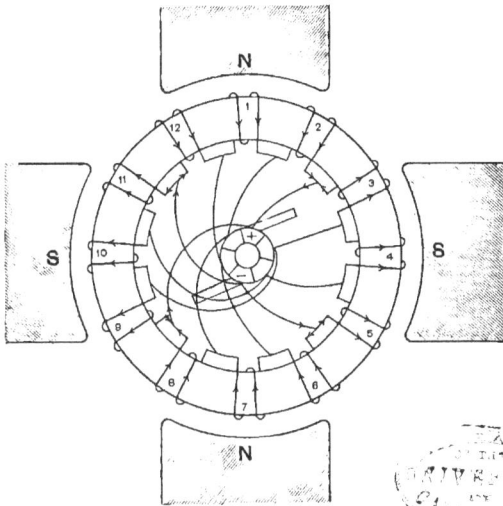

Fig. 5

FOUR CIRCUIT SINGLE WINDING.

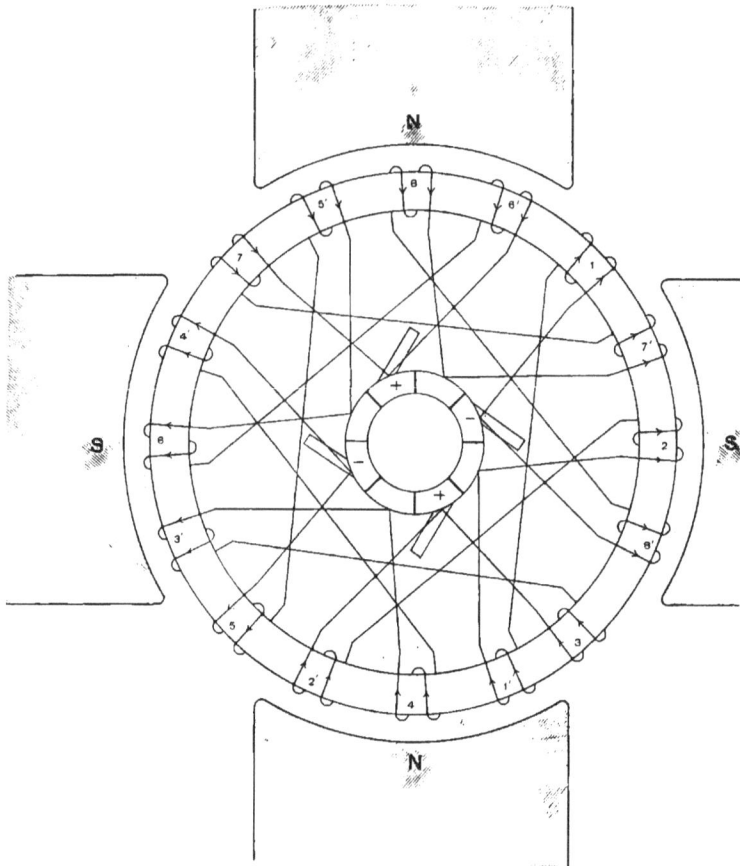

Fig. 6

FOUR CIRCUIT SINGLE WINDING.

In Fig. 6 the number of commutator segments is made
equal to half the number of coils by connecting two coils in
series between each pair of adjacent segments. The coils so
connected in series are situated in adjoining fields of opposite
polarity. This winding has the disadvantage that coils at
quite different potentials are adjacent, as may be seen by following through the various armature circuits from brush to
brush. This increases the difficulty of insulating. The volts
per bar also, for the same number of conductors per coil, are
twice as high as in the simple gramme ring. If it is necessary,
for any reason, to halve the number of bars, it would be preferable to combine two *adjacent* coils into one, and retain the
advantages of the simple gramme ring connection.

But in cases where the shape of the frame necessitates
somewhat unequal magnetic circuits, this connection averages
up the unequal induction in the various coils, and therefore
tends to diminish the sparking which might, with a simple
gramme ring in such an unbalanced magnetic system, be
considerable.

If s = number of coils, and n = number of poles, then any
coil is connected across to one $\left(\frac{s}{n} \pm 1\right)$ in advance of it, and
the two free ends of this pair of coils are connected to
adjacent commutator segments.

Figure 7 is merely a step in advance of Fig. 6, and the advantages and disadvantages pointed out in the discussion of Fig. 6 apply in still greater degree to Fig. 7.

It will be seen that the number of commutator segments is reduced to one-fourth of the number of coils by the connecting in series of four coils, one in each field, between two adjacent segments of the commutator.

As in the previous figure, the rule for connecting the coils is to connect each coil to one $\left(\dfrac{s}{n} \pm 1 \right)$ in advance.

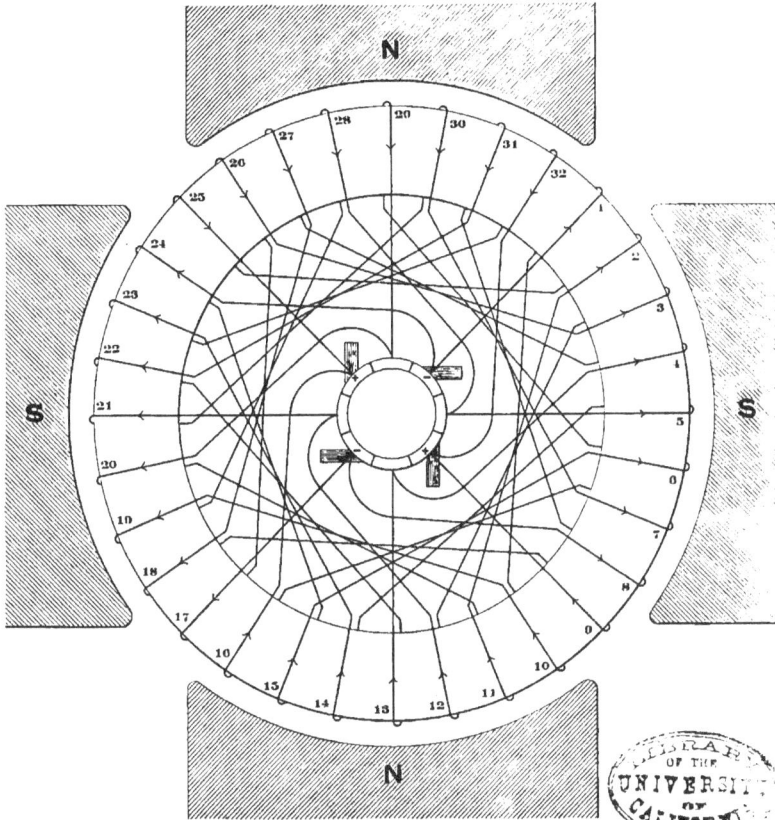

Fig. 7
FOUR CIRCUIT SINGLE WINDING.

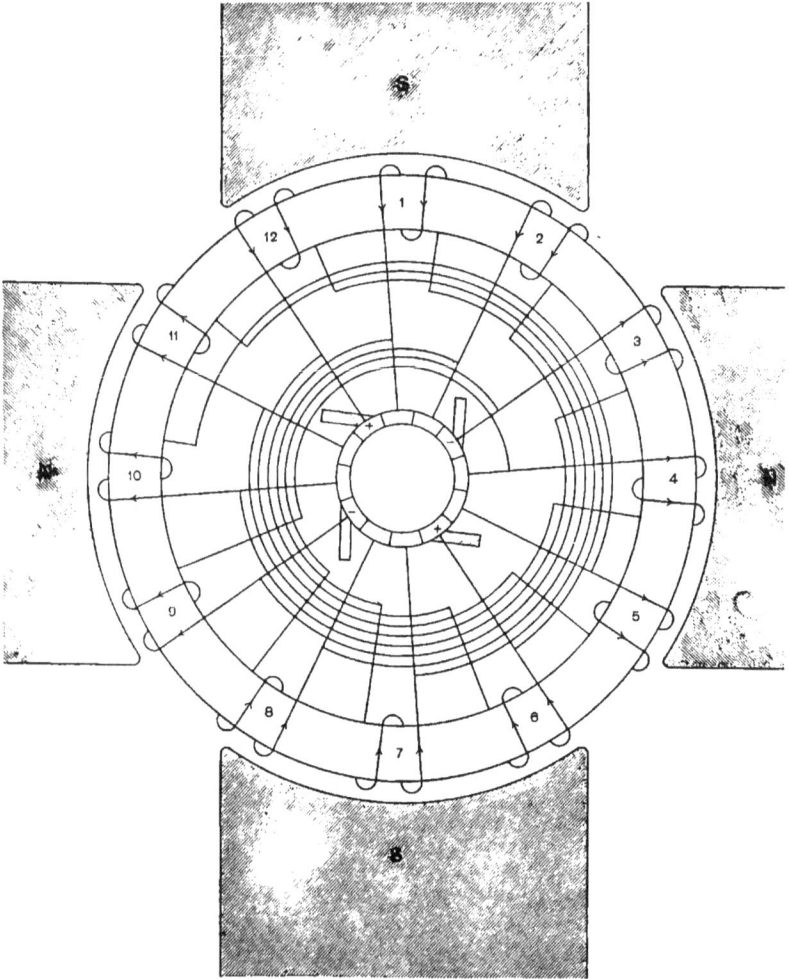

Fig. 8
FOUR CIRCUIT, SINGLE WINDING.

Figure 8 represents a winding in which the coils of one circuit, from
brush to brush, instead of being adjacent to each other, are situated in
different fields. For instance, the circuits through the armature in the
position shown are, —

$$\longrightarrow \quad \begin{aligned} -\left\{\begin{matrix} 3 & 10 & 5 \\ 8 & 1 & 6 \end{matrix}\right\} + \\ -\left\{\begin{matrix} 2 & 7 & 12 \\ 9 & 4 & 11 \end{matrix}\right\} + \end{aligned} \quad \longrightarrow$$

It is important to note that when the armature has entered the position in which four coils are short-circuited, the short-circuiting of any coil
occurs, not at any one brush, but through the pair of brushes of like polarity.
This would enable sparking to be diminished by connecting the two positive
brushes together through a suitable resistance (ohmic or inductive), and leading off to the load from the middle point of this resistance. The magnitude
of the resistance, if ohmic, would be limited only by the permissible loss
therein. High resistance leads to the commutator, and high-resistance
brushes have been used with considerable success; but in both of these
cases heat has to be developed in undesirable localities. But in the above
method of connection, the insertion of this resistance externally to the
brushes will not increase the heating of the machine. This resistance is also
so located that it could be adjusted in experimental work, and the difference in sparking noted by having a short-circuiting switch shunted around
the resistance.

Another advantage of this winding is that pointed out in the remarks on
Fig. 6, that in cases where the shape of the frame necessitates somewhat
unequal magnetic circuits, this connection will average up the unequal
induction in the various coils, and thereby diminish the sparking that
would otherwise occur.

CHAPTER II.

DOUBLE-WOUND GRAMME RINGS.

FIGURE 9 and the immediately following diagrams relate to a class of very great importance, which are known as double, triple, quadruple, etc., windings.

Very satisfactory results have been attained by the use of windings of this class. The most important advantage of the double winding is that the current is commutated at two different parts of the bearing surface of the brush; each independent volume of current being, therefore, only one-half of what it would be for a single winding. The importance of this feature has in practice been found to be very great.

Another important feature of this winding is that the successive commutator bars of one winding are not adjacent to each other, but alternate with the bars of the other winding; the two windings being put in parallel by the use of wide brushes. The result is that a section is very unlikely to be short-circuited by dirt or an arc. It also makes a very flexible winding, owing to the readiness with which any number of parallels may be arranged. Thus, in a six-pole field, we may have four, six, eight, etc., parallels.

It is necessary for a double winding that the brush should bear over a surface greater than the width of one segment (plus insulation); for a triple winding, greater than the width of two segments, etc.

In Fig. 9, which represents a two-circuit, doubly re-entrant, double-wound, simple gramme ring, the circuits through the armature are,—

$$\longrightarrow \quad -\begin{cases} 9 & 10 & 1 & 2 & 3 \\ 8 & 7 & 6 & 5 & 4 \end{cases} \vdash \quad \longrightarrow$$

$$\quad -\begin{cases} 9' & 10' & 1' & 2' & 3' \\ 8' & 7' & 6' & 5' & 4' \end{cases} +$$

After the armature has revolved through $\dfrac{360}{20 \times 2} = 9°$, coils 3 and 8 will be short-circuited, and the circuits through the armature will become,—

$$\longrightarrow \quad -\begin{cases} 9 & 10 & 1 & 2 \\ 7 & 6 & 5 & 4 \end{cases} + \quad \longrightarrow$$

$$\quad -\begin{cases} 9' & 10' & 1' & 2' & 3' \\ 8' & 7' & 6' & 5' & 4' \end{cases} +$$

Thus it will be seen that there will be a lack of balance between the two windings. First they will be of equal length; after 9° revolution, one will have one less section in series between the brushes; 9° later they will be equal again; and after still another 9° the other winding will have the smaller number of turns. This lack of symmetry will be less apparent as the number of sections is increased, and becomes of very little importance with the large numbers of conductors employed in practical work.

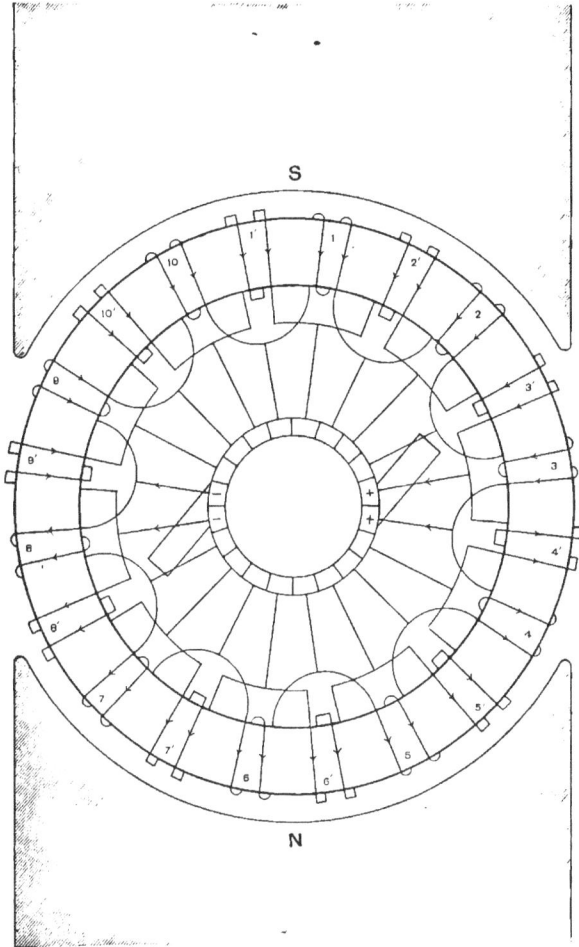

Fig. 9
TWO CIRCUIT DOUBLE WINDING

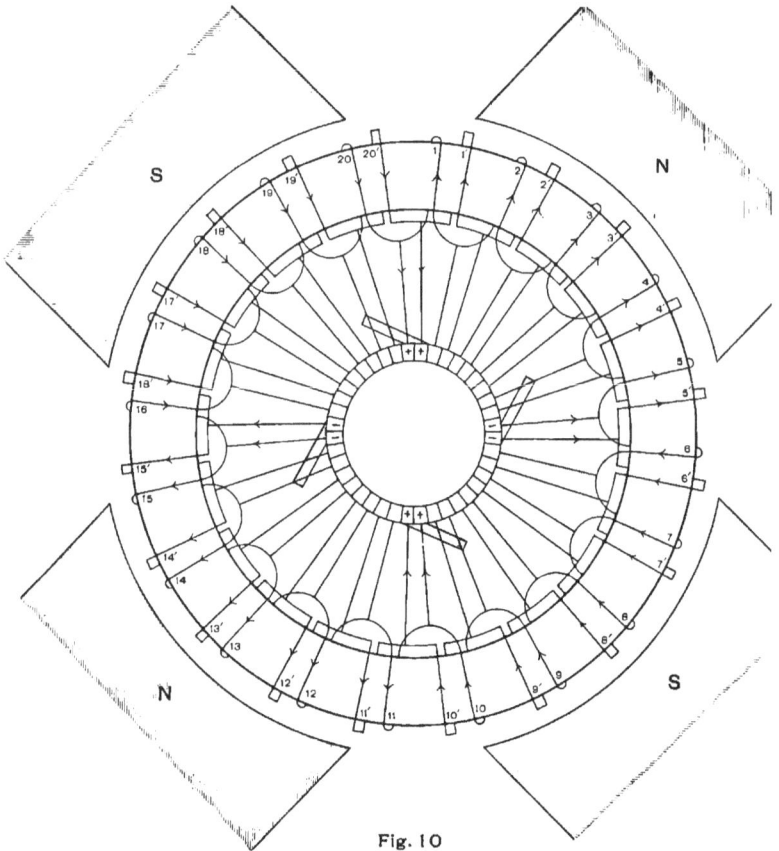

Fig. 10

FOUR CIRCUIT DOUBLE WINDING.

Figure 10 shows a similar winding in a four-pole field. The circuit through the armature in the position shown is, —

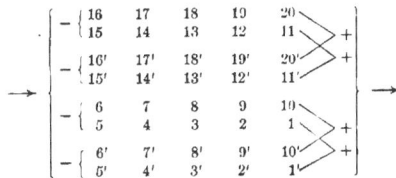

$$\longrightarrow \left[\begin{array}{l} - \begin{cases} 16 \quad 17 \quad 18 \quad 19 \quad 20 \\ 15 \quad 14 \quad 13 \quad 12 \quad 11 \end{cases} \!\!\!\!\searrow + \\ - \begin{cases} 16' \quad 17' \quad 18' \quad 19' \quad 20' \\ 15' \quad 14' \quad 13' \quad 12' \quad 11' \end{cases} \!\!\!\!\nearrow + \\ - \begin{cases} 6 \quad 7 \quad 8 \quad 9 \quad 10 \\ 5 \quad 4 \quad 3 \quad 2 \quad 1 \end{cases} \!\!\!\!\searrow + \\ - \begin{cases} 6' \quad 7' \quad 8' \quad 9' \quad 10' \\ 5' \quad 4' \quad 3' \quad 2' \quad 1' \end{cases} \!\!\!\!\nearrow + \end{array} \right] \longrightarrow$$

After turning through $\dfrac{360}{40 \times 2} = 4.5°$, coils 15', 20', 5', and 10' will be short-circuited, and the circuits through the armature will be, —

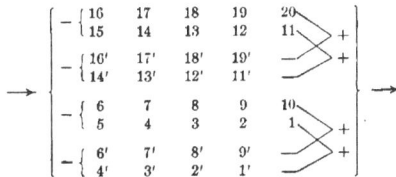

$$\longrightarrow \left[\begin{array}{l} - \begin{cases} 16 \quad 17 \quad 18 \quad 19 \quad 20 \\ 15 \quad 14 \quad 13 \quad 12 \quad 11 \end{cases} \!\!\!\!\searrow + \\ - \begin{cases} 16' \quad 17' \quad 18' \quad 19' \\ 14' \quad 13' \quad 12' \quad 11' \end{cases} \!\!\!\!\diagdown\!\!\!\diagup + \\ - \begin{cases} 6 \quad 7 \quad 8 \quad 9 \quad 10 \\ 5 \quad 4 \quad 3 \quad 2 \quad 1 \end{cases} \!\!\!\!\searrow + \\ - \begin{cases} 6' \quad 7' \quad 8' \quad 9' \\ 4' \quad 3' \quad 2' \quad 1' \end{cases} \!\!\!\!\diagdown\!\!\!\diagup + \end{array} \right] \longrightarrow$$

Here can be seen again the lack of symmetry noted in remarks on Fig. 9.

A very useful winding is that shown in Fig. 11. It, also, is a four-circuit double winding. It is one of a class with very interesting properties. It differs from the double winding shown in Fig. 10, in that the two windings are components of one re-entrant system. Any one section is no longer exclusively an element of one of two windings, but changes from one winding to the other four times per revolution, being short-circuited at the neutral point for a brief period at the occurrence of each of these transfers. These features are secured by adding one section to the doubly re-entrant double winding shown in Fig. 10, and, as in that figure, making the connections, not between adjacent sections, but always by passing over one section. The number of sections being odd, it will be seen that after having progressed twice around the ring, all sections will have been passed through, and the winding will have arrived at the other terminal of the section from which it started.

Triple, quadruple, and higher orders of windings may be treated analogously.[1]

The circuits through the armature in the position shown in Fig. 11 are,—

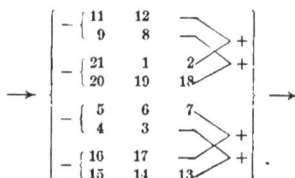

$$\rightarrow \begin{bmatrix} -\begin{cases} 11 & 12 \\ 9 & 8 \end{cases} \begin{array}{c} \diagdown \\ \diagdown \end{array} \Big\rangle + \\ -\begin{cases} 21 & 1 & 2 \\ 20 & 19 & 18 \end{cases} \begin{array}{c} \diagup \\ \diagup \end{array} \Big\rangle \begin{array}{c} + \\ + \end{array} \\ -\begin{cases} 5 & 6 & 7 \\ 4 & 3 \end{cases} \begin{array}{c} \diagdown \\ \diagdown \end{array} \Big\rangle + \\ -\begin{cases} 16 & 17 \\ 15 & 14 \end{cases} \begin{array}{c} \diagup \\ \diagup 13 \end{array} \Big\rangle \begin{array}{c} + \\ + \end{array} \end{bmatrix} \rightarrow .$$

Coil 10 is, at this instant, short-circuited. An instant later coil 10 becomes active, and coil 2 becomes short-circuited. The circuits through the armature then become,—

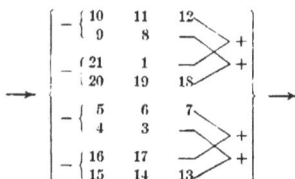

$$\rightarrow \begin{bmatrix} -\begin{cases} 10 & 11 & 12 \\ 9 & 8 \end{cases} \begin{array}{c} \diagdown \\ \diagdown \end{array} \Big\rangle + \\ -\begin{cases} 21 & 1 \\ 20 & 19 & 18 \end{cases} \begin{array}{c} \diagup \\ \diagup \end{array} \Big\rangle \begin{array}{c} + \\ + \end{array} \\ -\begin{cases} 5 & 6 & 7 \\ 4 & 3 \end{cases} \begin{array}{c} \diagdown \\ \diagdown \end{array} \Big\rangle + \\ -\begin{cases} 16 & 17 \\ 15 & 14 & 13 \end{cases} \begin{array}{c} \diagup \\ \diagup \end{array} \Big\rangle + \end{bmatrix} \rightarrow$$

The order in which the various coils will be short-circuited is 10, 2, 15, 7, 20, 12, 4, 17, etc., so that the 21 coils will each have been short-circuited once when the armature shall have revolved through $\frac{360°}{4} = 90°$. Therefore the angular interval between corresponding positions of two successive short circuits is $\frac{90°}{21} = 4.28°$.

[1] Such windings will be designated as singly re-entrant, to distinguish them from others, such as those of Figs. 9 and 10, which are doubly re-entrant.

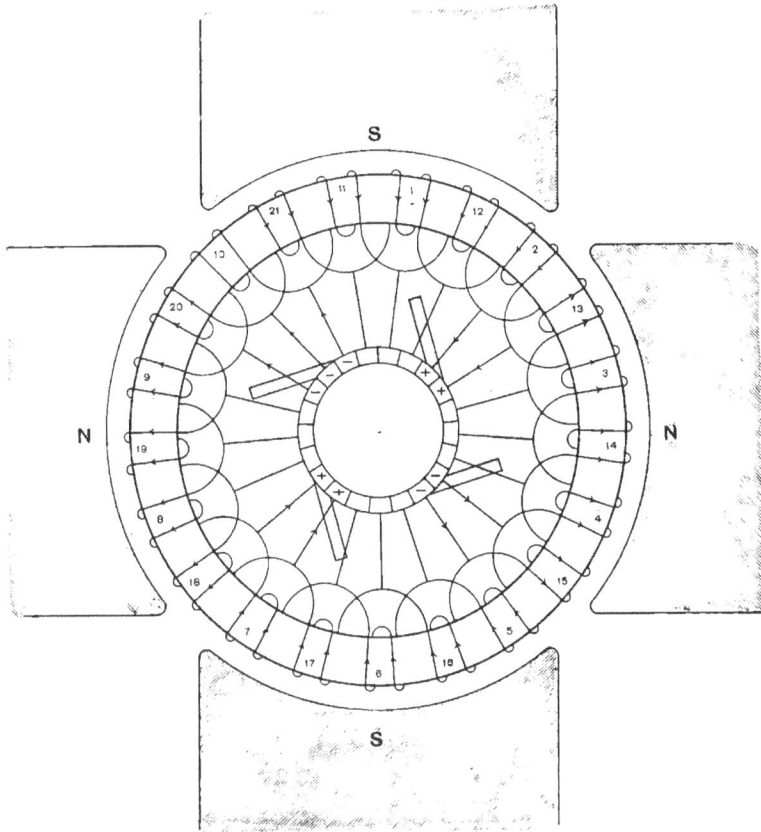

Fig. 11

FOUR CIRCUIT DOUBLE WINDING

All of the windings so far described have as many circuits through the armature as there are pole pieces, and form a class by themselves known as multiple-circuit windings. Four-pole fields have usually been considered, but the modifications of the diagrams and text to apply them to larger numbers of poles, are obvious.

In general, the number of sets of brushes equals the number of poles and the number of circuits through the armature. Different numbers of segments and brushes are due to modifications, and do not affect the underlying character of the windings as a class. Some of these modifications have been described. Others can be worked out as the occasion requires.

Too much importance cannot be attached to the general rule that interpolations and cross-connections are almost always very undesirable.

CHAPTER III.

TWO-CIRCUIT, SINGLE-WOUND, MULTIPOLAR RINGS.

THE next windings to be considered form a class which, independently of the number of poles, have only two circuits through the armature. These are known as two-circuit windings. Such windings possess the practical advantage that the number of conductors, as compared with multiple-circuit windings, is only $\frac{2}{N}$ as great, hence the space required for insulation is only $\frac{2}{N}$ as great as with the multiple-circuit windings, in consequence of which the diameter of the armature, or the depth of space occupied by the armature conductors, may be less than with the multiple-circuit windings, thereby diminishing the cost of material.

Further, on account of the lesser number of conductors, the cost of the labor of winding is correspondingly diminished.

In practice, the two-circuit gramme windings have been applied only to armatures of small output, under which condition lack of symmetry of the armature coils with respect to the points of commutation is not particularly objectionable. Only two sets of collecting brushes are necessary for the collection of current; in practice generally but two sets have been used.

In the "short-connection"[1] type of two-circuit gramme windings, the circuits from brush to brush consist of conductors influenced by all the poles, so that the electromotive forces generated in the two circuits are necessarily equal, a feature that may prove advantageous when the depth of air-gap is so small that any slight eccentricity of the armature affects the magnetic flux at the different poles.

In the "long-connection" type of two-circuit gramme winding, the two circuits from brush to brush consist of conductors influenced by only one-half of the poles, so that the electromotive forces generated in the two circuits are unequal, unless the sum of the lines at the poles of the same sign is equal to the sum of the lines at the poles of the opposite sign. In magnetic circuits of ordinarily good design this condition is fulfilled even though the fluxes at the different poles are unequal. So the winding is practically as good as the "short-connection" winding, and possesses certain other advantages stated in the text, that make its use preferable.

For armatures the outputs of which are so great that several sets of collecting brushes are required, these windings possess the same disadvantages as two-circuit drum windings, a discussion of which is to be found under that caption.

---

[1] Called "short-connection" type because coils in adjacent fields are connected together. This distinguishes it from the "long-connection" type, in which coils twice as far apart are connected together.

23

Figure 12 represents one of the most practicable two-circuit windings for multipolar-ring armatures. It may be designated as the long-connection type of the two-circuit gramme winding, and one of its chief advantages is, that no great differences of potential exist between adjacent coils.

In the figure is shown the case of a four-pole, two-circuit, single-wound, long-connection ring armature. In the position chosen, the circuits through the armature are, —

$$\longrightarrow \quad -\left\{ \begin{matrix} 11 & 4 & 12 & 5 & 13 & 6 & 14 \\ 2 & 9 & 1 & 8 & 15 & 7 & - \end{matrix} \right\} + \quad \longrightarrow$$

Coils 3 and 10, in series, are at this instant short-circuited by the negative brush. A little later, coils 7 and 15 will be short-circuited by the positive brush. When this occurs, the negative brush will bear upon the middle of a segment.

The number of commutator segments is equal to the number of coils, and must be odd for armatures with an even number of pairs of poles; but may be odd or even for armatures with an odd number of pairs of poles. The relation that must subsist in two-circuit, multipolar-ring, long-connection windings, between the number of coils (s) and the number of poles (n), is, —

$$s = \frac{n}{2} y \pm 1,$$

where $y =$ pitch. (The pitch is the number of coils to be advanced through in arranging the end connections. In the diagram, for instance, the pitch $y = 7$, and the end of coil 1 is joined to the beginning of coil $1 + 7 = 8$; the end of 8 to the beginning of $8 + 7 = 15$; the end of 15 to the beginning of $15 + 7 = 22$ (or 7), etc.) Mr. Gisbert Kapp has prepared the following table for two-circuit, multipolar-ring, long-connection windings by substituting numerical values for n in the above formula : —

TWO-CIRCUIT, MULTIPOLAR-RING, LONG-CONNECTION WINDINGS.

	MACHINE HAS					
	4 poles	6 poles	8 poles	10 poles	12 poles	14 poles
The number of coils must be equal to	$2 y \pm 1$	$3 y \pm 1$	$4 y \pm 1$	$5 y \pm 1$	$6 y \pm 1$	$7 y \pm 1$

For two-circuit, multipolar-*ring* machines with long-connection windings, y, the pitch, may be *any* integer. (Note that these conditions do not hold for *drum* windings.)

Mr. Kapp has also prepared the following table, showing the practicable choice of angular distances between brushes in these two-circuit, multipolar windings : —

NUMBER OF POLES.	ANGULAR DISTANCE BETWEEN BRUSHES.				
2	180				
4	90				
6	60	180			
8	45	135			
10	36	108	180		
12	30	90	150		
14	25.7	77	128	180	
16	22.5	67.5	112	158	
18	20	60	100	140	180
20	18	54	90	126	162

The smaller possible angles, namely, 20° for 18 poles, and 18° for 20 poles, are in practice too small to be admissible, and are, therefore, not given in the table.

Fig. 12

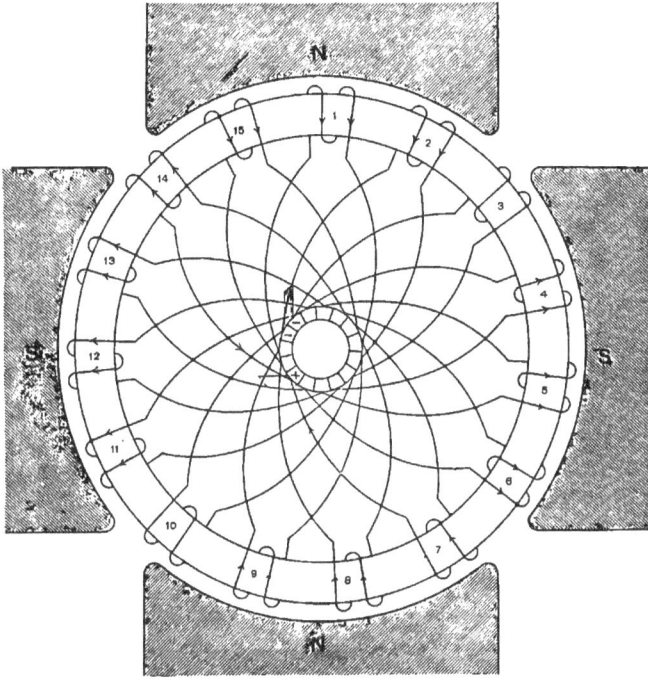

TWO CIRCUIT, SINGLE WINDING.

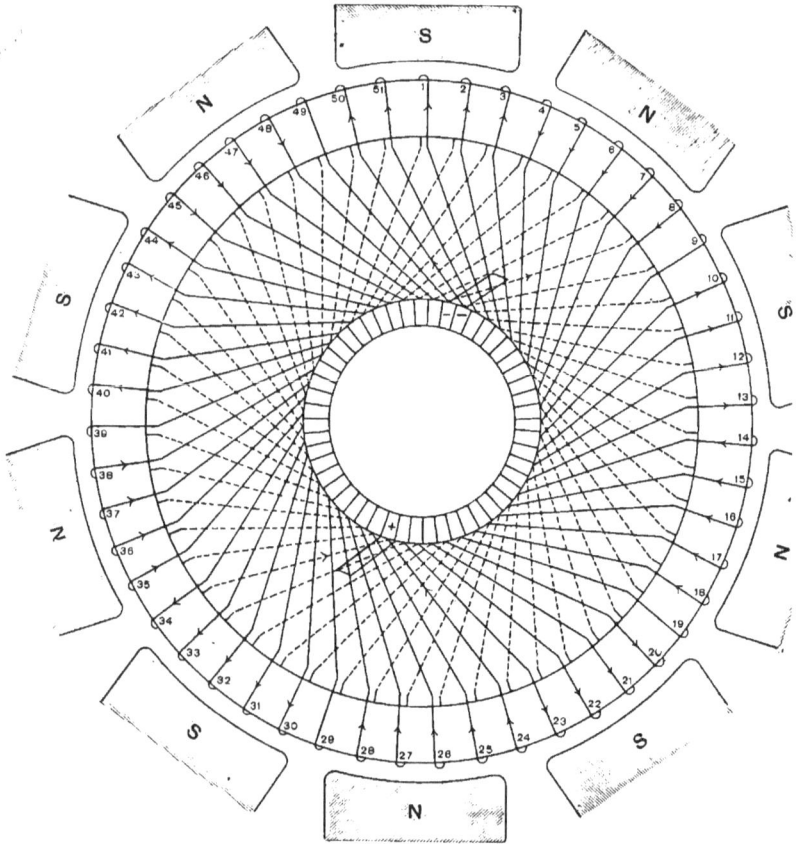

Fig. 13

TWO CIRCUIT, SINGLE WINDING.

Figure 13 represents a two-circuit, single-wound, long-connection, ten-pole ring armature. Substituting in the formula for the number of coils

$$s = \frac{n}{2}\, y \pm 1$$

the pitch, $y = 10$, and the number of poles, $n = 10$, gives $s = \frac{10}{2} \cdot 10 \pm 1 = 51$ or 49. 51 coils are taken in this case. The end of coil 1 is joined to the beginning of coil $1 + 10 = 11$; the end of 11 to the beginning of 21, etc.

The brushes are shown 180° apart, and at the position given the negative brush short-circuits the coils 9, 19, 29, 39, and 49. The circuits through the armature are, —

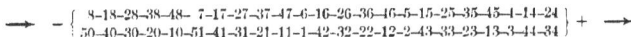

$$-\left\{ \begin{array}{l} 9\text{--}18\text{--}28\text{--}38\text{--}48\text{--} \ 7\text{--}17\text{--}27\text{--}37\text{--}47\text{--}6\text{--}16\text{--}26\text{--}36\text{--}46\text{--}5\text{--}15\text{--}25\text{--}35\text{--}45\text{--}4\text{--}14\text{--}24 \\ 50\text{--}40\text{--}30\text{--}20\text{--}10\text{--}51\text{--}41\text{--}31\text{--}21\text{--}11\text{--}1\text{--}42\text{--}32\text{--}22\text{--}12\text{--}2\text{--}43\text{--}33\text{--}23\text{--}13\text{--}3\text{--}44\text{--}34 \end{array} \right\} +$$

This diagram and table show very clearly that with an odd number of pairs of poles and an odd number of coils, an odd number of coils are short-circuited at one time, so that, as the total number of coils is odd, an even number is left to be divided between the two armature circuits, which are, therefore, equal. Referring back to Fig. 12, it will be seen that in the case of an even number of pairs of poles, an even number of coils are short-circuited, and as the total number of coils is necessarily odd, an odd number remains to be divided between the two armature circuits, so that these are necessarily unequal.

If, however, in Fig. 13 the brushes are put 108° apart
instead of 180°, coil 24 would be taken from the circuit given in the upper line of numbers and put in the
other circuit. There would then be 24 coils in one circuit,
and 22 in the other, instead of 23 in both. With the large
number of coils used in practice, however, these slight in-
equalities cause no trouble.

If y were chosen odd, 9 for instance, s would equal 46
or 44.

$$S = \frac{n}{2} \cdot y \pm 1 = \frac{10}{2} \cdot 9 \pm 1 = 46 \text{ or } 44.$$

This is in accordance with the observation made above,
that in the case of an odd number of pairs of poles the
number of coils may be even. The diagram for this case
is given in Fig. 14, where $s = 46$, $n = 10$, $y = 9$. In the position shown, coils 8, 17, 26, 35, and 44 are short-circuited
by the negative brush, and coils 31, 40, 3, 12, and 21 by the
positive brush. The circuits through the armature are, —

$$\longrightarrow \quad - \left\{ \begin{array}{l} 7\text{--}16\text{--}25\text{--}34\text{--}43\text{--}\ 6\text{--}15\text{--}24\text{--}33\text{--}42\text{--}5\text{--}14\text{--}23\text{--}32\text{--}41\text{--}4\text{--}13\text{--}22 \\ 45\text{--}36\text{--}27\text{--}18\text{--}\ 9\text{--}46\text{--}37\text{--}28\text{--}19\text{--}10\text{--}1\text{--}38\text{--}29\text{--}20\text{--}11\text{--}2\text{--}39\text{--}30 \end{array} \right\} + \quad \longrightarrow$$

giving, as in Fig. 13, two equal paths through the armature.

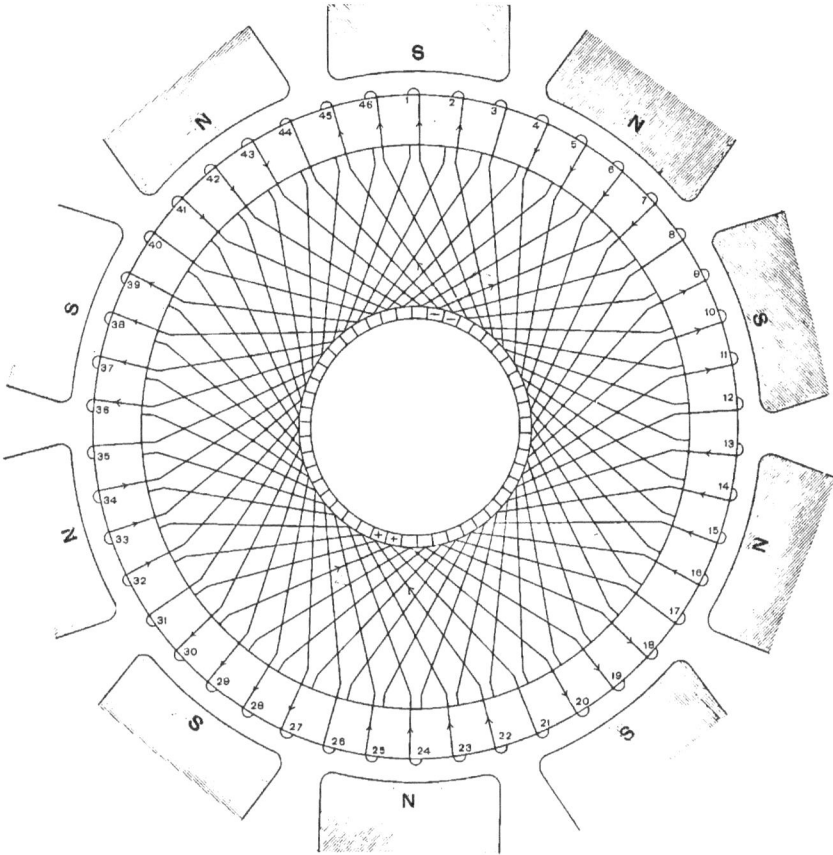

Fig. 14

TWO CIRCUIT, SINGLE WINDING.

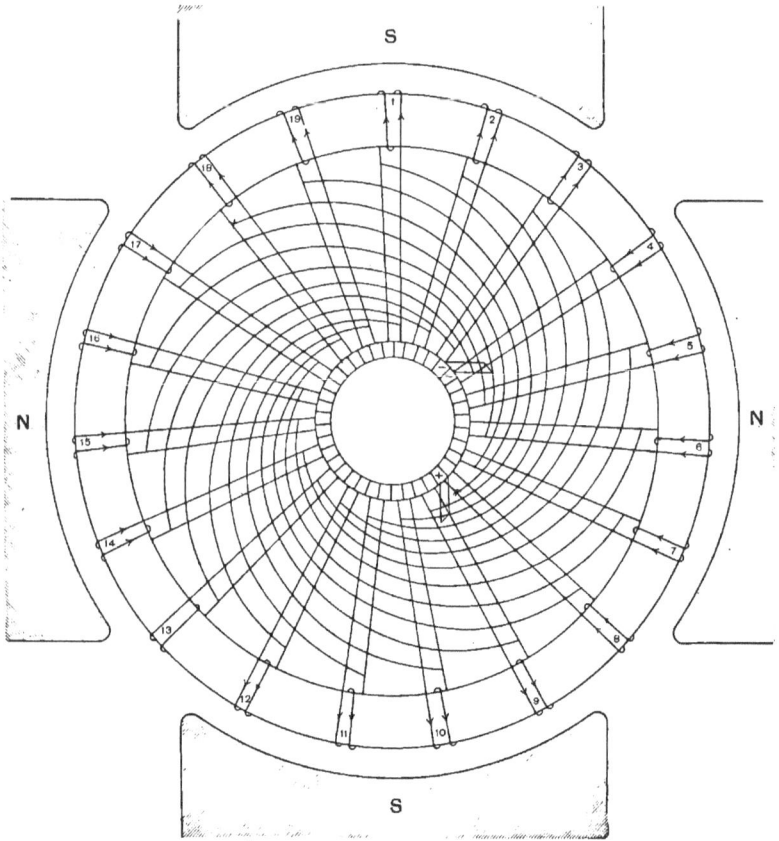

Fig. 15

TWO CIRCUIT, SINGLE WINDING.

In Fig. 15 is given a winding that has been used in practice with considerable success, owing partly to the extreme regularity of all connections, and still more to the fact that it involves the use of twice as many commutator segments as coils. Only one coil in series is short-circuited at each brush, and the volts per segment are one-half what they would be in the unmodified long-connection winding. The number of coils to be used is, as in the unmodified winding, $s = \frac{n}{2} \cdot y \pm 1$. Thus, in Fig. 15, $n = 4$, $y = 9$, $s = \frac{4}{2} \cdot 9 + 1$ $= 19$. Coil 1 is connected to coil 10, etc.

It will also be noted that those segments $\left[\frac{360}{\frac{n}{2}}\right]^\circ$ from each other are connected together. The number of segments $= \frac{n}{2} \cdot s$, of which $\frac{n}{2}$, at distances of $\left[\frac{360}{\frac{n}{2}}\right]^\circ$ from each other, are connected together. If every other one of the radial connections from the coils to the commutator are discarded, the winding becomes once more the plain, long-connection, two-circuit, grammē winding.

At the position shown, coil 13 is short-circuited by the negative brush, and the circuits through the armature are, —

$$\longrightarrow \quad - \left\{ \begin{array}{l} 3\text{-}12\text{-}2\text{-}11\text{-}1\text{-}10\text{-}19\text{-} \ 9\text{-}18 \\ 4\text{-}14\text{-}5\text{-}15\text{-}6\text{-}16\text{-} \ 7\text{-}17\text{-} \ 8 \end{array} \right\} + \quad \longrightarrow$$

Figure 16 is an application of the same type of winding to a *six*-pole gramme ring. $n=6$, $y=6$, $s=\frac{n}{2}y\pm 1=\frac{6}{2}\cdot 6+1=19$. There are $19\times\frac{6}{2}=57$ segments. All segments distant from each other by $\frac{360}{\frac{n}{2}}=120°$ should be connected together. Some of the cross-connections are shown inside the armature.

At the position shown, coil 12 is short-circuited by the positive brush. The circuits through the armature are—

$$\longrightarrow -\left\{\begin{array}{l} 9\text{-}3\text{-}16\text{-}10\text{-}4\text{-}17\text{-}11\text{-}\ 5\text{-}18 \\ 15\text{-}2\text{-}\ 8\text{-}14\text{-}1\text{-}\ 7\text{-}13\text{-}19\text{-}\ 6 \end{array}\right\} + \longrightarrow$$

If the connections shown inside the commutator, together with one-third of the segments, had been omitted, there would have been an unequal distribution of potential about the commutator. Between two segments would be found a certain voltage, V, and between the next two $2V$; then V again, etc.

If it should be desirable to diminish the number of commutator segments to one-half the number of coils, it may be done by the method of connection shown in Fig. 17, page 34, which will be recognized at once as the multipolar *ring* counterpart of the two-circuit winding as applied to multipolar *drums*. This winding will be referred to as a "short connection," two-circuit gramme winding. In the "long-connection" type, examples of which have just been given, connection has been made between coils situated in fields of like polarity. But in the "short-connection" type, connection is made between coils in adjacent fields. Both methods are feasible in ring windings, because the two ends of a coil located at a certain point of the periphery are accessible for connection at the commutator end if desired, but in drum windings only one end of a conductor located at a given point of the periphery is accessible at the commutator end, the other end of the conductor being necessarily connected across at the opposite end of the armature, and in consequence, also, must be connected over to a conductor in an adjacent field of unlike polarity, in order that the electromotive force, which is, say, from front to back in the first conductor, may add itself to that in the second conductor, which must therefore be from back to front; that is, the second conductor must be situated in a field of opposite polarity. Thus there are two sub-classes of two-circuit, multipolar ring windings, in the first of which (the "long-connection" winding) coils in fields of *like* polarity are connected in succession, and in the second of which, as in the two-circuit, multipolar drum winding, the conductors immediately succeeding each other are situated in fields of *opposite* polarity.

In this "short-connection" winding for two-circuit multipolar rings the formula for determining the proper number of coils, s, for any number of poles, n, is—

$$s = ny \pm 2,$$

where y, the pitch, may equal any integer, odd or even.

In connecting up this "short-connection" type of winding the following additional rule should be borne in mind in the interpretation and application of the meaning of the pitch, y: The number of coils in this winding, being from the formula always even, if y is also even, it is necessary in connecting up to use as the pitch, alternately, $(y-1)$ and $(y+1)$ instead of always y. Otherwise, if the coils are numbered successively

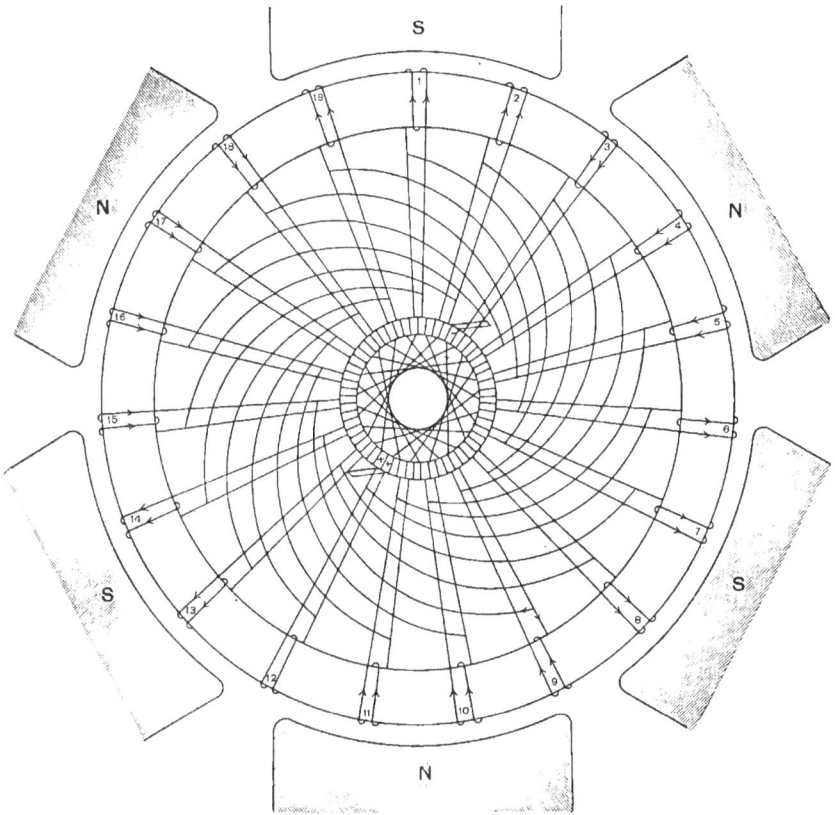

Fig. 16

TWO CIRCUIT, SINGLE WINDING.

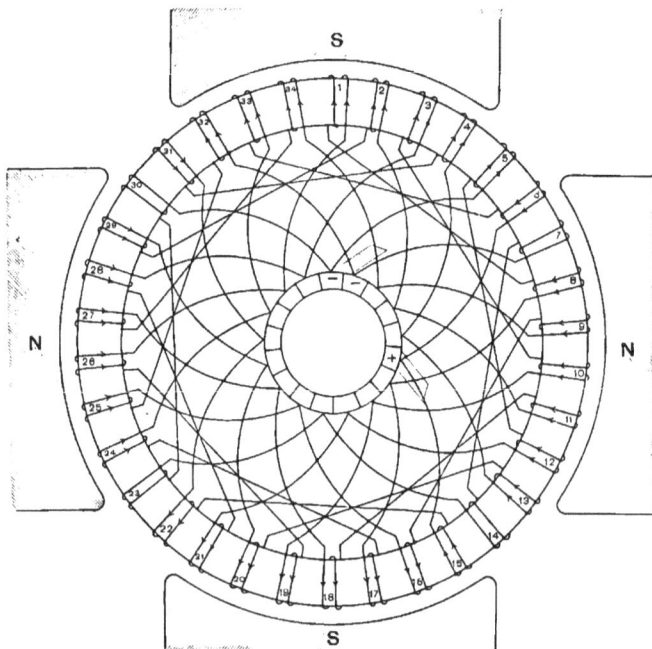

Fig. 17

TWO CIRCUIT, SINGLE WINDING.

from No. 1 on, the even-numbered coils would never be touched, if an odd-numbered conductor were started with, and *vice versa*. If y were used every time as the pitch, a double winding would be obtained. This case will be treated later.

It may also be well to note that $(y-3)$ and $(y+3)$ could be used alternately as the pitch. It is thought, however, that no advantages, and several disadvantages, would result from such a choice of pitches.

Figure 17 represents a two-circuit, single-wound, four-pole ring of the "short-connection" type just described.

$$n = 4, \; y = 8, \; s = ny \pm 2 = 4 \times 8 + 2 = 34.$$

This is the case referred to above, in which, s being even and also y, $(y-1)$ and $(y+1)$ must be used alternately as the pitch in connecting up. The sequence of connections will be seen in the figure to be $1, 1 + 7 = 8$, $8 + 9 = 17, 17 + 7 = 24$, etc.

Number of commutator segments $= \frac{34}{2} = 17$.

In the position shown, coils 7, 14, 23, and 30, in series, are short-circuited at the negative brush, and the circuits through the armature are, —

$$\longrightarrow \quad - \left\{ \begin{array}{l} 5\text{–}12\text{–}21\text{–}28\text{–} \; 3\text{–}10\text{–}19\text{–}26\text{–}1\text{–} \; 8\text{–}17\text{–}24\text{–}33\text{–} \; 6\text{———} \\ 32\text{–}25\text{–}16\text{–} \; 9\text{–}34\text{–}27\text{–}18\text{–}11\text{–}2\text{–}29\text{–}20\text{–}13\text{–} \; 4\text{–}31\text{–}22\text{–}15 \end{array} \right\} + \quad \longrightarrow$$

There are 14 coils in one path and 16 in the other. A little later, coils 6, 33, 24, and 17_f in series, will be short-circuited by the positive brush, and coils 7, 14, 23, and 30 will take their place, the circuits through the armature then becoming, —

$$\longrightarrow \quad - \left\{ \begin{array}{l} 7\text{–}14\text{–}23\text{–}30\text{–} \; 5\text{–}12\text{–}21\text{–}28\text{–}3\text{–}10\text{–}19\text{–}26\text{–}1\text{–} \; 8\text{———} \\ 32\text{–}25\text{–}16\text{–} \; 9\text{–}34\text{–}27\text{–}18\text{–}11\text{–}2\text{–}29\text{–}20\text{–}13\text{–}4\text{–}31\text{–}22\text{–}15 \end{array} \right\} + \quad \longrightarrow$$

A further inspection of the diagram will show the unsymmetrical arrangement of the short-circuited and adjacent coils, causing the induction in some coils to act in opposition to that in others with which it is in series. This is less marked with large numbers of coils.

The chief disadvantages of the "short-connection" winding are that adjacent coils have between them, periodically, the full E.M.F. of the armature, and that the end windings are complicated.

Figure 18 represents another two-circuit, single-wound, "short-connection" gramme winding, in which $s = ny \pm 2$ $= 4 \times 5 \pm 2 = 22$. In this case y, the pitch, is odd, and consequently the sequence of connections is 1, $1+5=6$, $6+5$ $=11$, $11+5=16$, etc., thus advancing each time by 5, and not, as in the case of Fig. 17, page 34, where y was even, alternately by $(y+1)$ and $(y-1)$. Corresponding ends of coils are connected together; thus, the end of 1 and the end of 6, the beginning of 6 and the beginning of 11, etc.

At the position shown, coils 5, 10, 15, and 20 are short-circuited by the negative brush, and the circuits through the armature are,—

$$\longrightarrow \quad - \left\{ \begin{array}{l} 22\text{--}17\text{--}12\text{--}\ 7\text{--}2\text{--}19\text{--}14\text{--}\ 9\text{------} \\ 3\text{--}\ 8\text{--}13\text{--}18\text{--}1\text{--}\ 6\text{--}11\text{--}16\text{--}21\text{--}4 \end{array} \right\} + \quad \longrightarrow$$

The winding is subject to the disadvantages noted in connection with Fig. 17, page 34.

Instead of having the objectionable crossings at the terminals of the coils, as shown in Fig. 18, page 37, alternate coils should be wound right and left handedly. This would only be useful in cases where all the connecting is done at one end, which should be avoided when possible.

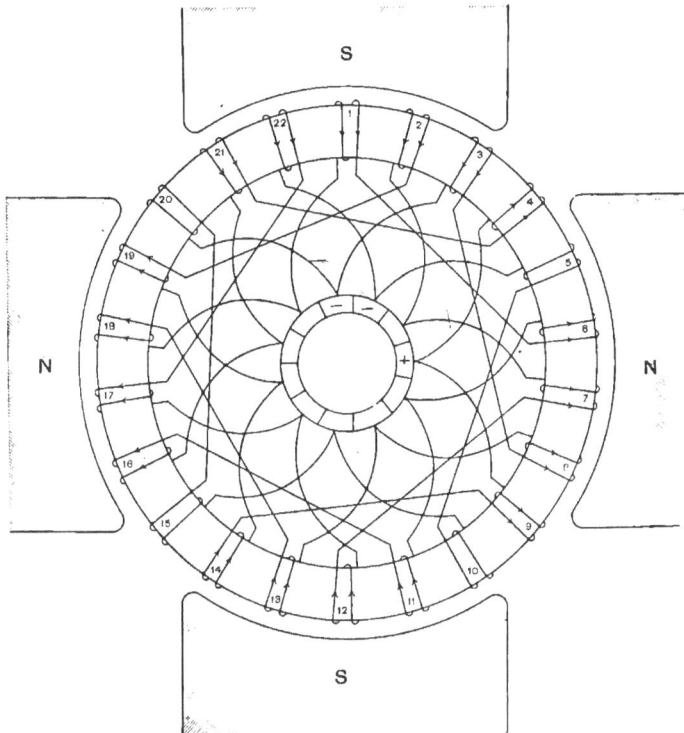

Fig. 18
TWO CIRCUIT, SINGLE WINDING.

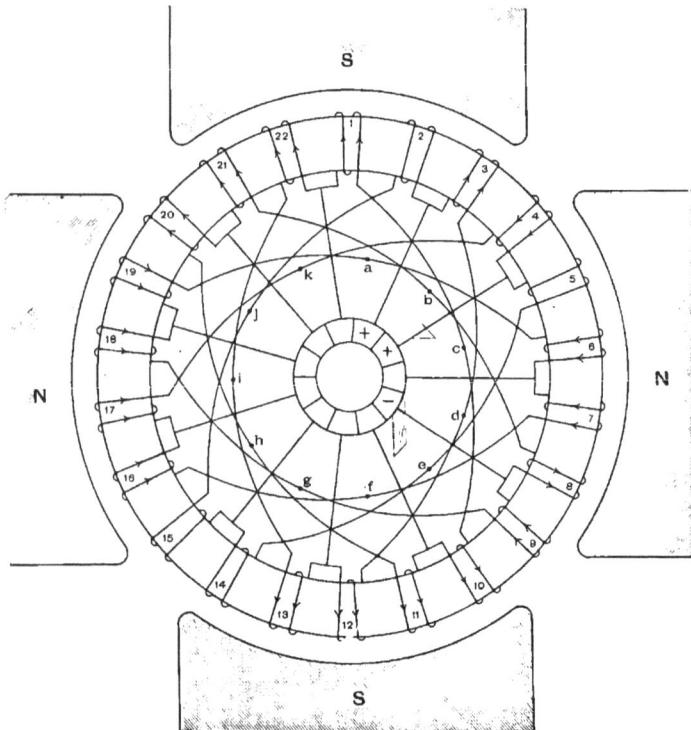

Fig. 19

TWO CIRCUIT, SINGLE WINDING.

Instead of connecting together in pairs coils lying in fields of opposite polarity, as in Figs. 17 and 18, adjacent coils may be connected together as shown in Fig. 19, and these connected across to coils in the nearest field of like polarity. The number of commutator segments is equal to one-half of the number of coils. The inherent identity of this and the "long-connection" winding may be seen by doing away with the leads to the commutator segments, and substituting leads from the eleven points lettered a, b, c, d, etc. The result will be a simple "long-connection" gramme winding, with half as many coils of twice as many turns each.

Therefore, the best way of laying out such a winding is to apply the rules for the "long-connection" winding, and make the connections shown in Fig. 19, instead of those of the regular "long-connection" gramme winding.

This winding gives half as many commutator segments as coils.

In the position shown, coils 5, 14, 15, and 2 are short-circuited by the positive brush, and the circuits through the armature are, —

$$\longrightarrow \quad - \left\{ \begin{array}{l} 8\text{--}21\text{--}20\text{--}11\text{--}10\text{--}\ 1\text{--}22\text{--}13\text{--}12\text{--}3 \\ 9\text{--}18\text{--}19\text{--}\ 6\text{--}\ 7\text{--}16\text{--}17\text{--}\ 4\text{------} \end{array} \right\} + \quad \longrightarrow$$

CHAPTER IV.

TWO-CIRCUIT, MULTIPLE-WOUND, MULTIPOLAR RINGS.

THE next class is that of the two-circuit, multiple-wound, long-connection ring windings. The general formula is, —

$$s = \frac{n}{2} \times y \pm m,$$

where

s = number of coils,

n = number of poles,

y = pitch,

m = number of windings.

The "m" windings will consist of a number of independently re-entrant windings equal to the greatest common factor of "y" and "m."

Therefore, when it is desired that the "m" windings shall combine to form *one re-entrant* system, it will be necessary that the G.C.F. of "y" and "m" shall be made equal to 1.

Figure 20 represents a two-circuit, doubly re-entrant, double-wound ring armature.

$$s = 26, \qquad n = 4, \qquad m = 2.$$

$$s = \frac{n}{2} \times y \pm m, \qquad 26 = \frac{4}{2} \times y + 2, \qquad \therefore y = 12.$$

Greatest common factor of y (12) and m (2) is 2. Therefore the winding will be doubly re-entrant.

At the position shown, coils 24 and 12, in series, are short-circuited by the negative brush. The circuits through the armature are, —

$$\longrightarrow \left[-\left\{ \begin{matrix} 25\text{--}13\text{--}1\text{--}15\text{--}3\text{--}17\text{---} \\ 26\text{--}14\text{--}2\text{--}16\text{--}4\text{--}18\text{---} \end{matrix} \right\} + \atop -\left\{ \begin{matrix} 10\text{--}22\text{--}8\text{--}20\text{--}6\text{---------} \\ 11\text{--}23\text{--}9\text{--}21\text{--}7\text{--}19\text{--}5 \end{matrix} \right\} + \right] \longrightarrow$$

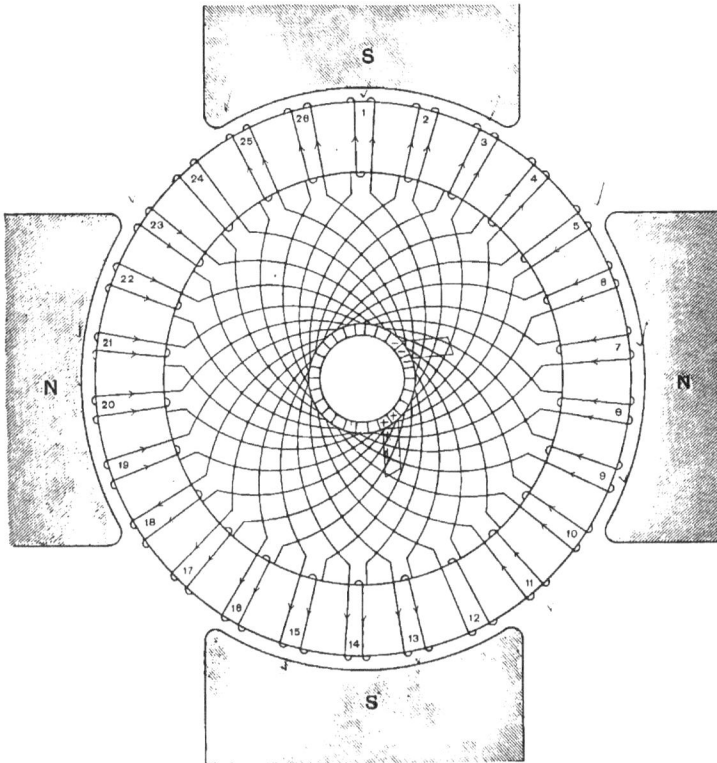

Fig. 20

TWO CIRCUIT, DOUBLE WINDING.

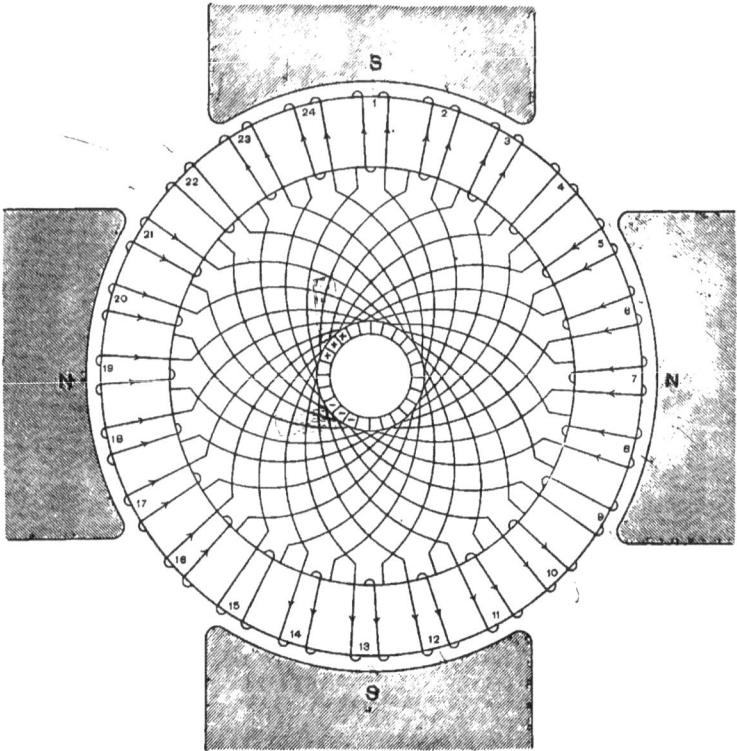

,Fig. 2 I

TWO CIRCUIT, DOUBLE WINDING.

Figure 21 represents a two-circuit, singly re-entrant, double-wound ring armature.

In this case $y=11$, $n=4$, and $m=2$. $s=\frac{1}{2}\times11\pm2=20$ or 24. 24 coils are taken. G.C.F. of "y" and "m" being 1, the winding is singly re-entrant.

In the position given, coils 9 and 22 are short-circuited at the negative brush, and 4 and 15 at the positive. The circuits through the armature are,—

$$\longrightarrow \left[\begin{array}{c} -\left\{\begin{array}{l}20\text{--}7\text{--}18\text{--}5\text{--}16\text{---}\\21\text{--}8\text{--}19\text{--}6\text{--}17\text{---}\end{array}\right\}+\\[1em] -\left\{\begin{array}{l}11\text{--}24\text{--}13\text{--}2\text{--------}\\10\text{--}23\text{--}12\text{--}1\text{--}14\text{--}3\end{array}\right\}+\end{array}\right] \longrightarrow$$

Figure 22 represents another two-circuit, singly re-entrant, double-wound ring armature.

$$m = 2, \quad n = 6, \quad y = 7, \quad s = \frac{n}{2}y \pm 2 = \tfrac{6}{2} \times 7 \pm 2 = 19 \text{ or } 23.$$

"y" and "m" being prime, the winding is singly re-entrant.

At the position shown, coils 4, 11, and 18 are short-circuited at the positive brush, and the circuits through the armature are: —

$$\longrightarrow \left\{ \begin{array}{l} - \left\{ \begin{array}{l} 15\text{--}22\text{--}6\text{--}13\text{--}20 \\ 14\text{--}21\text{--}5\text{--}12\text{--}19 \end{array} \right\} + \\[2mm] - \left\{ \begin{array}{l} 8\text{--} \ 1\text{--}17\text{--}10\text{--}3\text{--} \\ 7\text{--}23\text{--}16\text{--} \ 9\text{--}2\text{--} \end{array} \right\} + \end{array} \right\} \longrightarrow$$

Two two-circuit, singly re-entrant, triple windings for gramme rings are given below without diagrams: —

$$m = 3, \; n = 6, \; y = 7, \; s = \frac{n}{2} \times y \pm 3 = \frac{6}{2} \times 7 + 3 = 24.$$

The connections would be, —

1–8–15–22–5–12–19–2–9–16–23–6–13–20–3–10–17–24–7–14–21
–4–11–18–1

$$m = 3, \; n = 10, \; y = 10, \; s = \tfrac{10}{2} \times 10 - 3 = 47.$$

1–11–21–31–41–4–14–24–34–44–7–17–27–37–47–10–20–30–40–3
–13–23–33–43–6–16–26–36–46–9–19–29–39–2–12–22–32–42
–5–15–25–35–45–8–18–28–38–1

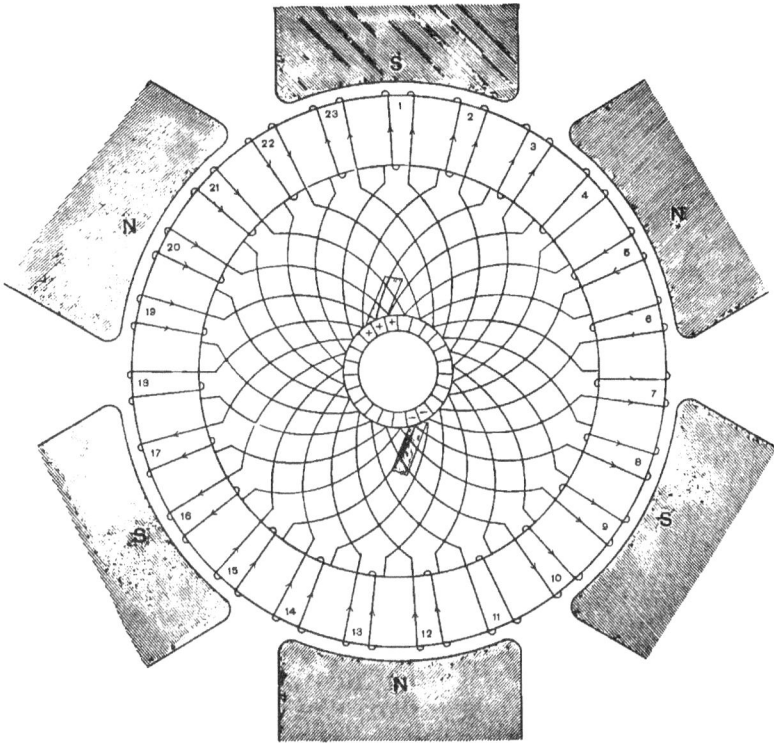

Fig. 22

TWO CIRCUIT, DOUBLE WINDING.

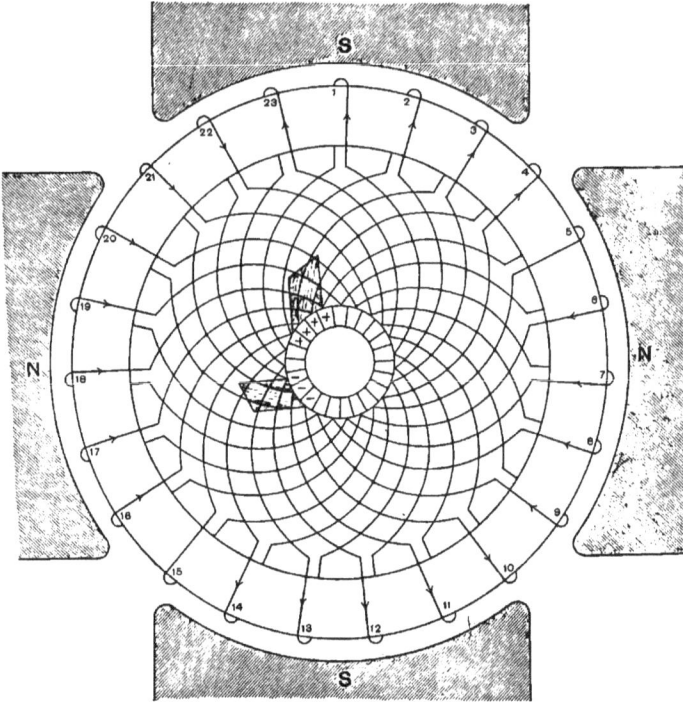

Fig. 23
TWO CIRCUIT, TRIPLE WINDING.

Figure 23 represents a two-circuit, singly re-entrant, triple winding.

$$m=3, \ n=4, \ y=10, \ s=\tfrac{1}{2}\times 10 \pm 3 = 23.$$

"m" and "y" being prime, the winding is singly re-entrant.

In the position shown, coils 5 and 15, in series, are short-circuited by the positive brush. The circuits through the armature are, —

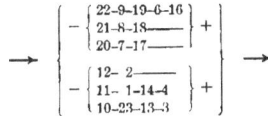

$$\longrightarrow \quad \left\{ \begin{array}{l} - \left\{ \begin{array}{l} 22\text{-}9\text{-}19\text{-}6\text{-}16 \\ 21\text{-}8\text{-}18\text{-------} \\ 20\text{-}7\text{-}17\text{-------} \end{array} \right\} + \\[2mm] - \left\{ \begin{array}{l} 12\text{-} 2\text{-------} \\ 11\text{-} 1\text{-}14\text{-}4 \\ 10\text{-}23\text{-}13\text{-}3 \end{array} \right\} + \end{array} \right\} \quad \longrightarrow$$

The extreme irregularity of the various circuits in multiple is not characteristic of the winding, but is merely due to the very small number of coils chosen. In practical cases it would be negligible.

From the formula and conditions of page 40, and from the examples just given, it will be seen that two-circuit, multiple-wound, ring windings may be divided into the three following cases: —

CASE I. — "y" and "m" are mutually prime. This gives a singly re-entrant winding of "m" multiple windings.

Illustration: — $n=4, \ y=7, \ m=4, \ s=\tfrac{1}{2}\times 7 + 4 = 18.$

Connections are, —   1-8-15-4-11-18-7-14-3-10-17-6-13-2-9-16-5-12-1.

May be expressed symbolically as ⓞⓞⓞ

CASE II. — "y" a multiple of "m." This gives "m" independently re-entrant windings.

Illustration : — $n=4, \ y=8, \ m=4, \ s=\tfrac{1}{2}\times 8 + 4 = 20.$

Connections are, —   1- 9-17-5-13-1
                          2-10-18-6-14-2
                          3-11-19-7-15-3
                          4-12-20-8-16-4

May be expressed symbolically as ◯ ◯ ◯ ◯.

CASE III. — "y" and "m" have common factors. This gives a number of independently re-entrant windings, equal to the greatest common factor of "y" and "m."

Illustration : — $n=4, \ y=6, \ m=4, \ s=\tfrac{1}{2}\times 6 + 4 = 16.$

The result is a two-circuit, quadruple winding with two independently re-entrant windings, because 2 is the greatest common factor of "y" and "m."

The connections are, —   1-7-13-3-9-15-5-11-1   and   2-8-14-4-10-16-6-12-2

May be expressed symbolically as ⓞ ⓞ.

Case II. is really a special instance of Case III.

The above formula and controlling conditions will be found to hold for all numbers of poles, coils, pitches, and windings of the two-circuit, long-connection type of gramme-ring armature windings.

Figure 24 is a two-circuit, singly re-entrant triple winding of the type described in connection with Figs. 15 and 16, which, it should be remembered, is only a modification of the long-connection type.

$$n = 4, \ y = 10, \ m = 3, \ s = \frac{n}{2} \times y \pm m = \frac{4}{2} \times 10 + 3 = 23.$$

At the position shown, coil 21 is short-circuited at the negative brush, and coils 3 and 4 at the positive brush. The circuits through the armature are, —

$$\longrightarrow \left\{ \begin{array}{c} - \left\{ \begin{array}{l} 8\text{--}18\text{--}\ 5\text{----} \\ 9\text{--}19\text{--}\ 6\text{--}16 \\ \text{----}20\text{--}\ 7\text{--}17 \end{array} \right\} + \\ - \left\{ \begin{array}{l} 22\text{--}12\text{--}\ 2\text{--}15 \\ \text{----}11\text{--}\ 1\text{--}14 \\ \text{----}10\text{--}23\text{--}13 \end{array} \right\} + \end{array} \right\} \longrightarrow$$

Figure 24 should be compared with Figs. 15 and 16.

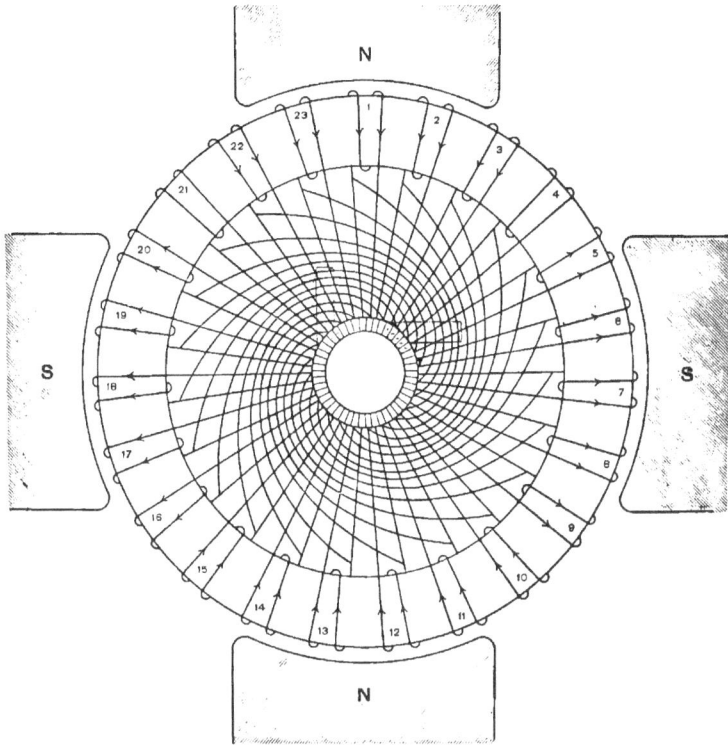

Fig. 24

TWO CIRCUIT, TRIPLE WINDING.

CHAPTER V.

DRUM ARMATURE WINDINGS.

In drum windings, all connections from bar to bar must be made upon the rear and front ends exclusively, it not being practicable to bring connections through inside from back to front as is the case with rings. Consideration of this limitation will show that the two sides of any one coil must be situated in fields of opposite polarity, so that the electromotive forces, generated in the active conductors of a coil by their passage through their respective fields, shall be in the same direction.

In the case of a drum, it should also be noted that a coil is linked with the whole or nearly the whole flux from one pole piece, instead of, as in the ring armature, with only one-half of the flux.

BIPOLAR DRUM WINDINGS.

The winding of bipolar armatures is much less simple in the case of drums than in that of rings, and it will therefore be necessary to give considerable attention to the various methods in which such windings may be carried out.

Figure 25 represents essentially the winding devised by von Hefner-Alteneck. It is used chiefly for small, smooth-core, wire-wound armatures, and the element of the winding, represented in the diagram by a pair of face conductors, and a back connection consists usually, in practice, of a coil of several turns, comparable in some respects to the coil of the ring windings ; but in the diagram only one turn per coil will be shown. This will also be advantageous, inasmuch as large, iron-clad, bar-wound, multipolar drum armatures are derived from, and diagrammatically are very analogous to, the wire-wound, smooth-core armatures now under consideration.

An examination of Fig. 25 shows that, starting from a commutator segment, the winding proceeds over the front end to conductor No. 1; down No. 1 over the back to conductor No. 8, which, it should be noted, is adjacent to the conductor diametrically opposite No. 1. From No. 8 the winding returns to the next commutator segment, and is then carried to conductor No. 3 (skipping No. 2, which will later be joined over the back to a conductor almost diametrically opposite to it), down No. 3, over the back to No. 10, etc. From this it is seen that the "pitch" on the back end is 7 and on the front end is −5.

In the position shown, the circuits through the armature are, —

$$\longrightarrow \quad -\left\{ \begin{matrix} 10- & 3-8- & 1- & 6-15 \\ 7-14-9-16-11- & 2 \end{matrix} \right\} + \quad \longrightarrow$$

The coil represented by the conductors 13 and 4 is short-circuited at the positive brush, and coil 12-5 at the negative brush.

The customary convention is adopted in the diagram, ⊗ indicating a current from the observer into the paper, and ● a current up out of the surface of the paper toward the observer.

A serious fault of this winding is that large differences of potential exist between adjacent conductors (or, usually, groups of conductors). This would be of no importance with the small numbers of conductors represented in these diagrams, but in actual cases, large numbers of conductors are used, and are placed close together in order to waste no available space.

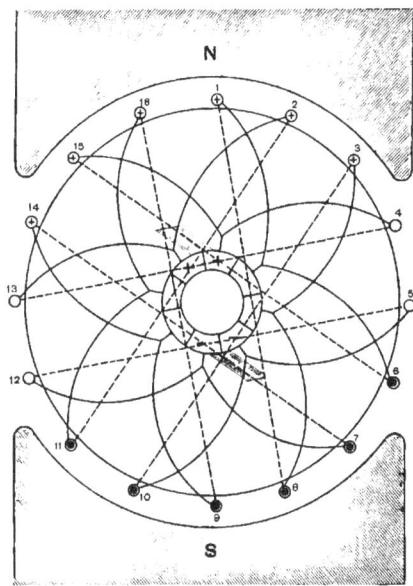

Fig. 25

TWO CIRCUIT, SINGLE WINDING.

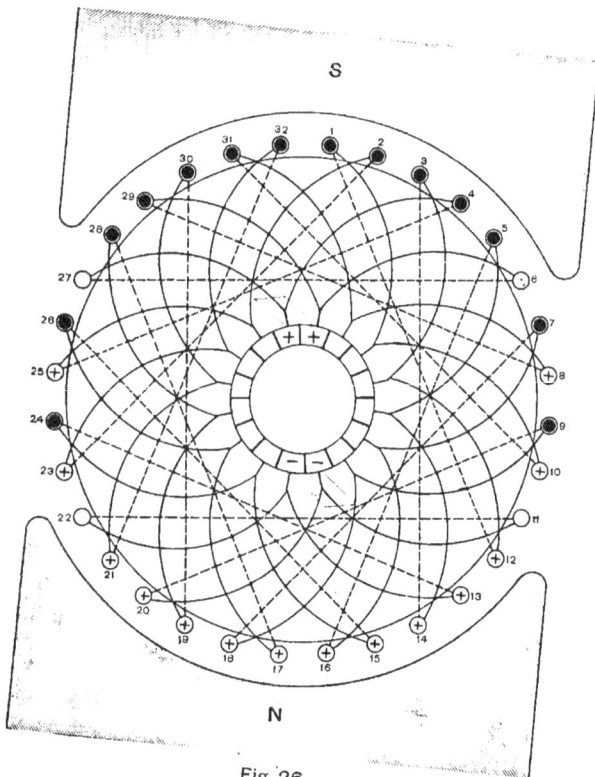

Fig. 26

TWO CIRCUIT, SINGLE WINDING.

Figure 26 gives the diagram of a winding discussed by Swinburne. Its characteristic feature is the use of a small pitch (in the figure the pitch at the back end is 11, and at the front end it is −9), whereby the turns consist of conductors separated by a much smaller angular distance than in the von Hefner-Alteneck winding.

An advantage of this winding is that there is much less crossing of the end connections than is the case where the pitch is taken larger. Thus the difficult question of insulation at the ends of the armature is greatly simplified.

Still further, it has been pointed out by Swinburne that the demagnetizing effect of the armature on the field is reduced, as may be seen from the fact that the currents in the conductors in the demagnetizing belt between the pole tips, namely, 23, 24, 25, and 26, and in 7, 8, 9, and 10, are alternately in opposite directions, and thus neutralize each other.

A serious disadvantage is that the short-circuited coils, 6–27 and 11–22, are considerably removed from the neutral line. This, together with the fact that the counter-electromotive forces present in several conductors of the circuit between brushes detract from the volts per unit of length of armature wire, reduces to rather small limits the extent to which such connecting over short chords should be carried.

In the position shown, the circuits through the armature are, —

$$\longrightarrow \quad -\left\{ \begin{matrix} 20\text{–}\ 9\text{–}18\text{–}\ 7\text{–}16\text{–}\ 5\text{–}14\text{–}\ 3\text{–}12\text{–}\ 1\text{–}10\text{–}31\text{–}\ 8\text{–}29 \\ 13\text{–}24\text{–}15\text{–}26\text{–}17\text{–}28\text{–}19\text{–}30\text{–}21\text{–}32\text{–}23\text{–}\ 2\text{–}25\text{–}\ 4 \end{matrix} \right\} + \quad \longrightarrow$$

In Fig. 27 it will be seen that the number of coils is odd (in the two preceding diagrams it was even), with the result that the two active sides of such coils may now be diametrically opposite.

This would not, however, usually be advisable, as it makes many more crossings at the ends, and therefore increases the difficulty of insulating.

Some advantage results from bringing the short-circuited coil (in the figure, coil 24–9 is short-circuited by the negative brush), exactly in the neutral line, this being, of course, only possible when the conductors forming its active sides are diametrically opposite.

The circuits through the armature in the position shown are, —

$$\longrightarrow \quad - \begin{cases} 22-\ 7-20-\ 5-18-\ 3-16-1-14-29-12-27-10-25 \\ 11-26-13-28-15-30-17-2-19-\ 4-21-\ 6-23-\ 8 \end{cases} + \quad \longrightarrow$$

The pitch on the back end is 15, and on the front end it is −13.

Owing to the number of segments being odd, only one coil is short-circuited at once, unless wide brushes are used.

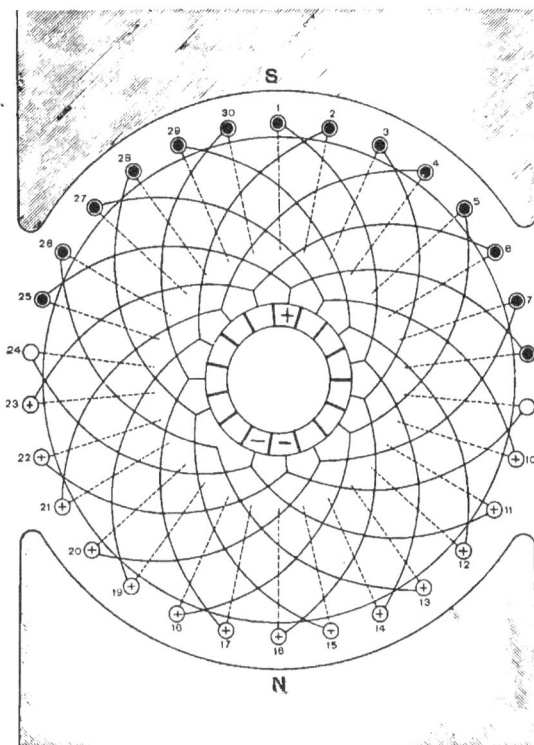

Fig. 27

TWO CIRCUIT, SINGLE WINDING.

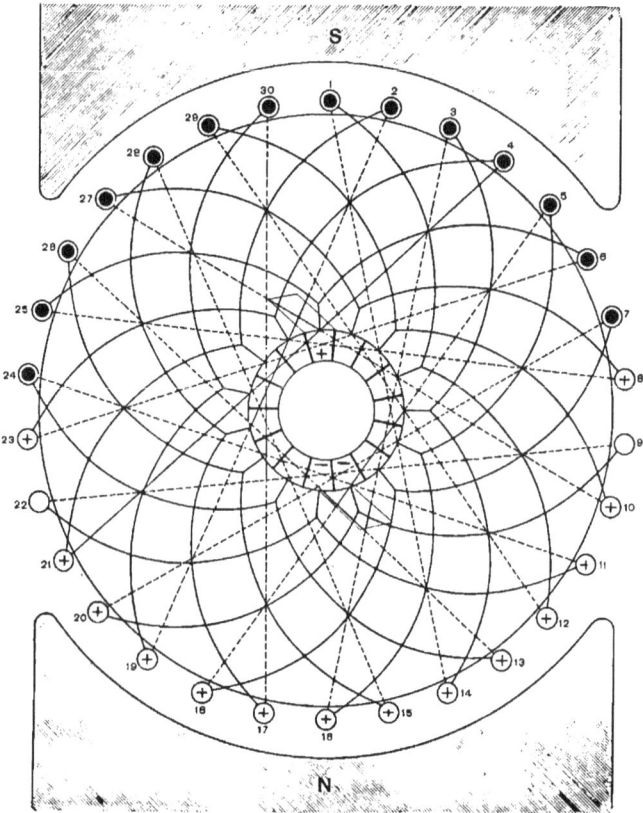

Fig. 28

TWO CIRCUIT, SINGLE WINDING.

In Fig. 28 there is also an odd number of coils (and therefore an odd number of commutator segments). But instead of connecting over the back from No. 1 to No. 16 (the conductor diametrically opposite No. 1) as in Fig. 17, connection is made over the back from No. 1 to No. 14, then over the front to No 3, etc , the pitch at the back end being 13, and on the front end −11. It is, therefore, a mild form of the Swinburne chord winding, as described in connection with Fig. 26. The end connections are better distributed and have fewer crossings than was the case in Fig. 27, where diametrically opposite conductors were connected into coils.

In the position shown, coil 22–9 is short-circuited at the negative brush, and the circuits through the armature are, —

$$\longrightarrow \quad - \begin{cases} 11\text{--}24\text{--}13\text{--}26\text{--}15\text{--}28\text{--}17\text{--}30\text{--}19\text{--} \ 2\text{--}21\text{--} \ 4\text{--}23\text{--} \ 6 \\ 20\text{--} \ 7\text{--}18\text{--} \ 5\text{--}16\text{--} \ 3\text{--}14\text{--} \ 1\text{--}12\text{--}29\text{--}10\text{--}27\text{--} \ 8\text{--}25 \end{cases} + \quad \longrightarrow$$

In Fig. 29 the winding is carried on over a still shorter chord, the pitch at the back end being 11 and at the front end −9.

It is very instructive to compare Figs. 27, 28, and 29, all of which have 30 face conductors (15 coils). But in Fig. 27 diametrically opposite conductors are connected over the back, the back pitch being 15. Figure 28 is a weak chord winding, the back pitch being 13. Figure 29 is a decided chord winding, the back pitch being 11. The points to be compared are the positions of the short-circuited conductors with reference to the neutral line ; the amount of neutralizing of the effect of the demagnetizing belt between pole tips, and the comparative amount of crossing of connectors at the ends.

In Fig. 27 it was shown that diametrically opposite conductors could be connected into coils if the number of coils were chosen *odd*.

The same object may be attained with an *even* number of coils by winding them in two layers instead of in one layer, as has been the case in all the heretofore described bipolar drum armatures.

It should be again noted that the term "conductors" is used in these explanations, although "groups of conductors" could often be substituted therefor in small, smooth-core, wire-wound armatures.

Thus the set of "one-layer windings," just described, are those in which "conductors" or "groups of conductors" are, in the completed winding, arranged in one layer, although the individual wires of such a group may optionally occupy one or several layers. In the same way, the two-layer windings now to be described are those in which the completed winding consists of "conductors" or "groups of conductors" arranged in two layers, although the actual depth of individual wires may, when desirable, be greater than two.

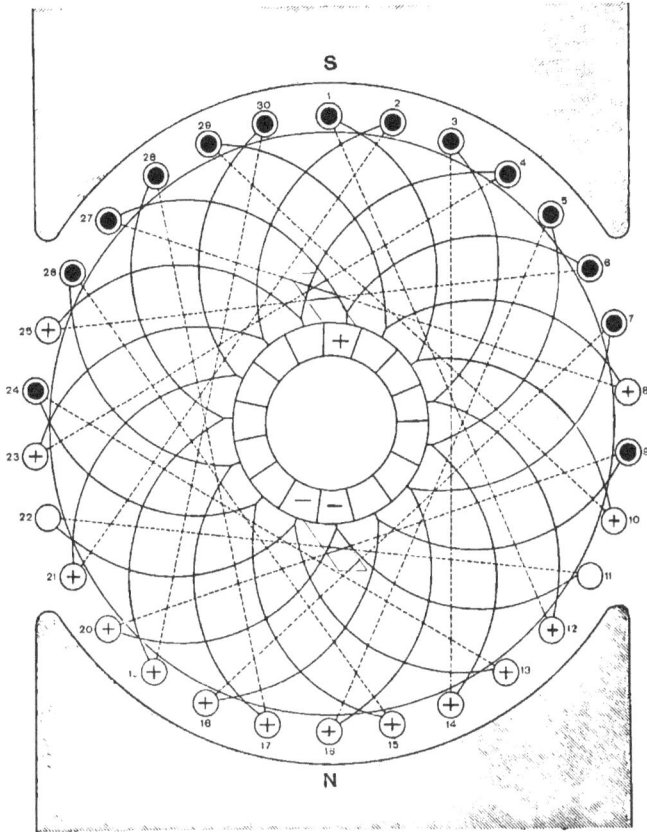

Fig. 29
TWO CIRCUIT, SINGLE WINDING.

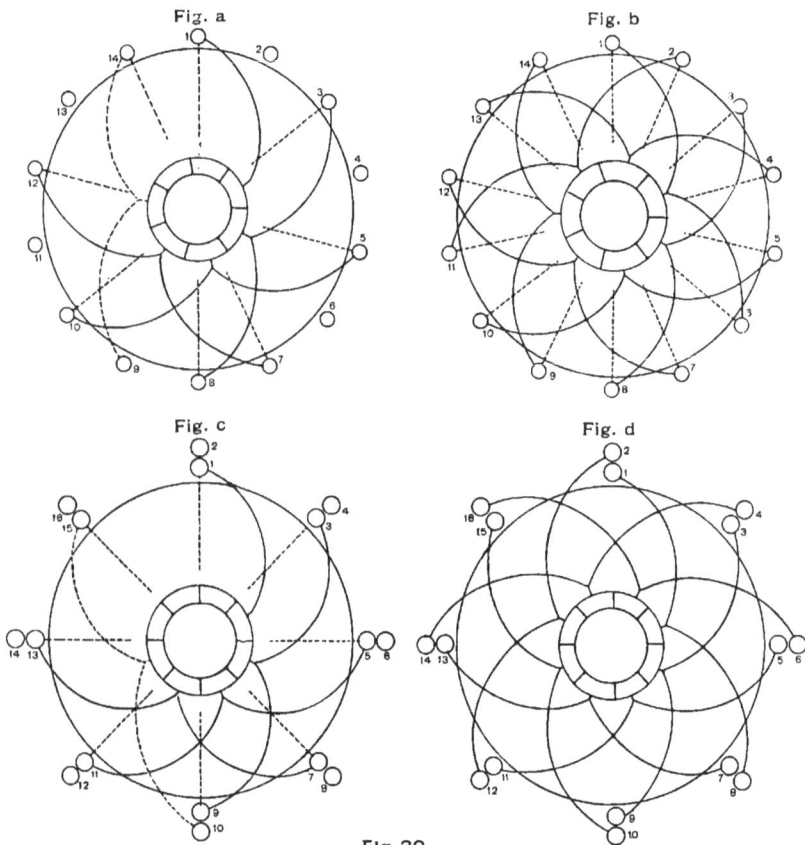

Fig. a

Fig. b

Fig. c

Fig. d

Fig. 30
a, b, c and d.

In Fig. 30, diagrams a and b represent a single-layer bipolar drum winding with an odd number of coils, in which diametrically opposite conductors are connected together into coils. In diagram a the first half of the winding is carried out and proceeds from a commutator bar to conductor No. 1, to 8, to 3, to 10, to 5, to 12, to 7, to 14, and is then ready for the second half. It will be seen that at this stage only every other coil is connected up, and that only one-half of the commutator segments are utilized. Diagram b shows the winding completed. This winding, which is of the type shown in Fig. 27, is given here for comparison with the two-layer winding shown in diagrams c and d. In Fig. c it will be seen that the first half is exactly the same as the first half of the one-layer winding (except that it contains eight conductors instead of seven), and at the completion of the first half all the conductors of the lower layer are connected up in the order 1–9–3–11–5–13–7–15, and only one-half of the commutator segments are connected in. The coils remaining for the second half, instead of lying between those of the first half, occupy an outer layer. Diagram d shows the completed winding, with all the coils and commutator segments utilized.

Figure 31 represents a two-layer winding with thirty-two conductors, with diametrically opposite conductors connected into coils over the back end.

These back-end connèctions are not shown, because they would interfere with the clearness of the diagram. The connections are 1–17–3–19–5–21, etc. In the position shown, coil 25–9 is short-circuited at the negative brush and 26–10 at the positive brush, and the circuits through the armature are, —

$$\longrightarrow \quad - \begin{cases} 23\text{-} \ 7\text{-}21\text{-} \ 5\text{-}19\text{-} \ 3\text{-}17\text{-}1\text{-}16\text{-}32\text{-}14\text{-}30\text{-}12\text{-}28 \\ 11\text{-}27\text{-}13\text{-}29\text{-}15\text{-}31\text{-}18\text{-}2\text{-}20\text{-} \ 4\text{-}22\text{-} \ 6\text{-}24\text{-} \ 8 \end{cases} + \quad \longrightarrow$$

It will be seen from this table that maximum difference of potential exists between conductors lying directly over each other in different layers, such as 27 and 28, or 7 and 8. But *adjacent* conductors have only small differences of potential; therefore, the two layers should be carefully insulated from each other.

It is an advantage to have the conductors 25–9 and 26–10 of the two short-circuited coils all situated on one diameter, as they may therefore be brought diametrical, and therefore are capable of being short-circuited more nearly in the neutral position.

A disadvantage of the winding is that, one-half being wound exclusively in the lower layer and the other half in the upper, they have unequal lengths and different peripheral speeds, and in those recurring positions in which the two circuits through the armature consist respectively of the lower and the upper layer, the condition will be unbalanced.

In practice, however, it is frequently found expedient to use this connection because of the ease of winding, the inequality being made as small as possible. It will be shown later how this inequality may be obviated; the winding will be, however, less easy to execute.

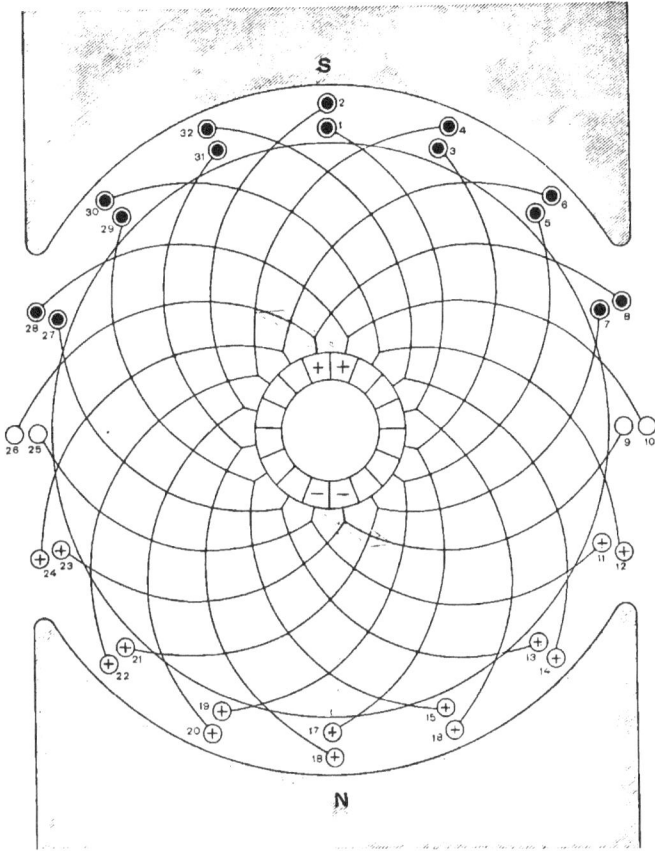

Fig. 31

TWO CIRCUIT, SINGLE WINDING.

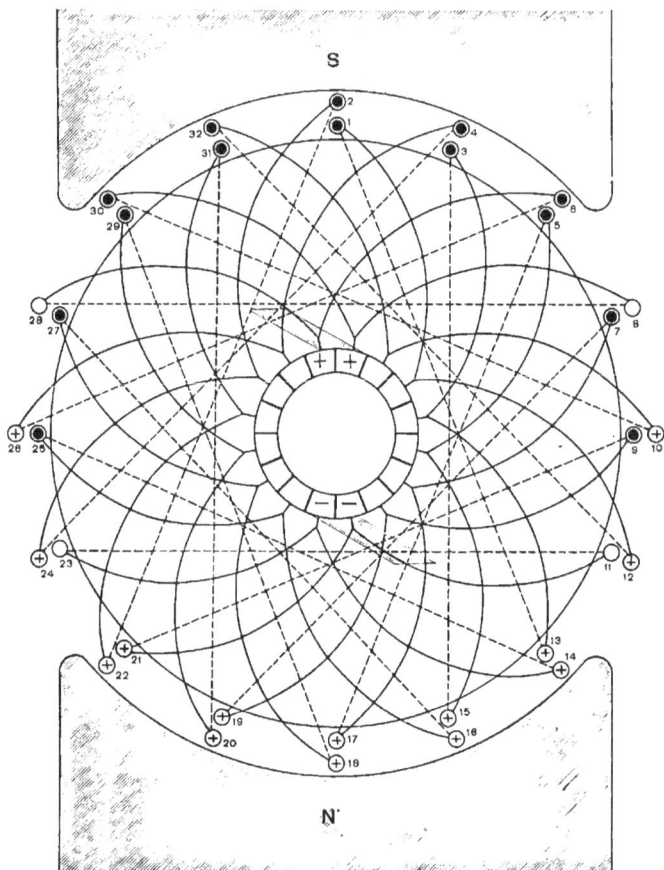

Fig. 32
TWO CIRCUIT, SINGLE WINDING

In Fig. 32 the winding is of the Swinburne type, being connected over the ends along a short chord. Thus, starting from a commutator segment, it passes down No. 1, over the back to No. 13, over the front to No. 3, and so on through 3, 15, 5, 17, 7, 19, 9, 21, 11, 23 ; but coming over the front from 23 it would naturally go to 13 of the lower layer. This, however, is already used, so the winding continues by No. 14, which is directly over No. 13 in the top layer, and then on through 25–16–27–18–29–20–31–22. From 22 it would naturally go to No. 1, but, as the winding is not yet completed, it must go instead to No. 2, which is directly over No. 1, and then proceed from 2 through 24–4–26–6–28–8–30–10–32–12, and then it closes on itself at No. 1. This winding is not at all difficult, because, although the lower layer is not entirely completed before beginning to wind the upper layer, yet in that part of the armature on which it is desired to wind the upper layer, the lower layer is entirely completed, and for quite a distance beyond, so that there would be no trouble in inserting the necessary insulation, etc.

In the position shown, coil 28–8 is short-circuited at the positive brush, and coil 23–11 at the negative brush. It is a disadvantage to have the short-circuited coils so far from the neutral line.

The circuits through the armature in the given position are, —

$$\longrightarrow \quad - \left\{ \begin{array}{l} 21\text{--} \ 9\text{--}19\text{--} \ 7\text{--}17\text{--} \ 5\text{--}15\text{--} \ 3\text{--}13\text{--}1\text{--}12\text{--}32\text{--}10\text{--}30 \\ 14\text{--}25\text{--}16\text{--}27\text{--}18\text{--}29\text{--}20\text{--}31\text{--}22\text{--}2\text{--}24\text{--} \ 4\text{--}26\text{--} \ 6 \end{array} \right\} \ + \quad \longrightarrow$$

It will be seen that in this armature there can be no position in which one layer belongs exclusively to one circuit and the other to the other circuit. Therefore the discrepancy in lengths and peripheral speeds of the two circuits through the armature will, at the most unfavorable moment, be less than when diametrically opposite conductors are connected into coils. The winding has, in common with all chord windings, the advantage of less crossings of the end connections. The diagram shows particularly well the absence of demagnetizing action in the zone of conductors between pole tips.

If, in Fig. 32, page 66, conductor No. 1 had been connected over the back to No. 15 instead of to No. 13, it would still have been a chord winding, but with somewhat less marked characteristics than that of Fig. 32. All the advantages and disadvantages would have been on a smaller scale.

Figure 33 represents a winding in which coils of the outer and inner layer are alternately connected. The rear-end connections are not drawn, but are diametrical. Thus the series is 1–15–4–18–5–19–8–22–9–23–12–26–13–27–16–2–17–3–20–6–21–7–24–10–25–11–28–14–1. This makes both circuits through the armature of very nearly equal length and of very nearly equal average peripheral speed.

In the position shown, coil 21–7 is short-circuited by the positive, and 22–8 by the negative brush. The circuits through the armature are, —

$$\longrightarrow \quad -\left\{ \begin{array}{l} 19-\ 5-18-\ 4-15-\ 1-14-28-11-25-10-24 \\ 9-23-12-26-13-27-16-\ 2-17-\ 3-20-\ 6 \end{array} \right\} + \quad \longrightarrow$$

For this winding to be regular, the number of conductors must be an *odd* multiple of **4**.

Other bipolar drum windings have been proposed by Hering, Western Electric Company, and others, each of which possesses certain special advantages. It might be well especially to consult an article by Hering in " Electrician and Electrical Engineer," Vol. 4, 1885, p. 423, and Vol. 5, 1886, p. 84.

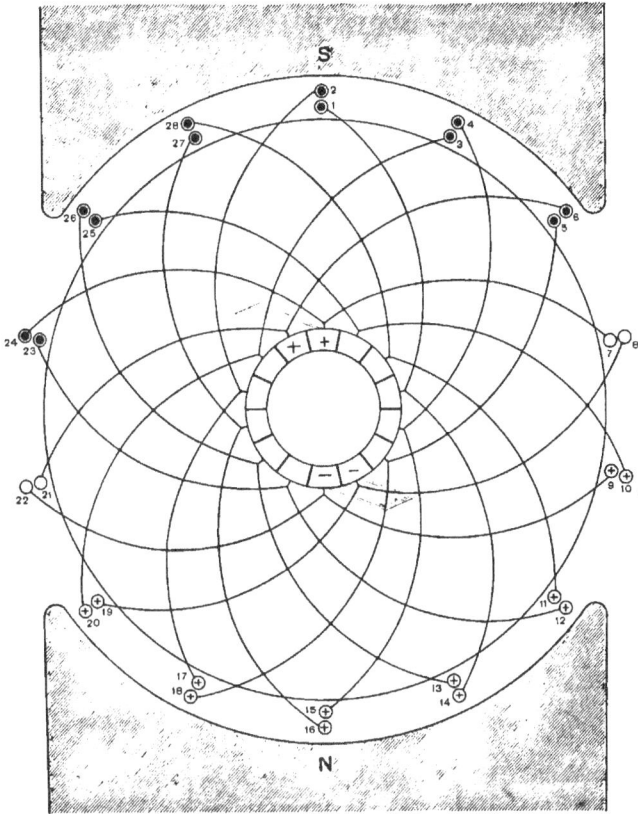

Fig. 33

TWO CIRCUIT, SINGLE WINDING.

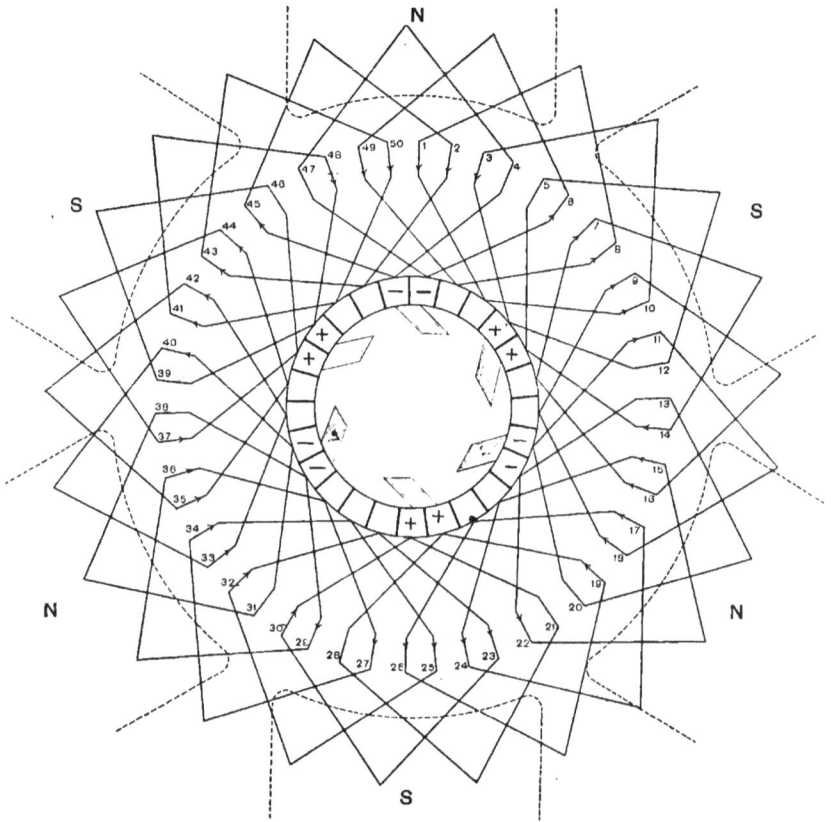

Fig. 34

SIX CIRCUIT, SINGLE WINDING.

CHAPTER VI.

MULTIPLE–CIRCUIT, SINGLE–WOUND, MULTIPOLAR DRUMS.

FOR multiple-circuit, multipolar drums, the condition to be fulfilled to make the winding re-entrant is that there shall be an even number of bars. The pitch at one end of the armature must exceed that at the other end by 2 (for single windings), each of these pitches being odd. If n is the number of poles and C the number of face conductors, the average pitch should not differ much from $\frac{C}{n}$; for if it is much less, two successive conductors will often lie under the same pole piece, and their induced electromotive forces will be in opposition to each other, whereas they should be additive. If the average pitch is much greater than $\frac{C}{n}$, the cross-connections will be unnecessarily long, and the armature resistance and cost of copper unnecessarily high. Suppose a preliminary calculation for a single-layer, six-pole machine shows that about 49 conductors are required, it will be seen that $\frac{C}{n} = \frac{49}{6} = 8.17$. The two-end pitches must both be odd numbers, and must differ by 2. Therefore, take 7 and 9. The mean pitch is 8. The condition to be fulfilled by the total number of conductors is that it shall be an even number. Let it be 50.

This case is shown in Fig. 34. In this diagram the radial lines represent the face conductors. The connecting lines on the inside represent the end connections at the commutator end, and those on the outside represent the end connections at the pulley end. The brushes are placed inside the commutator for convenience.

At the position shown, the conductors without arrow-heads are short-circuited. The circuits through the armature are,—

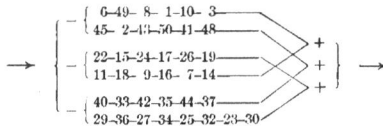

$$\rightarrow \quad \begin{cases} - \begin{cases} 6\text{--}49\text{--} 8\text{--} 1\text{--}10\text{--} 3\text{---} \\ 45\text{--} 2\text{--}43\text{--}50\text{--}41\text{--}48\text{---} \end{cases} \\ - \begin{cases} 22\text{--}15\text{--}24\text{--}17\text{--}26\text{--}19\text{---} \\ 11\text{--}18\text{--} 9\text{--}16\text{--} 7\text{--}14\text{---} \end{cases} \\ - \begin{cases} 40\text{--}33\text{--}42\text{--}35\text{--}44\text{--}37\text{---} \\ 29\text{--}36\text{--}27\text{--}34\text{--}25\text{--}32\text{--}23\text{--}30 \end{cases} \end{cases} \quad \begin{matrix} + \\ + \\ + \end{matrix} \quad \rightarrow$$

The front-end pitch is $y=9$, and the back-end pitch is $y=-7$.

If the pitches had been taken 7 and −5 instead of 9 and −7, retaining the same number (50) of face conductors, the diagram given in Fig. 35 would have been the result. This, it will be seen, is an application of the chord winding to a multipolar armature. The current in the conductors in the neutral zone is alternately in opposite directions, so that the demagnetizing action of the armature is small. The end connections are shorter, occupying less room and reducing the armature resistance and cost of copper. The short-circuited conductors are, however, at some distance from the neutral lines, and, although the electromotive forces in each pair will partly neutralize each other, it would be advisable, in cases where such chord windings are adopted, to have as great distances between pole tips as other circumstances permit.

In the given position, the short-circuited conductors are 4–49, 12–7, 20–15 28–23, 38–33, 46–41. The armature circuits are, —

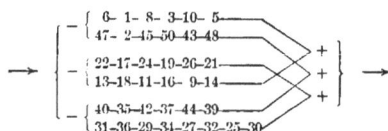

The front-end pitch is $y = 7$, and the back-end pitch $y = -5$.

If it should be considered desirable to have all the paths through the armature contain *exactly* the same number of conductors, then the number of face conductors should be chosen a multiple of the number of poles. But with a large number of conductors this would generally not be an important consideration.

In modern practice the conductors in large multipolar machines frequently consist of bars arranged in slots. The end connections then become strips arranged in two or more spiral layers at each end. If there were only one conductor per slot, two layers at each end would still be necessary, as it would be the same as if the lower conductors were brought up side of the upper conductors, and every other conductor would, therefore, as before, be connected over in an opposite direction from its neighbor.

For multiple-circuit, *single-wound* armatures there may be any even number of conductors per slot, and any number of slots. No new diagrams are necessary to show the cases of two or more conductors per slot, as Figs. 34 and 35 may be interpreted as having twenty-five slots and two conductors per slot, in which case odd-numbered conductors may be considered to belong to the upper layer, and even-numbered conductors to the lower layer. Connection is always made between odd and even numbered conductors, the pitch being always odd. The front-end and back-end pitches must differ by 2, and must have opposite signs.

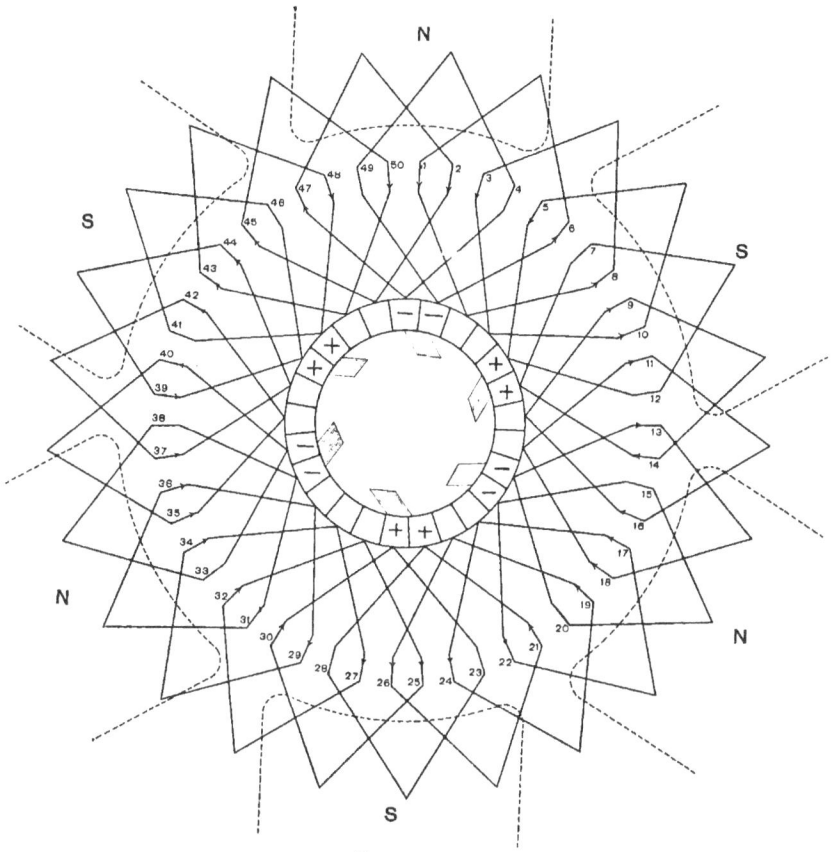

Fig. 35
SIX CIRCUIT, SINGLE WINDING.

S

Fig. 36

SIX CIRCUIT, SINGLE WINDING.

Figure 36 represents a six-circuit, single-wound, drum winding with eighty conductors. The number of conductors is purposely taken large, so that a study of the diagram and winding table may show the magnitude of the differences of potential in neighboring conductors.

At the given position, conductors 75–6, 9–20, 21–32, 35–46, 49–60, and 61–72 are short-circuited at the brushes.   The circuits through the armature are,—

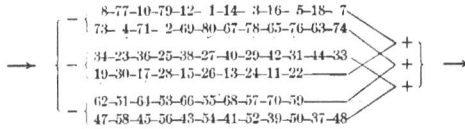

An inspection of the above table will show that the full difference of potential exists at recurring intervals between each pair of sequential conductors, such as 7 and 8, or 47 and 48. In practice, such conductors will often consist of two bars lying one above the other in the same slot. This shows that such upper and lower layers in a slot should be carefully insulated.   On the other hand, alternately sequential conductors, as 5 and 7, or 47 and 45, have between them only the small difference of potential of two conductors in series; so that, in practice, where such conductors usually belong both to the upper or both to the lower layer of the same slot, comparatively thin layers of insulation suffice.   For instance, it is often the case in multiple-circuit windings that there are four conductors per slot, arranged two wide and two deep.   This case would require that the horizontal layer of insulation between conductors should be much thicker than the vertical layer.

For this class of windings (multiple-circuit, single-wound drums) a formula is superfluous, and the following summary of conditions will suffice : —

There may be any even number of conductors, except that in ironclad windings the number of conductors must also be a multiple of the number of conductors per slot.

The front and back pitches must both be odd, and must differ by 2; therefore the average pitch is even.

The average pitch "y" should not be very different from $\frac{c}{n}$, *where c = number of conductors, and* n = *number of poles.   For chord windings, "y" should be smaller than* $\frac{c}{n}$ *by as great an amount as other conditions will permit.*

CHAPTER VII.

MULTIPLE-CIRCUIT, MULTIPLE-WOUND, MULTIPOLAR DRUMS.

THE next windings to be considered are multiple-circuit, *multiple-wound*, multipolar drums.

The following rules control these windings : —

The number of conductors, C, must be an even number. The pitches must be odd. If $y =$ front-end pitch, then $-(y-2m) =$ back-end pitch, where $m =$ number of windings (double, triple, quadruple. etc.).

These "m" windings may form one re-entrant winding, "m" independent re-entrant windings, or a number of re-entrant windings equal to some factor of "m," each of which re-entrant windings is composed of two or more components.

To determine the proper number of conductors for any of the above cases, the following rule should be observed : —

If "m" equals the number of windings, and "C" equals the number of face conductors, then the number of independently re-entrant windings will be equal to the greatest common factor of $\frac{C}{2}$ *and* m.

For instance, if a quadruple winding has 28 conductors, then the greatest common factor of $(m=4)$ and $\left(\frac{C}{2} = \frac{28}{2} = 14\right)$ is 2, and the quadruple winding will consist of *two* independent double windings, each of the two being re-entrant. This may be represented symbolically as ⓐ ⓐ.

If $C = 24$, and $m = 4$, the greatest common factor of $\left(\frac{C}{2} = \frac{24}{2} = 12\right)$ and $(m = 4)$ is 4, and the quadruple winding will be made up of *four* independent single windings. This may be represented symbolically as ○ ○ ○ ○.

If $C = 26$ and $m = 4$, the greatest common factor of $\left(\frac{C}{2} = \frac{26}{2} = 13\right)$ and $(m = 4)$ is 1, and the quadruple winding will consist of *one* singly re-entrant quadruple winding. This may be represented symbolically as ⓐ.

The above rule applies to any winding (double, triple, quadruple, etc.).

It is interesting to note that, for "multiple-*circuit*" windings, the rule for the number of multiple windings is independent of the number of poles and of the pitch.

The number of conductors. "C," the average pitch, "y," and the number of poles, "n," should be so chosen that $n \times y$ shall be somewhere nearly equal to C, being preferably a little smaller than C.

77

Figure 37 which, like Figs. 34 and 35, has six poles and fifty conductors, is a singly re-entrant triple winding. $C = 50$; $m = 3$. Greatest common factor of $\frac{C}{2}$ and m is 1. Therefore, by the preceding rule, the result is one singly re-entrant triple winding. The winding may be represented symbolically as ⓞ.

The average pitch should be a little less than $\frac{C}{n} = \frac{50}{6} = 8.33$, and the forward and backward pitches must differ by $(2\,m = 6)$. Therefore the front end pitch is taken $y = 11$, and the back-end pitch $y = -5$.

In the given position, conductors 49 and 4 are short-circuited at a negative brush, and 12 and 7 at a positive brush. The circuits through the armature are,—

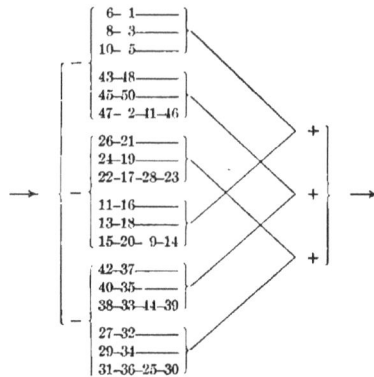

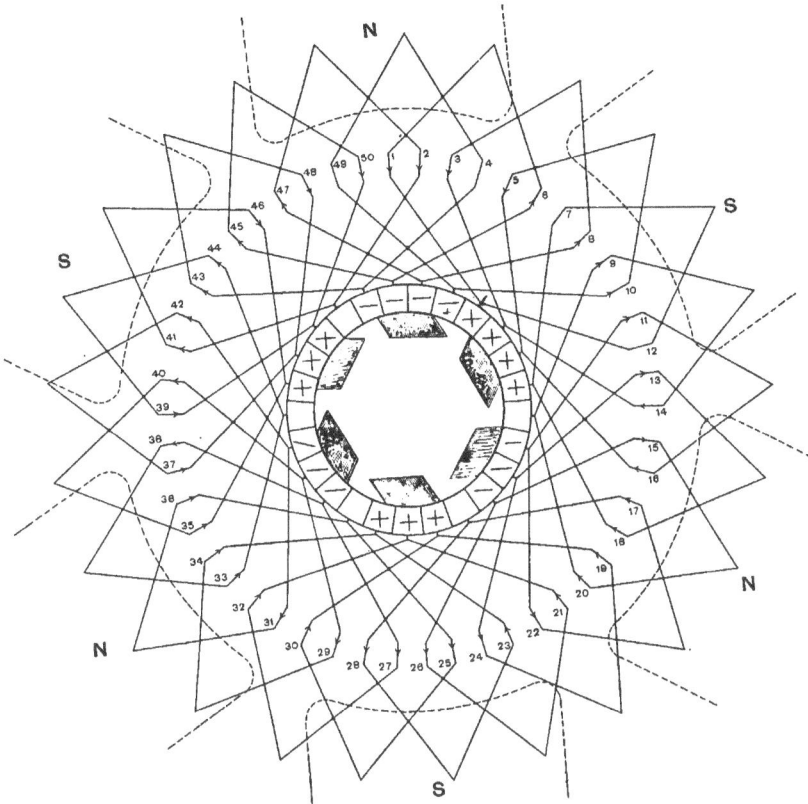

Fig. 37

SIX CIRCUIT, TRIPLE WINDING.

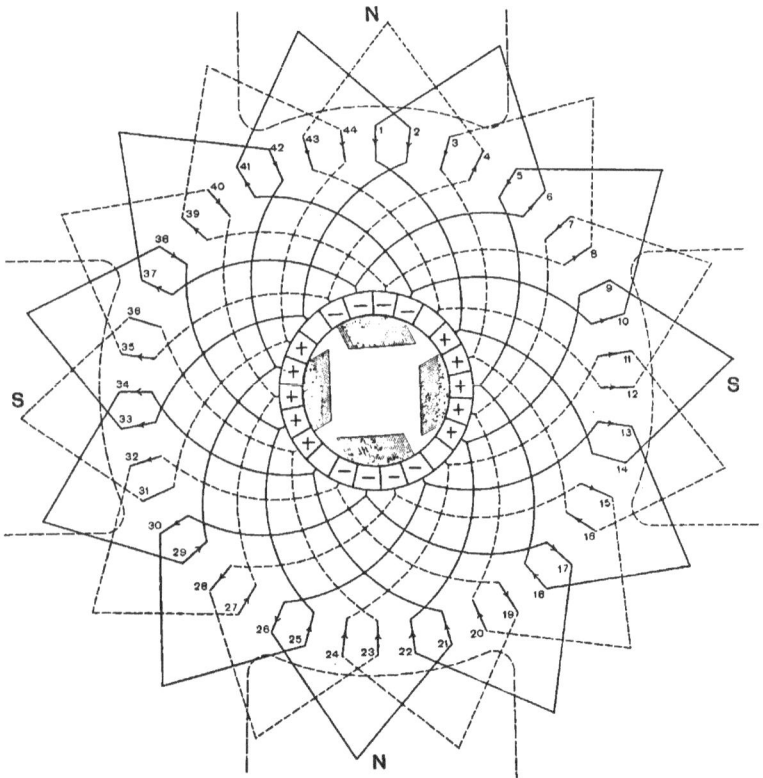

Fig. 38

FOUR CIRCUIT, QUADRUPLE WINDING.

Figure 88 is a four-circuit, doubly re-entrant quadruple winding in which $n=4$, $C=44$, and $m=4$. The greatest common factor of $\frac{C}{2}$ and "m," $i.e.$, of 22 and 4, is $\mathbf{2}$; therefore there are two independent, singly re-entrant, double windings. The winding may be represented symbolically by ⓐⓐ. These two windings are represented on the diagram by full and dotted lines. The front-end pitch has been taken 13, and the back-end pitch -5, the difference being necessarily $2\,m=8$. Inspection will show that the two windings are,—

1-14- 9-22-17-30-23-38-33-2-41-10-5-18-13-26-21-34-29-42-37-6-1
and
3-16-11-24-19-32-27-40-35-4-43-12-7-20-15-28-23-36-31-44-39-8-3

In the given position, 9-14 and 31-36 are short-circuited at the positive brushes. The circuits through the armature are,—

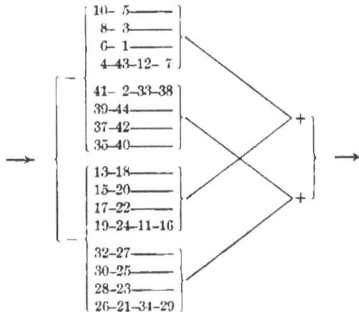

```
              ┌ 10- 5──── ⎤
              │  8- 3──── │
              │  6- 1──── │
              │  4-43-12- 7 ⎦
              ┌ 41- 2-33-38 ⎤
              │ 39-44──── │                +
  ─→   ⎡─     │ 37-42──── │  ⟍        ⟋   ⎤    ─→
       ⎣      ⎦ 35-40──── ⎦    ⟍    ⟋
              ┌ 13-18──── ⎤      ⟋⟍
              │ 15-20──── │    ⟋    ⟍
              │ 17-22──── │  ⟋        ⟍    +
       ⎡─     ⎦ 19-24-11-16 ⎦
       ⎣      ┌ 32-27──── ⎤
              │ 30-25──── │
              │ 28-23──── │
              └ 26-21-34-29 ⎦
```

The extreme irregularity exhibited in the diagrams and tables of the multiple windings is due to the necessarily small numbers of conductors chosen. With the magnitudes taken in practical work, everything will be sufficiently regular.

Figure 39 is the same quadruple winding as Fig. 38, except that the pitches are taken 15 and −7 instead of 13 and −5. This was drawn to emphasize the fact that there is nothing absolute in the choice of the pitch in these multiple circuit armatures, except that in the case of the *multiple windings*, the numerical differences between the forward and backward pitches must be equal to 2 m, where " m " is the number of windings. As before stated, the average pitch should not differ much from $\frac{C}{n}$, and should be somewhat less, rather than greater.

Figure 38, which partakes in a small degree of the nature of the short chord windings (as compared with Fig. 39), has a very much larger percentage of the conductors subjected to counter-induction than would be the case in actual practice with large numbers of conductors.

For instance, the average pitch might often be represented by some such number as 75. If it were to be a quadruple winding, the two pitches should differ by 2 m or 8. Therefore the forward pitch would be taken 79, and the backward pitch −71, so that the order of the winding would be 1–80–9–88, etc., whereas in the case of small numbers of conductors, such as in Fig. 38, the order of the winding was 1–14–9–22–17–30, etc. It will be evident that the distinction between these two cases is, that with the larger number of conductors there are many forward and backward steps before the original loop is crossed, thus : —

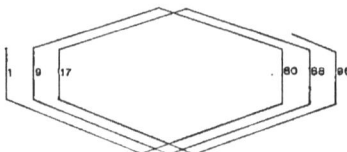

But in the case of the small number of conductors the loop is crossed almost at once, thus : —

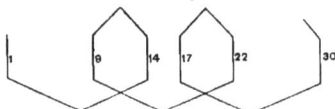

In other words, with multiple windings and small numbers of conductors, the numerical differences between the forward and backward pitches is a large percentage of the average pitch, whereas with the large numbers of conductors used in practice, it is a very small percentage of the average pitch.

The fact that irregularities are much exaggerated by the necessary choice of rather small numbers of conductors should be borne in mind in the study of these diagrams, particularly those of multiple windings.

If, instead of the quadruple windings consisting of two independent doubly re-entrant windings of Figs. 38 and 39, *one* singly re-entrant quadruple winding is desired, a number of conductors must be

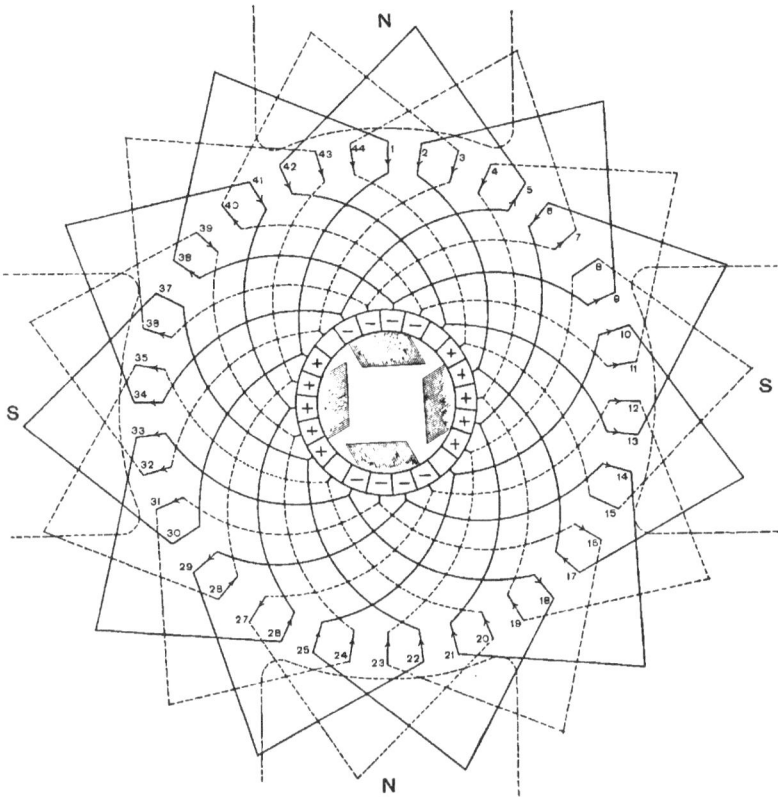

Fig. 39

FOUR CIRCUIT, QUADRUPLE WINDING.

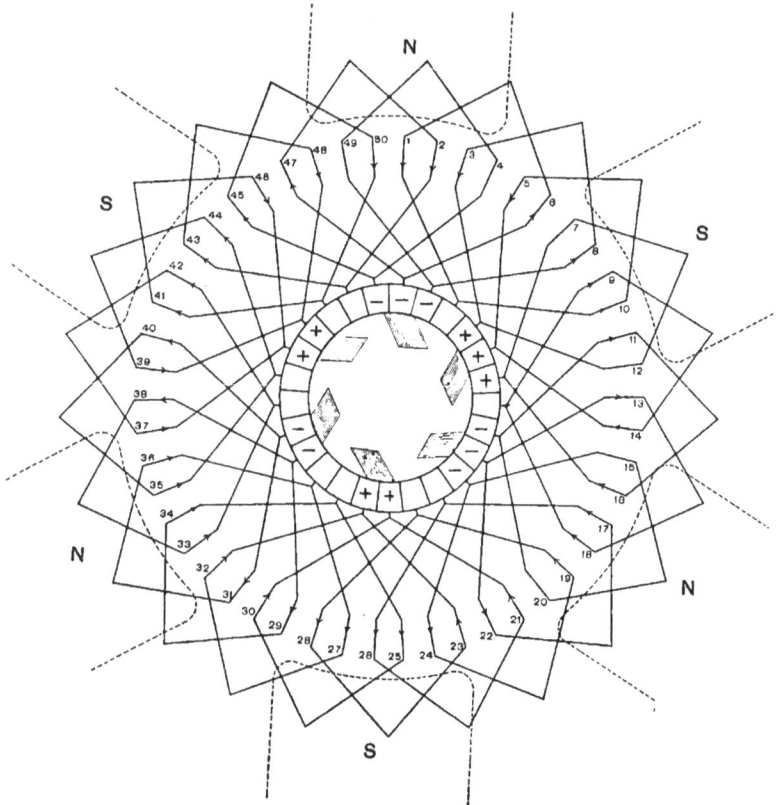

Fig. 40

SIX CIRCUIT, DOUBLE WINDING.

chosen such that $\frac{C}{2}$ and "m" (4) shall be mutually prime. Take $C=42$. Then $\frac{C}{2}=21$, and $m=4$, which are mutually prime. If the forward pitch is taken $y=13$, and the backward pitch $y=-5$, the winding will be, —

1–14–9–22–17–30–25–38–33–4–41–12,7–20–15–28–23–36–31–2–39–10–5–18–13–26
–21–34–29–42–37–8–3–16–11–24–19–32–27–40–35–6–1

This would be represented symbolically as (ooo) and would be a singly re-entrant quadruple winding.

If *four* entirely independent windings are desired, $\frac{C}{2}$ and "m" must have 4 for their greatest common factor. Taking $C=40$, and making the front and back pitches respectively $y=13$ and $y=-5$, the winding would be, —

1–14– 9–22–17–30–25–38–33– 6–1
3–16–11–24–19–32–27–40–35– 8–3
5–18–13–26–21–34–29– 2–37–10–5
7–20–15–28–23–36–31– 4–39–12–7

This could be represented symbolically as ○ ○ ○ ○, and would be a quadruply re-entrant, quadruple winding.

In Fig. 40 is shown a six-circuit, singly re-entrant, double winding. $C=50$, $n=6$, $m=2$. The greatest common factor of $\frac{C}{2}$ and "m" being 1, the winding is singly re-entrant, and may be represented symbolically as ⊙.

The forward pitch is $y=9$, and the backward pitch is $y=-5$.

In the given position, conductors 49–4, 7–12, and 15–20 are short-circuited. The circuits through the armature are, —

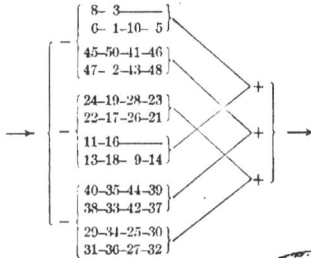

CHAPTER VIII.

TWO-CIRCUIT, SINGLE-WOUND, DRUM ARMATURES.

THE " two-circuit " windings now to be considered are distinguished by the fact that the *pitch is always forward*, instead of alternately forward and backward, as in the "multiple-circuit" windings, just described.

The sequence of connections leads the winding from a certain bar opposite one pole piece to a bar similarly situated opposite the next pole piece, and so on, so that as many bars as pole pieces are passed through before another bar in the original field is reached. Such progression around the armature is continued until all the bars are connected in, and the winding returns on itself.

Two-circuit, drum windings, like the two-circuit, gramme-ring windings, have for a given voltage the fraction $\frac{2}{n}$ as many conductors as multiple-circuit windings, with the attendant advantages, stated for the two-circuit, gramme-ring windings. The advantages, that the circuits from brush to brush consist of conductors influenced by all the poles, are — when there is but one turn in each coil — the same as in the two-circuit, short-connection ring winding. When there are several turns in the coil, the advantages are subject to the same reservations as in the two-circuit, long-connection, ring winding. The advantages, due to such arrangements of the conductors, have been confined to machines of small electrical output. In machines of large electrical output, in which there are a number of sets of brushes of the same sign (otherwise the cost of the commutator is excessive), the advantages possible from equal currents in the circuits have been over-balanced by the increased sparking due to unequal division of the current between the different sets of brushes of the same sign.

An examination of the diagrams will show that in the two-circuit windings the drop in the armature, likewise the armature reaction, is independent of any manner in which the current may be subdivided among the different sets of brushes, but depends only upon the sum of the currents at all the sets of brushes of the same sign. There are, in the two-circuit windings, no features that tend to cause the current to subdivide equally between the different sets of brushes of the same sign, and, in consequence, if there is any difference in contact resistance between the different sets of brushes, or if the brushes are not set with the proper lead with respect to each other, there will be an unequal division of the current.

When there are as many sets of brushes as poles, the density at each pole must be the same, otherwise the position of the different sets of brushes must be shifted with respect to each other to correspond to the different intensities, the same as in the multiple-circuit windings.

In practice it has been found difficult to prevent the shifting of the current from one set of brushes to another. The possible excess of current at any one set of brushes increases with the number of sets; likewise the possibility of excessive sparking. For this reason the statement has been sometimes made that the disadvantages of the two-circuit windings increase with the number of poles.

From the above, it may be concluded that any change of the armature with respect to the poles will in the case of two-circuit windings be accompanied by shifting of the current between the different sets of brushes; therefore to maintain a proper subdivision of the current the armature must be maintained in one position, with respect to the poles, and with exactness, since there is no counter action in the armature to prevent the unequal division of the current.

In the case of multiple-circuit windings, it will be noted that the drop in any circuit, likewise the armature reaction in the field in which the current is generated, tends to prevent the excessive flow of current from the corresponding set of brushes. On account of these features, together with the consideration that when there are as many brushes as poles the two-circuit armatures require the same nicety of adjustment with respect to the poles as the multiple-circuit windings, the multiple-circuit windings are generally preferable, even when the additional cost is taken into consideration.

Denoting the number of face conductors by "C," the number of poles by "n," and the average pitch by "y," the formula controlling the two-circuit, single-wound, multipolar drum, is, —

$$C = ny \pm 2.$$

It is preferable to have the pitch "y" the same at the two ends, because the two sets of end connections will then be of the same length, but the choice of the number of conductors "C" for any particular case is less restricted (when the number of poles is greater than four) if the front and back pitches are permitted to differ by 2. Each pitch, must, moreover, be an odd number, as, in order that the winding may pass through *all* the conductors before returning upon itself, it must pass alternately through odd and even numbered conductors. Also when, as is usually the case, the bars occupy two layers, it is necessary to connect from a conductor of the upper to one of the lower layer so as to obviate interference in the positions of the spiral end connections. Where different pitches are used at the front and back ends, each being odd, the *average* "y" appearing in the formula will be even.

In Fig. 41 is given a two-circuit, single winding for a four-pole drum. The pitch is $y = 9$ at both ends.

$$C = ny \pm 2 = 4 \times 9 \pm 2 = 34 \text{ or } 38.$$

Thirty-four conductors were taken. If it is necessary to have thirty-four conductors, it would be better to take the average "y" equal to eight, and then to use $y = 9$ at one end and $y = 7$ at the other. It is thus possible to shorten the end connections at the end at which the shorter pitch is used, and thus avoid using an unnecessary amount of copper. This will also make the armature resistance less, and will give more room for the end connections.

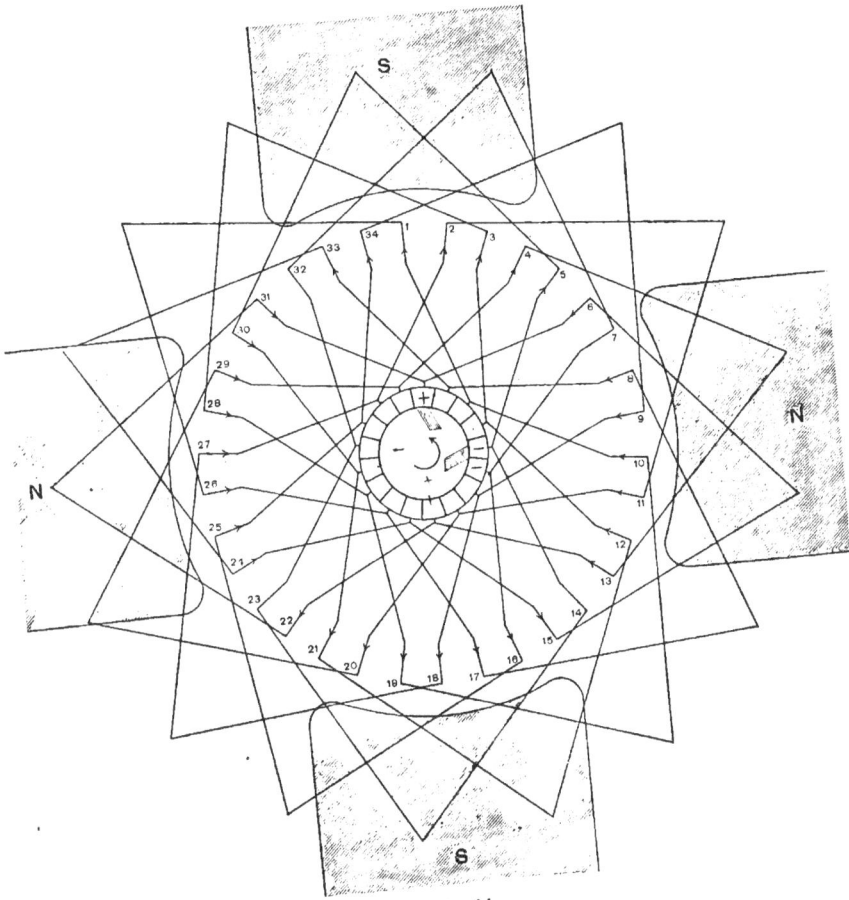

Fig. 41.

TWO CIRCUIT, SINGLE WINDING.

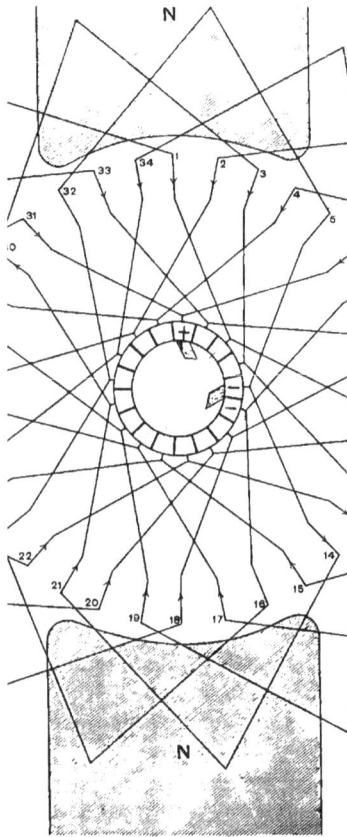

Fig. 42
TWO CIRCUIT, SINGLE WINDING.

In Fig. 42 this has been done, the front-end pitch being $y=9$ as before, but the back-end pitch being $y=7$. The average pitch is $y=8$.

$$C = ny \pm 2 = 4 \times 8 \pm 2 = 30 \text{ or } 34.$$

Thirty-four conductors have been taken.

If thirty-eight conductors should be preferable to thirty-four, then the best arrangement would be to use $y=9$ at both ends.

$$C = ny \pm 2 = 4 \times 9 \pm 2 = 34 \text{ or } 38.$$

This case has not been drawn, but it would be the proper method for thirty-eight conductors, as the only other way would be to have a front-end pitch $y=11$ and a back-end pitch $y=9$, giving an average pitch $y=10$.

$$C = ny \pm 2 = 4 \times 10 \pm 2 = 38 \text{ or } 42.$$

This last choice, i.e. pitches of 9 and 11, would be undesirable, as the connections at the end with a pitch of 11 would be unnecessarily long. Therefore, as a general rule, the pitch should be chosen a little less than $\frac{C}{n}$, and when this would result in an even pitch, the pitch at one end may be made $(y+1)$ and at the other end $(y-1)$. Of course, the advantage of having both sets of end connections exactly equal might offset the small saving in material. This would have to be determined for the case in hand. Often, however, even where the same pitch is used at both ends, other considerations make it necessary to use two differently proportioned sets of connecting strips.

This matter of the possibility of using two different pitches, so that the "y" of the equation $C = ny \pm 2$ may be any integer, odd or even, is not so very important in the case of four-pole armatures, as it does not increase the range of choice of conductors. But for six, eight, and higher numbers of poles the introduction of even integers for "y" gives many more possible numbers of conductors than if it were necessary to be confined to odd integers.

Thus, for the case of six-pole windings, the formula $C = ny \pm 2$ becomes $C = 6y \pm 2$. If "y" is put successively equal to 10, 11, 12, 13, 14, and 15, the possible numbers of bars will become as follows : —

$$y = 10 \quad C = 60 \pm 2 = 58 \text{ or } 62$$
$$y = 11 \quad C = 66 \pm 2 = 64 \text{ or } 68$$
$$y = 12 \quad C = 72 \pm 2 = 70 \text{ or } 74$$
$$y = 13 \quad C = 78 \pm 2 = 76 \text{ or } 80$$
$$y = 14 \quad C = 84 \pm 2 = 82 \text{ or } 86$$
$$y = 15 \quad C = 90 \pm 2 = 88 \text{ or } 92.$$

Thus it may be seen that if it were only permissible to use odd integers for "y," the possible conductors for this range would be limited to 64, 68, 76, 80, 88, and 92; but by using unequal pitches at the two ends, the average "y" becomes even, and the possible numbers of conductors to which the choice is limited is doubled. It is very important that this point should be borne in mind, as the rule often used for four-pole machines that C must equal number of poles times an odd number, plus or minus two, is sometimes mistakenly extended to larger numbers of poles, and a number of conductors is chosen either larger or smaller than is desired; whereas, if different pitches at the two ends had been used, a much more suitable choice might have been made.

Another limiting consideration is, that the numbers of conductors _per slot_ is governed largely by the capacity and voltage of the machine, so that sometimes two, sometimes four, and in exceptional cases even six or eight, bars might be desired per slot, therefore, the total number of conductors "C" must be a multiple of 2, 4, 6, or 8, as the case may be. If, in the case of a six-pole armature, only two conductors per slot are desired, the pitch may be either odd or even; but it will be found that where four conductors per slot are wanted, and where, therefore, "C" must be a multiple of 4, that only the numbers of conductors obtained by making "y" an odd integer meet the requirement. And if six conductors per slot should be wanted (and it seldom would be, because the mechanical fitting of the connections would be so troublesome), neither the use of an odd nor of an even integer would (in the case of a six-pole armature) give a possible number for "C."

In the following illustrative diagrams it will not be necessary to take pains to show how many conductors there are per slot. They will be drawn with the conductors spaced at equal intervals, and one, two, four, or more, as desired, may be supposed to be brought together in a slot.

In Fig. 43 is given a diagram for a six-pole, two-circuit, single-wound, drum armature. The pitch is $y = 11$ at both ends.

$$C = ny \pm 2 = 6 \times 11 \pm 2 = 66 \pm 2 = 64 \text{ or } 68.$$

Sixty-eight conductors were taken, and they could be arranged one, two, or four per slot, as other conditions might determine.

In the position shown, the positive brush short-circuits the group of conductors 5–62–51–40–29–18, all in series. The circuits through the armature are, —

→ –{ 6–17–28–39–50–61– 4–15–26–37–48–59– 2–13–24–35–46–57–68–11–22–33–41–55–66– 9–20–31–42–53–64–7 } + →
      { 63–52–41–30–19– 8–65–54–43–32–21–10–67–56–45–34–23–12– 1–58–17–36–25–11– 3–60–10–38–27 10—— }

An examination of the preceding table will show that immediately sequential conductors, such as 6 and 7, have between them, at recurring periods, the full difference of potential of the winding. But _alternately_ sequential pairs of conductors, as 6 and 4, or 63 and 65, have between them only the difference of potential of "n" bars.

For the above analysis, only the two full-lined brushes were supposed to be in service. If, however, the four brushes shown by the dotted lines were added, the short-circuited bars would consist of groups of two each, in series between _different_ brushes of like sign. In the given position, these groups would be 5–62, 51–40, and 29–18 at the positive brushes, and 63–52, 41–30, and 17–6 at the negative brushes. The circuits through the armature would be the same, with the exception that the bars short-circuited by the negative brushes would now disappear from the list. These six conductors, 6, 17, 63, 52, 41, 30, have been underlined in the table, and are marked on the diagram by small circles.

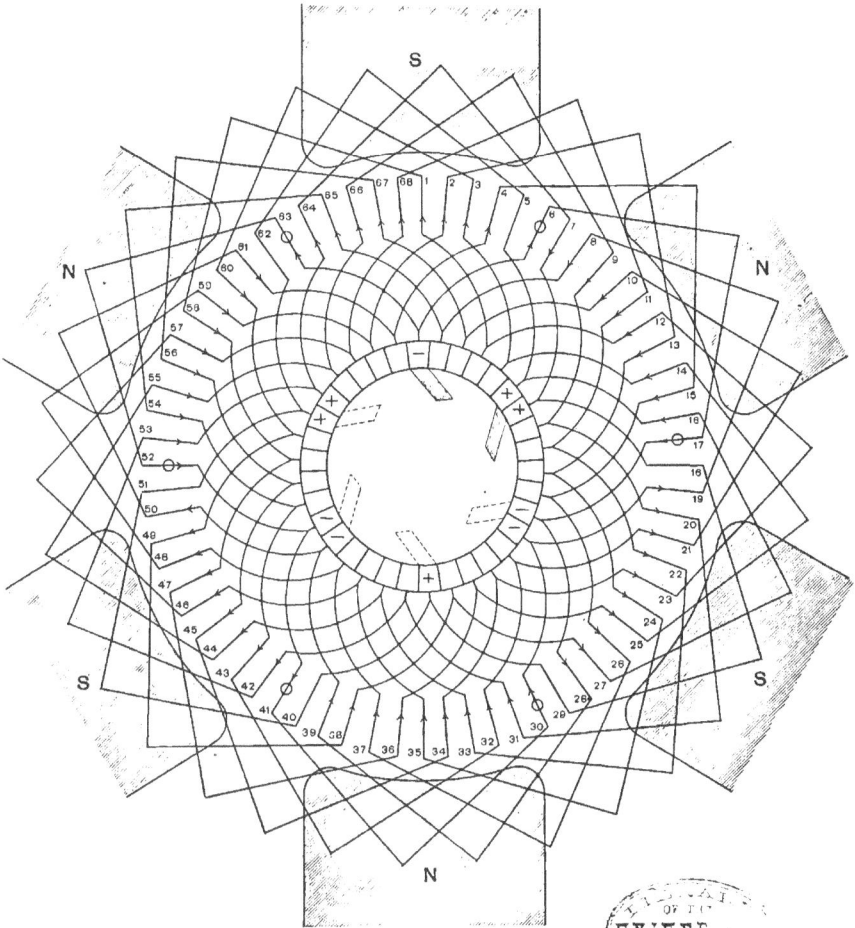

Fig. 43

TWO CIRCUIT, SINGLE WINDING.

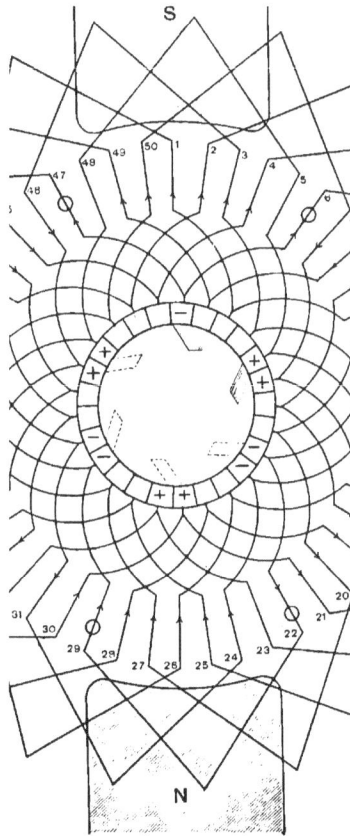

Fig. 44
TWO CIRCUIT, SINGLE WINDING.

In Fig. 44 is given a diagram for a two-circuit, six-pole armature. The back-end pitch is $y=7$, and the front-end pitch is $y=9$. Therefore the average pitch is $y=8$.

$$C=ny\pm2=6\times8\pm2=46 \text{ or } 50.$$

Fifty conductors are taken. As in the preceding diagram, only the six conductors without arrow-heads are short-circuited when the two full-line brushes alone are active. But when all six brushes bear on the commutator, the conductors designated by small circles are also short-circuited.

TWO-CIRCUIT WINDINGS WITH CROSS-CONNECTED COMMUTATORS.

Figures 45, 46, 47, and 48 are illustrative of a class of two-circuit windings that possess the distinctive feature that the number of coils may bear a relation to the number of poles not possible with the other two-circuit windings described. An examination of the diagrams will show that the different coils of a winding may be subdivided in groups, each group having either as many coils as there are pairs of poles, or half as many, these different groups being connected in series by a cross-connected commutator.

Figure 45 is an example of this class. As will be seen, it consists of an eight-pole drum armature, with fifty-six conductors connected up as a two-circuit, single winding.

The underlying principle is best understood by noting one "element" of the winding, such as the eight polar conductors drawn with very heavy lines. It starts from a certain commutator segment, and after proceeding under each of the eight pole pieces, it returns to the adjacent segment. It should be further observed that, unlike the heretofore described two-circuit drum armatures, the conductors of this element are separated from each other by an angular distance equal exactly to $\frac{360}{8} = 45°$, instead of, as in the ordinary two-circuit drum windings, being separated by an angular distance a little greater or less than this.

$$C = 56, \quad n = 8, \quad y \text{ (the "pitch")} = \tfrac{56}{8} = 7.$$

It should be particularly noted that, with this winding, a number of conductors is used which is an exact multiple of the number of poles. This, of course, is not possible with the ordinary two-circuit drum windings, which are controlled by the formula —

$$C = ny \pm 2.$$

As will be seen from the diagram, this winding requires cross-connection of the commutator, but in many machines this disadvantage might be offset by the fact that, owing to the symmetrical arrangement of the conductors with reference to the pole pieces, the objectionable "selective commutation" of the ordinary type would probably be avoided.

To return to a study of the diagram, it will be seen that there are $\dfrac{C}{n} = \dfrac{56}{8} = 7$ sets of "elements" exactly the same as that above described, except that each is located at an angular distance of $\frac{360}{7}$ from the preceding one. To facilitate comprehension of the diagram, these seven "elements" have been drawn in with different styles of lines, and are readily distinguishable.

It is therefore obvious that, if it were not for the commutator cross-connections, the winding would consist of seven sets of eight conductors each, and that each such set has its two terminals at a pair of adjacent segments. These individual coils are put in the proper series relation between brushes by the commutator cross-connection. The resultant design is perfectly symmetrical.

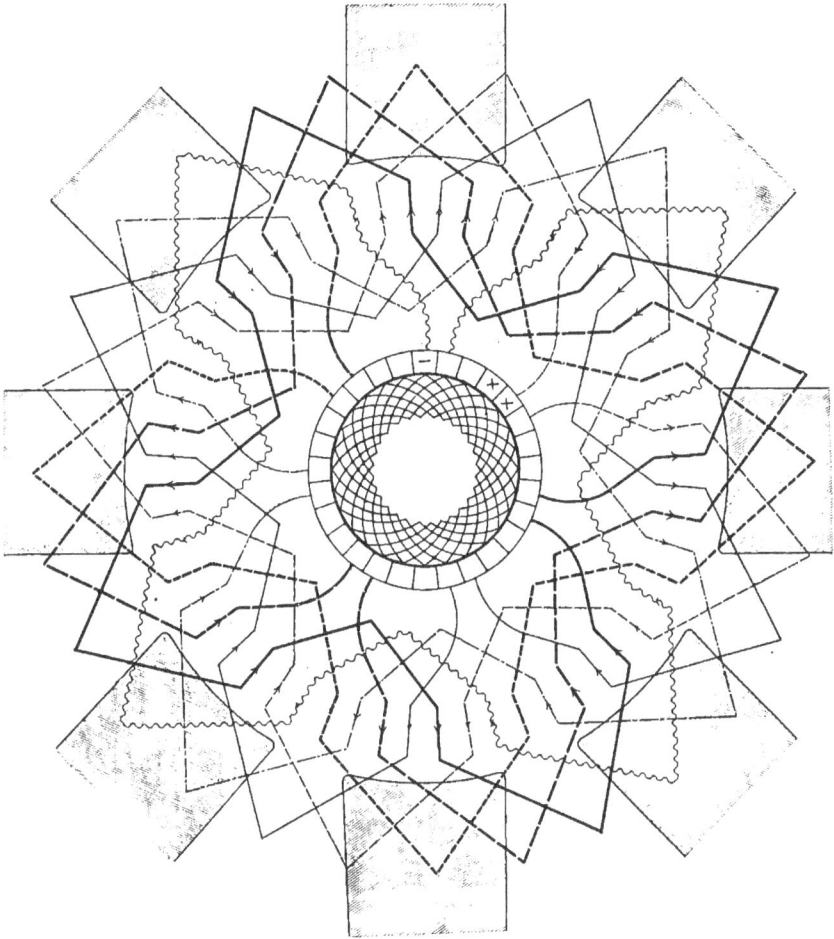

Fig. 45

Figure 46 differs only in having forty-eight conductors, with the necessary consequence that, the pitch being even ($\frac{48}{8} = 6$), it has to be different at the front and back. It is seven at the commutator end, and five at the other end. This slight irregularity makes the wording of the description of Fig. 45 not absolutely applicable to this diagram, the chief difference being that, although every pair of successive conductors are exactly similarly located with respect to a pair of poles as every other pair, the same cannot be said of every individual conductor of an element, the distance between them being successively greater and less than ($\frac{360°}{48}$).

Figure 47 represents a two-circuit single-winding, identical with Fig. 45, except that the connecting leads at the front end are twice as long.

This is used in some "form" windings, where the two ends of a coil are brought out in front at a point half-way between the two slots holding the wires of a coil. The long front connections would never be used in bar windings, where each face conductor of the diagram represents only one conductor, for it would be a waste of copper. Short leads such as those of Fig. 45 would, for such bar windings, always be used.

An "element" of the winding may be readily seen from the heavy lining in the diagram.

Windings of same type as Fig. 47 could be made corresponding to Fig. 46, as well as to Fig. 45. In fact, the underlying principle of this winding is identical with that of the type illustrated by Figs. 45 and 46.

Fig. 47

Figure 48 represents a two-circuit single winding for an eight-pole machine, in which four conductors constitute an element. The number of conductors is here taken to be fifty-two. There are therefore $\frac{52}{4} = 13$ elements. It is a condition of this winding that the number of elements *must* be an odd number. From this it follows that the total number of conductors cannot be a multiple of the number of poles.

It serves, therefore, for numbers of conductors with which the previously described winding (where C is a multiple of n) could not be used. It probably, however, would not be so well balanced as in the case where C is a multiple of n. The commutator requires cross-connecting, as shown in the diagram. The cross-connections at the front end are of twice the usual length.

WENSTRÖM TWO-CIRCUIT, WIRE-WOUND ARMATURE.

Figure 49 represents a winding devised by Wenström to lessen the depth of the end windings of wire wound armatures.

The particular case represented by the diagram had thirty-five lozenge-shaped slots, each containing four conductors. For the sake of clearness only the connections of the wires between two adjacent commutator segments are shown, and no difficulty will be found in completing the winding, by continuing on through the remaining segments.

This method is, of course, only suitable for wire-wound armatures and like most such wire windings, it is difficult to repair.

It is to be noted that these armatures, which have been quite extensively used, were completely ironclad, there being no slot opening.

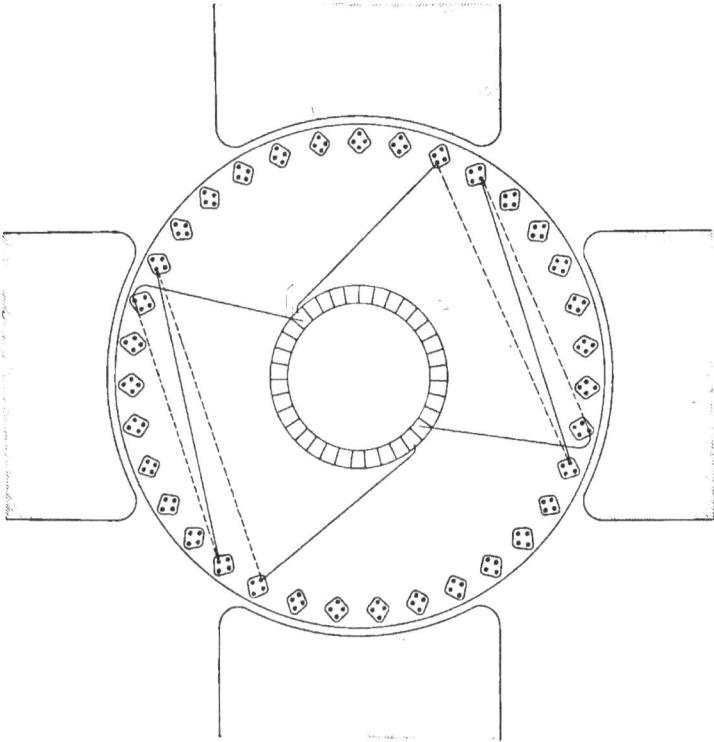

Fig. 49

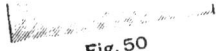

Fig. 50

TWO CIRCUIT, SINGLE WINDING.

CHAPTER IX.

INTERPOLATED COMMUTATOR SEGMENTS.

In Fig. 50 is given a two-circuit single winding. $n=6$, $y=13$, $C=ny\pm2=6\times13\pm2=76$ or 80. Eighty conductors have been taken. This would naturally give forty commutator segments. Suppose speed, strength of field, and active length of conductors to be of such magnitudes as to generate one volt per conductor. Noting that, as shown in the figure, twelve conductors are short-circuited, there will be $\frac{80-12}{2}=34$ active conductors in series between brushes. Therefore the total E.M.F. will be 34 volts. There would be (before interpolating) $\frac{40-6}{6}=5.67$ segments between every two neutral points of the commutator. Therefore there would be $\frac{34}{5.67}=6$ volts between every two adjacent segments.

Suppose this to be higher than is desired. It might then be proposed to double the number of segments by the method of cross-connecting shown in Fig. 50. This will increase the number of segments to eighty. Following the circuit through from the negative to the positive brush, the conductors have been labeled 1 volt, 2 volts, 3 volts, etc., adding one volt for each conductor. Taking the potential of the negative brush as zero, this gives the potential of each conductor. Following down from each conductor to its attached segments, they have been numbered in a corresponding manner; thus the four segments connected to the two bars at 20 volts potential have been marked 20, etc.

An examination of the figure will now make it apparent that proceeding from the neutral points (at zero potential) the voltage increases alternately by two and by four volts per segment, the average being three volts per segment. Therefore, although the average volts per segment have been decreased to one-half of what they were for forty segments, half of the segments have between them only one-third, and the remainder, two-thirds, of the original volts per segment. Therefore, for a six-pole armature, the volts per segment cannot be halved by interpolation. And in order to reduce them to one-third throughout, it is not sufficient to cross-connect as shown in the figure, but it is necessary to triple the natural number of segments and cross-connect every three corresponding segments. This would be far from simple.

107

A fairly large number of conductors was taken in Fig. 50, in order to give a thorough explanation of the principles involved in interpolating segments. The further study of the subject can, however, be more satisfactorily carried on with small numbers of conductors.

In Fig. 51 is shown another two-circuit, single winding, with $n = 6$, $y = 7$, $C = ny \pm 2 = 6 \times 7 \pm 2 = 40$ or 44. Forty-four conductors are taken. Without interpolation, twenty-two segments would be used. Here $3 \times 22 = 66$ segments are used. This is arrived at by connecting together every three corresponding commutator segments.

If, as in the preceding figure, only two segments had been cross-connected, the connections shown by the full lines would have sufficed. Cross-connecting every three corresponding segments involved the addition of the dotted line connections. This, as the diagram shows, doubles the total number of commutator cross-connections, and is therefore mechanically objectionable.

But the volts between bars are now everywhere equal instead of being alternately V and $2V$ as in Fig. 50. This may be seen by an examination of the numbers on the conductors and segments, which have been arranged according to the conventional method described.

Thus, proceeding from the segments under the negative brush, the voltage would increase regularly by two volts per segment up to the positive brush, so that whereas, in the former cases, the order was 2, 4, 8, 10, 14, 16, etc., it is now 2, 4, 6, 8, 10, 12, 14, 16, etc.

Fig. 51

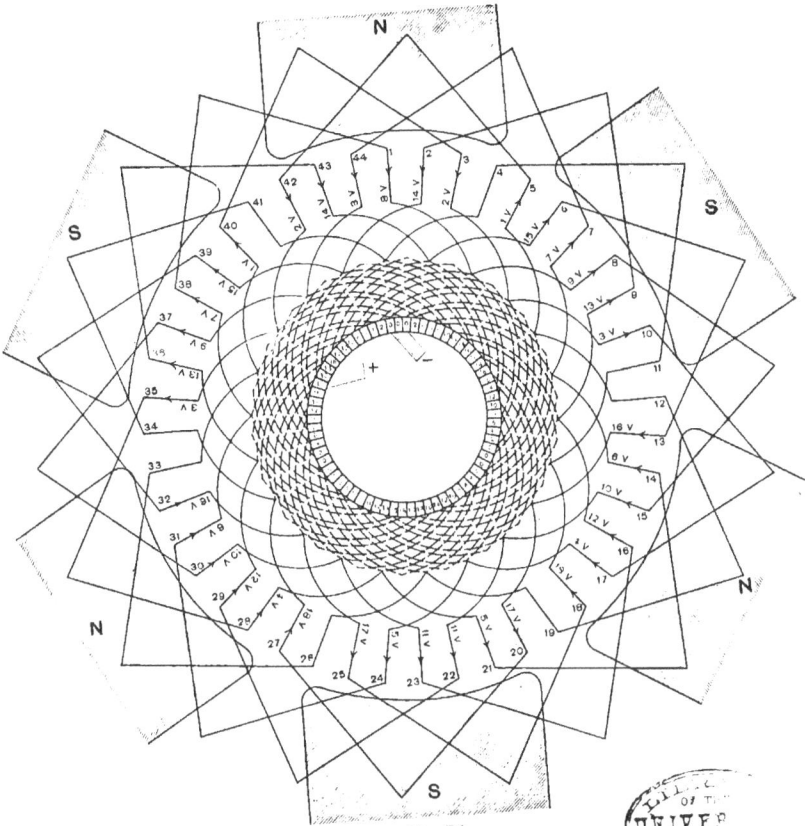

TWO CIRCUIT, SINGLE WINDING.

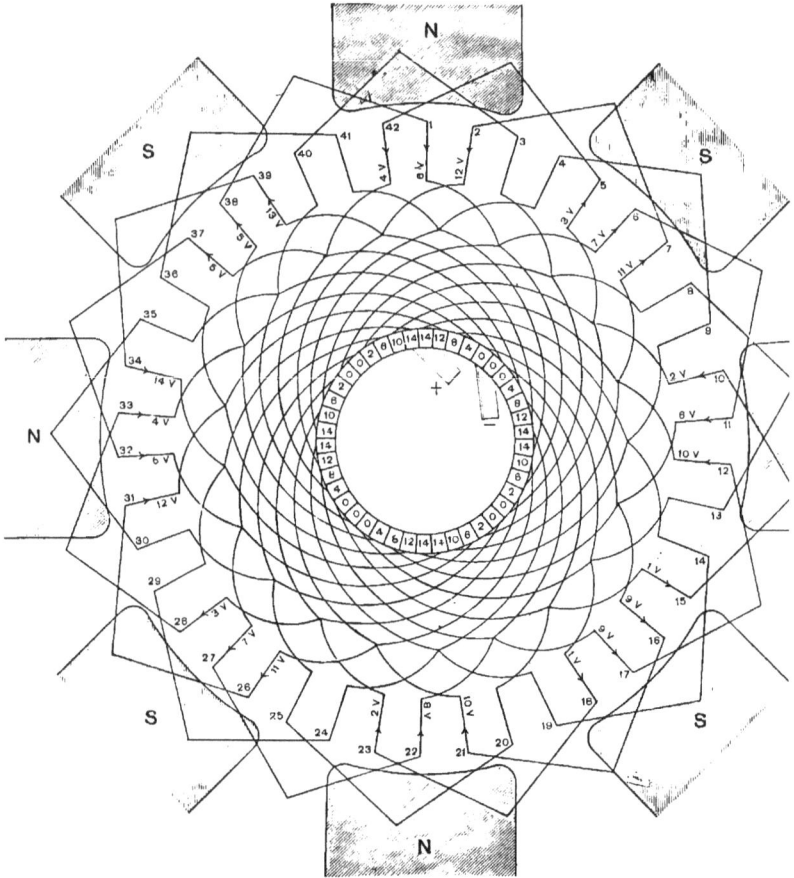

Fig. 52
TWO CIRCUIT, SINGLE WINDING.

In Fig. 52 is given the diagram of a two-circuit, single-wound, eight-pole armature with forty-two conductors. $C = ny \pm 2$; $8 \times 5 + 2 = 42$. It is given to show that, with even numbers of pairs of poles, the number of commutator bars may be doubled by interpolation, and that the result will be to halve the volts between every two segments instead of producing the unsymmetrical result observed in the case of an odd number of pairs of poles.

An examination of Fig. 52 will show that commutator segments 180° apart are cross-connected. The scheme of studying the relative potential of conductors and commutator segments is the same as that used in the case of the two preceding figures, and can be followed through without trouble. Some confusion may result from the fact that owing to the small number of conductors taken, the length of the two circuits through the armature are quite unequal, one path consisting of twelve conductors, and the other of fourteen. As the positive neutral points where these two paths meet must be at the same potential, all the segments at these positions have been indicated as being at a potential of fourteen volts, so that the sequence of figures giving the potentials of the segments is, in four of the eight cases, 0, 4, 8, 12, 14; increasing regularly by four volts until the very end, where the increase is but two volts.

In the other four cases, for the same reason, the sequence is 0, 2, 6, 10, 14, showing the irregularity at the negative neutral points. With the large number of conductors used in practice no misunderstanding would result.

With an even number of pairs of poles it is not necessary to be confined to using only twice the natural number of commutator segments. Thus in Fig. 53 is given the same eight-pole winding as in Fig. 52, with the exception that eighty-four segments are used instead of forty-two. The natural number of segments would be twenty-one.

As the conventions used in the previous descriptions are followed in mapping out the relative potentials of the various parts, no further explanations will be necessary.

Fig. 53

T,WO CIRCUIT, SINGLE WINDING.

CHAPTER X.

TWO-CIRCUIT, MULTIPLE-WOUND, DRUM ARMATURES.

THE next class is that of the two-circuit, *multiple-wound*, drum armature.
The general formula is :—

$$C = ny \pm 2\,m,$$

where
$C =$ number of face conductors,
$n =$ number of poles,
$y =$ average pitch,
$m =$ number of windings.

The " m " windings will consist of a number of *independently* re-entrant windings, equal to the greatest common factor of " y " and " m." Therefore, where it is desired that the " m " windings shall combine to form *one re-entrant* system, it will be necessary that the greatest common factor of " y " and " m " be made equal to 1.

Also, when " y " is an *even* integer, the pitch must be taken alternately as $(y-1)$ and $(y+1)$.

In Fig. 54 is reproduced a winding described by E. Arnold (" Die Ankerwicklungen der Gleichstrom-Dynamomaschinen," p. 70, Fig. 80), and by Dr. Kittler (" Handbuch der Elektrotechnik," 2d ed., p. 535, Fig. 403, b). It is classified by them as a four-circuit, single winding. They show four narrow brushes, and point out that the winding has the peculiarities that, in connecting up, the pitch is always taken forward, and that the short-circuiting of a coil occurs between opposite brushes of like polarity, instead of entirely at one brush, as is usually the case. They give no further instances of the application of this winding, except that Herr Arnold proposes for it the formula :—

$$C = n(y \pm 1),$$

and adds that if $\frac{C}{2}$ and " y " have a common factor, a singly re-entrant winding is not obtained, several independently re-entrant windings being the result. He follows this statement with a diagram having $C = 28$, $n = 4$, and $y = 6$.

$$[28 = 4(6+1)],$$

which gives two independently re entrant windings, and shows, as before, four points of commutation.

Returning to a consideration of Fig. 54, it may be seen that at the given position, conductors 5–12 and 21–28 are short circuited at the negative brushes, and 13–20 and 29–4 at the positive.

The circuits through the armature are, —

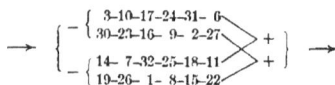

$$\longrightarrow \quad \left\{ \begin{array}{l} - \left\{ \begin{array}{l} 3\text{-}10\text{-}17\text{-}24\text{-}31\text{-} \ 6 \\ 30\text{-}23\text{-}16\text{-} \ 9\text{-} \ 2\text{-}27 \end{array} \right\rangle + \\ - \left\{ \begin{array}{l} 14\text{-} \ 7\text{-}32\text{-}25\text{-}18\text{-}11 \\ 19\text{-}26\text{-} \ 1\text{-} \ 8\text{-}15\text{-}22 \end{array} \right\rangle + \end{array} \right\} \quad \longrightarrow$$

114

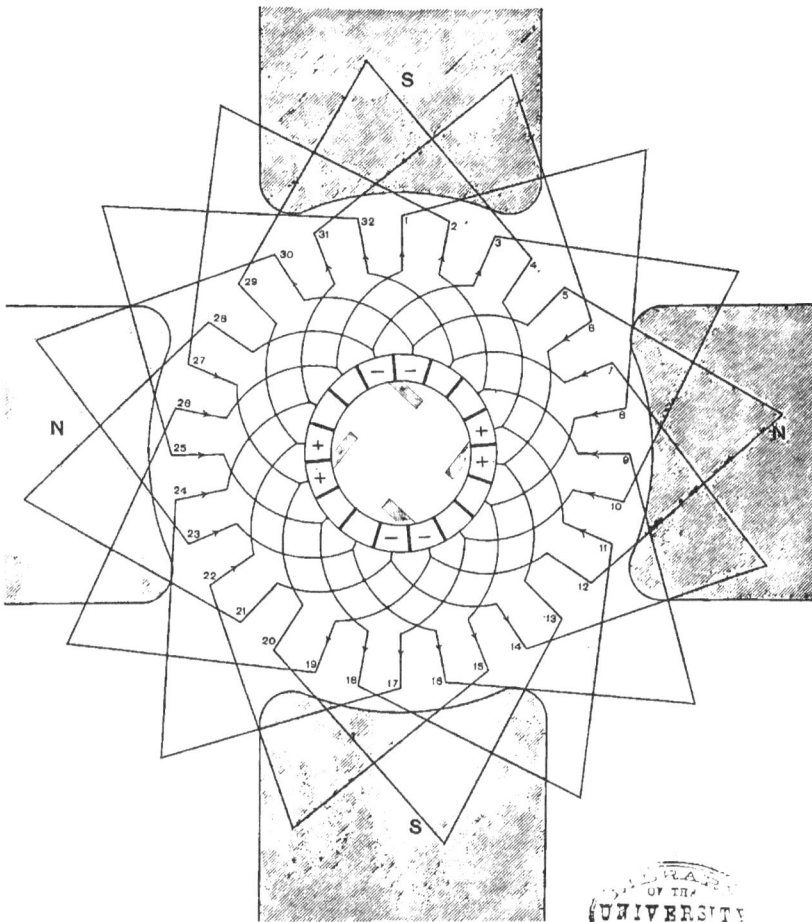

Fig. 54

TWO CIRCUIT, DOUBLE WINDING.

Fig. 55

TWO CIRCUIT, DOUBLE WINDING.

Now in Fig. 55 will be found the very same winding as in Fig. 54, with the exception that two wide brushes are shown instead of four narrow ones. Short-circuiting of a coil now necessarily occurs at one brush, and a study of the winding shows that it is one of the singly re-entrant multiple-wound type, this particular one being a two-circuit, singly re-entrant, double winding.

At the position shown, conductors 7–14–21–28 are short-circuited at the negative brush, and 15–22–29–4 at the positive. The circuits through the armature are : —

$$\longrightarrow \quad - \left[\begin{array}{l} 3\text{--}10\text{--}17\text{--}24\text{--}31\text{--}\ 6\text{------} \\ 30\text{--}23\text{--}16\text{--}\ 9\text{--}\ 2\text{--}27\text{--}20\text{--}13 \\ 32\text{--}25\text{--}18\text{--}11\text{------} \\ 5\text{--}12\text{--}19\text{--}26\text{--}\ 1\text{--}\ 8\text{------} \end{array} \right] \ + \quad \longrightarrow$$

It will be seen that, owing to the very small number of conductors, the winding is extremely irregular, but it will not be difficult to perceive that the nature of the course taken by the current through the armature remains essentially unaltered from that of Fig. 54, consisting, as there, of four paths with an average of six conductors in series per path. The current, however, enters the armature from one wide brush, which always spans more than one segment, and departs from a similar wide brush $\left(\dfrac{360}{n}\right)^\circ$ removed. But in the former case (Fig. 54), it entered two of the paths by one narrow negative brush, and the other two by another, situated $\left[\dfrac{360}{\frac{n}{2}}\right]^\circ$ distant.

It appears, therefore, conclusive that Fig. 54 is in all essential respects identical with a two-circuit, singly re-entrant, double winding, but this was probably not perceived by the above-mentioned authors: otherwise they would undoubtedly have extended the principle to higher orders of multiples and other numbers of poles. An *eight*-pole, two-circuit, singly re-entrant, triple winding (which would, of course, follow *six* paths through the conductors of the armature) would probably not have been considered possible, their conception of the winding apparently being that it was a multiple winding with as many paths through the conductors of the armature as the machine had poles. The formula and rules enunciated in this investigation follow naturally from the true conception of this winding, whereas the formula and condition stated by Herr Arnold may be seen, by a few attempts to apply it, to be entirely inadequate for the purpose of obtaining the necessary data for constructing such windings.

The two preceding figures (54 and 55) were given for the purpose of showing in how far the two-circuit, multiple windings have been understood in the past. The numbers of conductors were, however, entirely inadequate to fully illustrate the nature of the windings.

As this class promises to have a somewhat wide application, it is proposed to give a good many examples, selecting for the purpose various values of "C," "n," "y" and "m," and briefly analyzing each case on the basis of the rules given on page 114.

The symbolical representations heretofore used will be continued, thus:—

      ○   will represent a singly re-entrant single winding,

      ◎   will represent a singly re-entrant double winding,

      ○○   will represent a doubly re-entrant double winding,

      ⊙⊙   will represent a singly re-entrant triple winding,

      ○○○   will represent a triply re-entrant triple winding,

      ⊙⊙⊙   will represent a singly re-entrant quadruple winding,

      ◎◎   will represent a doubly re-entrant quadruple winding,

      ○○○○   will represent a quadruply re-entrant quadruple winding.

According to the above nomenclature, Fig. 40 would be a six-circuit, singly re-entrant, double winding [◎]; Fig. 37 would be a six-circuit, singly re-entrant, triple winding [⊙⊙]; and Fig. 38 a four-circuit, doubly re-entrant, quadruple winding [◎◎]. The use of the middle expression, "singly, doubly, etc., re-entrant," is unavoidable for absolute definiteness, but it will in most cases be sufficiently definite to speak, for example, of a "six-circuit, triple winding " and a "two-circuit, quadruple winding," where absolute exactness would require them to be spoken of respectively as a "six-circuit, singly re-entrant, triple winding " and a "two-circuit, doubly re-entrant, quadruple winding."

Figure 56 is a four-pole, two-circuit, singly re-entrant, triple winding. It is represented symbolically thus: ⊙⊙. $n=4$, and $m=3$. In order that it should be singly re-entrant, it was necessary for the greatest common factor of "m" and "y" to be 1. Therefore "y" was taken equal to 16.

$$C = ny \pm 2\,m = 4 \times 16 \pm 2 \times 3 = 58 \text{ or } 70.$$

Seventy conductors have been taken, and "y" is alternately 15 and 17, it being, of course, impossible to use 16.

In the position shown, the conductors without arrowheads are short-circuited, and the circuits through the armature are:—

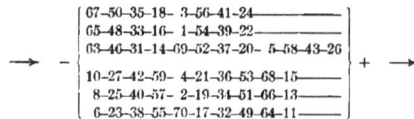

$$\longrightarrow - \begin{cases} 67\text{--}50\text{--}35\text{--}18\text{- }3\text{--}56\text{--}41\text{--}24\text{------} \\ 65\text{--}48\text{--}33\text{--}16\text{- }1\text{--}54\text{--}39\text{--}22\text{------} \\ 63\text{--}46\text{--}31\text{--}14\text{--}69\text{--}52\text{--}37\text{--}20\text{- }5\text{--}58\text{--}43\text{--}26 \\ 10\text{--}27\text{--}42\text{--}59\text{- }4\text{--}21\text{--}36\text{--}53\text{--}68\text{--}15\text{------} \\ 8\text{--}25\text{--}40\text{--}57\text{- }2\text{--}19\text{--}34\text{--}51\text{--}66\text{--}13\text{------} \\ 6\text{--}23\text{--}38\text{--}55\text{--}70\text{--}17\text{--}32\text{--}49\text{--}64\text{--}11\text{------} \end{cases} + \longrightarrow$$

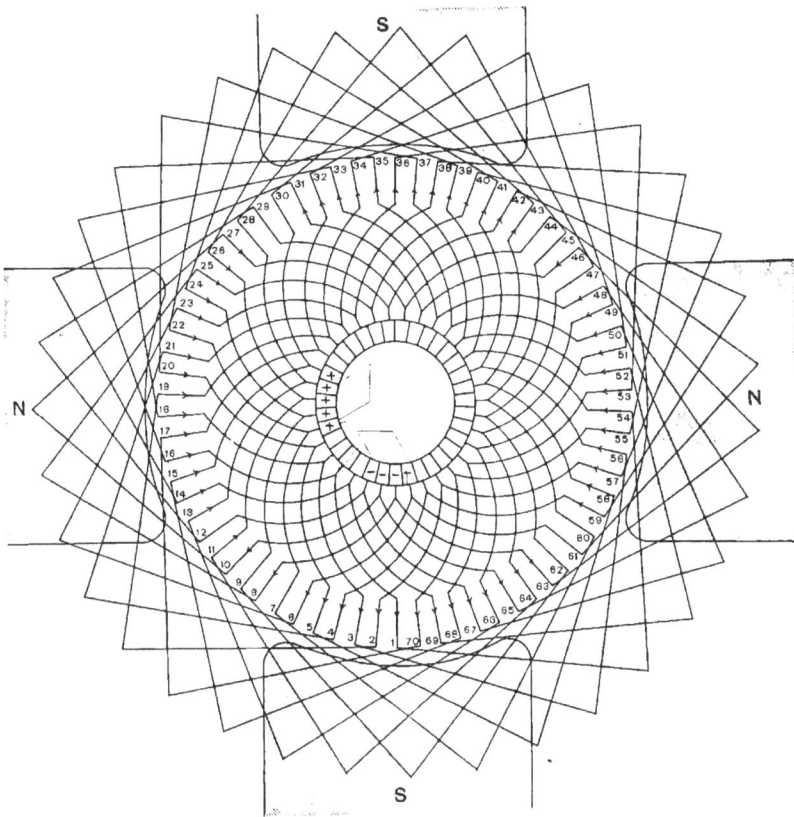

Fig. 56

TWO CIRCUIT, TRIPLE WINDING.

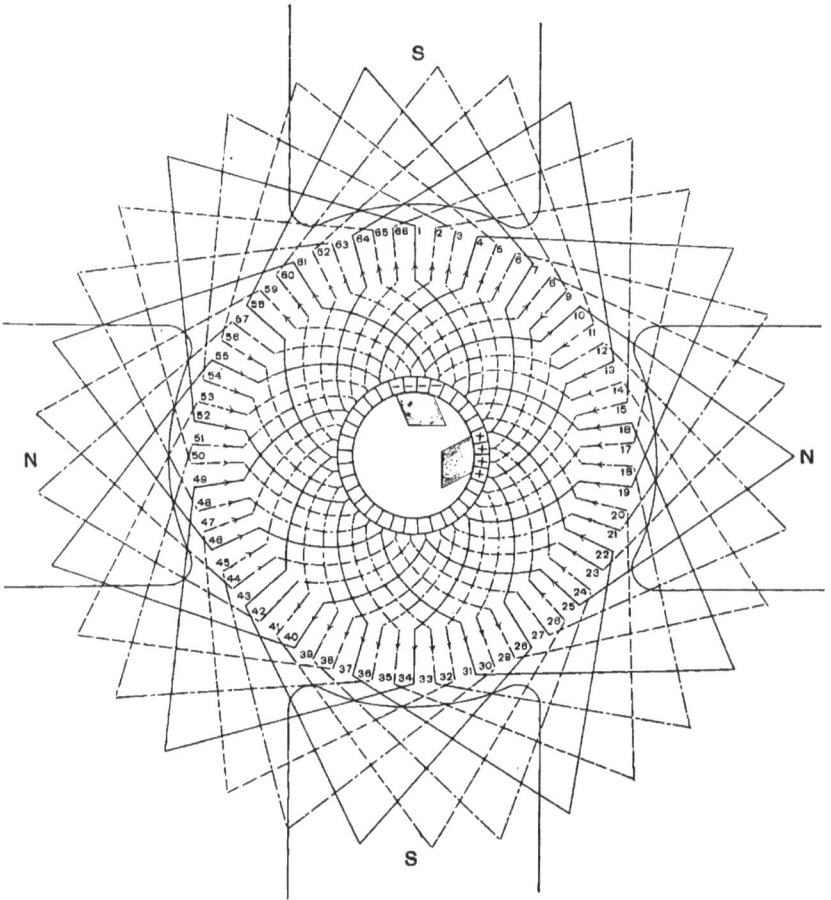

Fig. 57
TWO CIRCUIT, TRIPLE WINDING.

Figure 57 is a four-pole, two-circuit, triply re-entrant, triple winding. It would be represented symbolically as ○○○. $n = 4$, and $m = 3$. In order that it should be triply re-entrant, it was necessary for the greatest common factor of "m" and "y" to be 3. Therefore "y" was taken equal to 15.

$$C = ny \pm 2\,m = 4 \times 15 \pm 2 \times 3 = 54 \text{ or } 66.$$

Sixty-six conductors have been taken. The three independently re-entrant windings have been shown by three different styles of lines.

In the position shown, the conductors without arrow-heads are short-circuited, and the circuits through the armature are : —

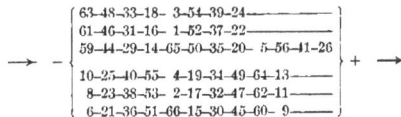

$$\longrightarrow \quad - \left\{ \begin{array}{l} 63\text{--}48\text{--}33\text{--}18\text{--}\ 3\text{--}54\text{--}39\text{--}24\text{------} \\ 61\text{--}46\text{--}31\text{--}16\text{--}\ 1\text{--}52\text{--}37\text{--}22\text{------} \\ 59\text{--}44\text{--}29\text{--}14\text{--}65\text{--}50\text{--}35\text{--}20\text{--}\ 5\text{--}56\text{--}41\text{--}26 \\ 10\text{--}25\text{--}40\text{--}55\text{--}\ 4\text{--}19\text{--}34\text{--}49\text{--}64\text{--}13\text{------} \\ 8\text{--}23\text{--}38\text{--}53\text{--}\ 2\text{--}17\text{--}32\text{--}47\text{--}62\text{--}11\text{------} \\ 6\text{--}21\text{--}36\text{--}51\text{--}66\text{--}15\text{--}30\text{--}45\text{--}60\text{--}\ 9\text{------} \end{array} \right\} + \quad \longrightarrow$$

It is interesting to compare this winding and table with the preceding, and to notice how very slightly they differ.

Figure 58 is a six-pole, two-circuit, singly re-entrant, double winding. It would be represented symbolically as ⓖ. $n=6$, and $m=2$.

In order that it should be singly re-entrant, it was necessary for the greatest common factor of "m" and "y" to be 1. Therefore "y" was taken equal to 9.

$$C = ny \pm 2\,m = 6 \times 9 \pm 2 \times 2 = 50 \text{ or } 58.$$

Fifty-eight conductors have been taken.

In the position shown, the circuits through the armature are : —

$$\longrightarrow \quad - \left\{ \begin{array}{l} 57\text{--}48\text{--}39\text{--}30\text{--}21\text{--}12\text{--}\ 3\text{--}52\text{--}43\text{--}34\text{--}25\text{--}16\text{------} \\ 55\text{--}46\text{--}37\text{--}28\text{--}19\text{--}10\text{--}\ 1\text{--}50\text{--}41\text{--}32\text{--}23\text{--}14\text{------} \\ 6\text{--}15\text{--}24\text{--}33\text{--}42\text{--}51\text{--}\ 2\text{--}11\text{--}20\text{--}29\text{--}38\text{--}47\text{--}56\text{--}7 \\ 4\text{--}13\text{--}22\text{--}31\text{--}40\text{--}49\text{--}58\text{--}\ 9\text{----------} \end{array} \right\} + \quad \longrightarrow$$

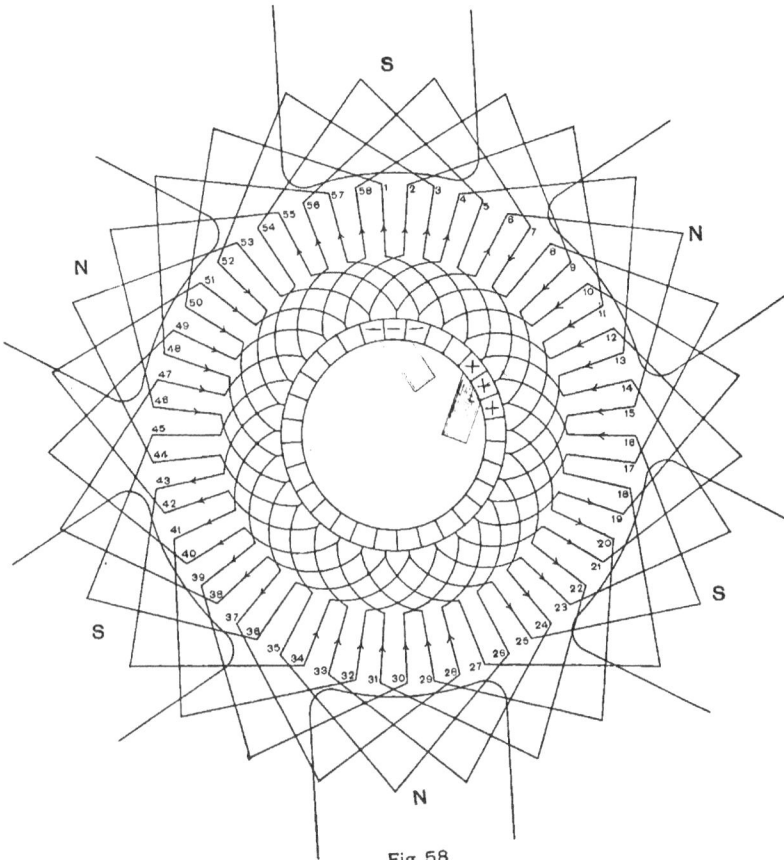

Fig. 58

TWO CIRCUIT, DOUBLE WINDING.

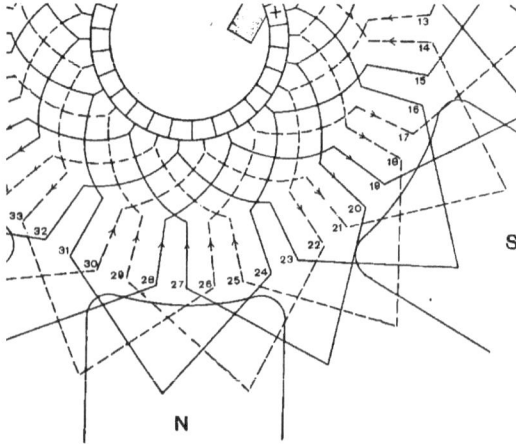

Fig. 59

TWO CIRCUIT, DOUBLE WINDING.

Figure 59 is a six-pole, two-circuit, doubly re-entrant, double winding, the symbolical representation being ◯ ◯. $n = 6$, and $m = 2$. In order that it should be doubly re-entrant, it was necessary for the greatest common factor of "m" and "y" to be 2. Therefore "y" was taken equal to 8.

$$C = ny \pm 2\,m = 6 \times 8 \pm 2 \times 2 = 44 \text{ or } 52.$$

Fifty-two conductors have been taken, and "y" is alternately 7 and 9, it being, of course, impossible to use $y = 8$.

In the position shown, the conductors without arrow-heads are short-circuited, and the circuits through the armature are:—

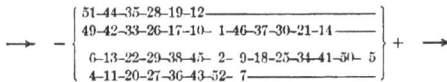

$$\longrightarrow \quad - \left\{ \begin{array}{l} 51\text{--}44\text{--}35\text{--}28\text{--}19\text{--}12 \text{————————} \\ 49\text{--}42\text{--}33\text{--}26\text{--}17\text{--}10\text{--} 1\text{--}46\text{--}37\text{--}30\text{--}21\text{--}14 \text{————} \\ 6\text{--}13\text{--}22\text{--}29\text{--}38\text{--}45\text{--} 2\text{--} 9\text{--}18\text{--}25\text{--}34\text{--}41\text{--}50\text{--} 5 \text{————} \\ 4\text{--}11\text{--}20\text{--}27\text{--}36\text{--}43\text{--}52\text{--} 7 \text{——————} \end{array} \right\} + \quad \longrightarrow$$

As frequently remarked in connection with other diagrams having small numbers of conductors, the very unequal lengths of the different paths through the armature is entirely caused by this choice of a small number of conductors, and would, to a large extent, disappear with all practicable numbers of conductors.

The two independently re-entrant windings are drawn respectively with full and with dotted lines.

Figure 60 is a six-pole, two-circuit, triply re-entrant, triple
winding.   It would be represented symbolically as $\bigcirc \bigcirc \bigcirc$.
$n = 6$, and $m = 3$.   In order that it should be triply re-entrant,
it was necessary for the greatest common factor of "m" and
"y" to be 3.   Therefore "y" was taken equal to 9.

$$C = ny \pm 2m = 6 \times 9 \pm 2 \times 3 = 48 \text{ or } 60.$$

Sixty conductors have been taken.

The three independently re-entrant windings have been
represented by three different styles of lines.

In the position shown, the circuits through the armature
are : —

$$\longrightarrow \quad - \left\{ \begin{array}{l} 59\text{--}50\text{--}41\text{--}32\text{--}23\text{--}14\text{---} \\ 57\text{--}48\text{--}39\text{--}30\text{--}21\text{--}12\text{---} \\ 55\text{--}46\text{--}37\text{--}28\text{--}19\text{--}10\text{--} \ 1\text{--}52\text{--}43\text{--}34\text{--}25\text{--}16 \\ \\ 6\text{--}15\text{--}24\text{--}33\text{--}42\text{--}51\text{--}60\text{--} \ 9\text{---} \\ 4\text{--}13\text{--}22\text{--}31\text{--}40\text{--}49\text{--}58\text{--} \ 7\text{---} \\ 2\text{--}11\text{--}20\text{--}29\text{--}38\text{--}47\text{--}56\text{--} \ 5\text{---} \end{array} \right\} \ + \ \longrightarrow$$

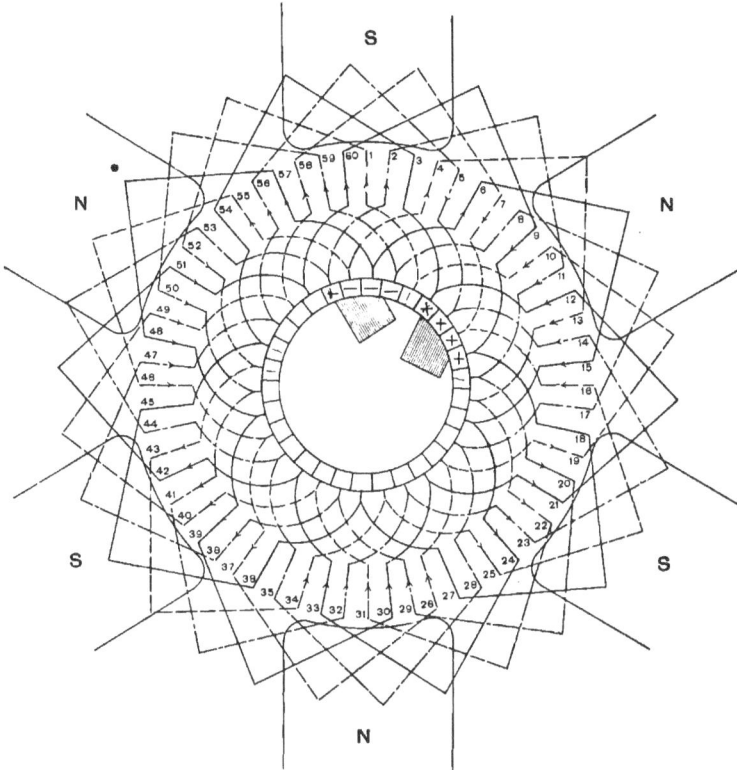

Fig. 60

TWO CIRCUIT, TRIPLE WINDING.

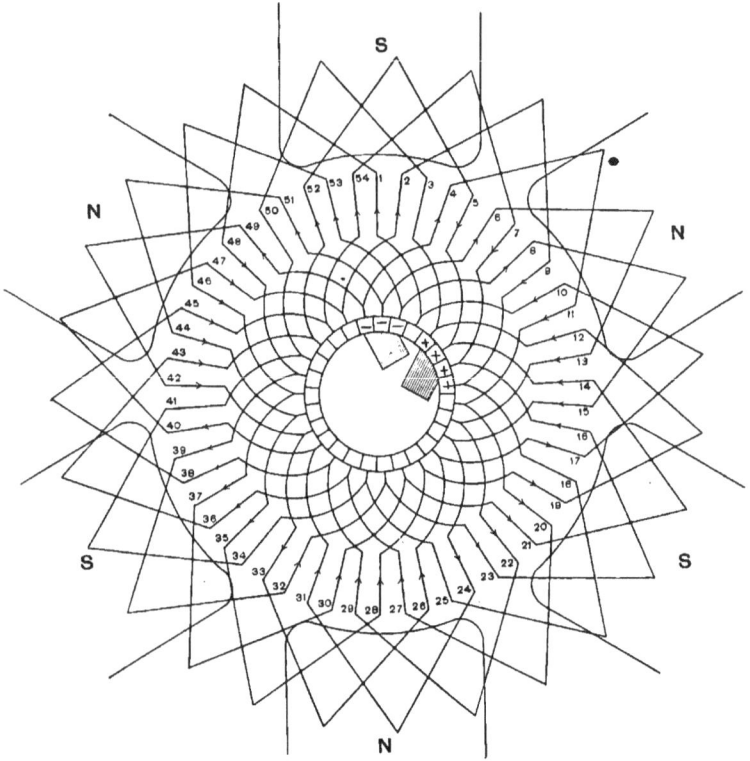

Fig. 61

TWO CIRCUIT, TRIPLE WINDING.

Figure 61 is a six-pole, two-circuit, singly re-entrant, triple winding. It may be symbolically expressed as ⊚. $n = 6$, and $m = 3$. In order that it should be singly re-entrant, it was necessary for the greatest common factor of "m" and "y" to be 1. Therefore "y" was taken equal to 8.

$$C = ny \pm 2m = 6 \times 8 \pm 2 \times 3 = 42 \text{ or } 54.$$

Fifty-four conductors have been taken, "y" is alternately 7 and 9, as it would, of course, be impossible to let $y = 8$.

In the position shown, the circuits through the armature are : —

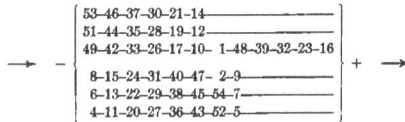

$$\rightarrow \quad - \left\{ \begin{array}{l} 53\text{--}46\text{--}37\text{--}30\text{--}21\text{--}14\text{---------} \\ 51\text{--}44\text{--}35\text{--}28\text{--}19\text{--}12\text{---------} \\ 49\text{--}42\text{--}33\text{--}26\text{--}17\text{--}10\text{--}\ 1\text{--}48\text{--}39\text{--}32\text{--}23\text{--}16 \\ \ 8\text{--}15\text{--}24\text{--}31\text{--}40\text{--}47\text{--}\ 2\text{--}9\text{---------} \\ \ 6\text{--}13\text{--}22\text{--}29\text{--}38\text{--}45\text{--}54\text{--}7\text{---------} \\ \ 4\text{--}11\text{--}20\text{--}27\text{--}36\text{--}43\text{--}52\text{--}5\text{---------} \end{array} \right\} \quad + \quad \rightarrow$$

Figure 62 is a six-pole, two circuit, triply re-entrant, triple winding. It would be represented symbolically as ◯◯◯. $n = 6$, $m = 3$. In order that it should be triply re-entrant, it was necessary for the greatest common factor of "y" and "m" to be 3. Therefore "y" was taken equal to 12.

$$C = ny \pm 2\,m = 6 \times 12 \pm 2 \times 3 = 66 \text{ or } 78.$$

Seventy-eight conductors have been taken, and "y" is alternately 11 and 13, as it would not be possible to let "y" = 12.

The three independently re-entrant windings have been represented by three different styles of lines.

In the position shown, the short-circuited conductors are those without arrow-heads. The circuits through the armature are : —

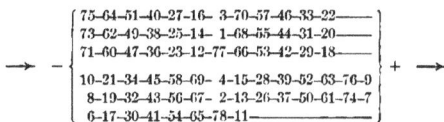

$$\longrightarrow \quad - \left\{ \begin{array}{l} 75\text{–}64\text{–}51\text{–}40\text{–}27\text{–}16\text{–}\ 3\text{–}70\text{–}57\text{–}46\text{–}33\text{–}22\text{———} \\ 73\text{–}62\text{–}49\text{–}38\text{–}25\text{–}14\text{–}\ 1\text{–}68\text{–}55\text{–}44\text{–}31\text{–}20\text{———} \\ 71\text{–}60\text{–}47\text{–}36\text{–}23\text{–}12\text{–}77\text{–}66\text{–}53\text{–}42\text{–}29\text{–}18\text{———} \\ 10\text{–}21\text{–}34\text{–}45\text{–}58\text{–}69\text{–}\ 4\text{–}15\text{–}28\text{–}39\text{–}52\text{–}63\text{–}76\text{–}9 \\ 8\text{–}19\text{–}32\text{–}43\text{–}56\text{–}67\text{–}\ 2\text{–}13\text{–}26\text{–}37\text{–}50\text{–}61\text{–}74\text{–}7 \\ 6\text{–}17\text{–}30\text{–}41\text{–}54\text{–}65\text{–}78\text{–}11\text{———————} \end{array} \right\} + \longrightarrow$$

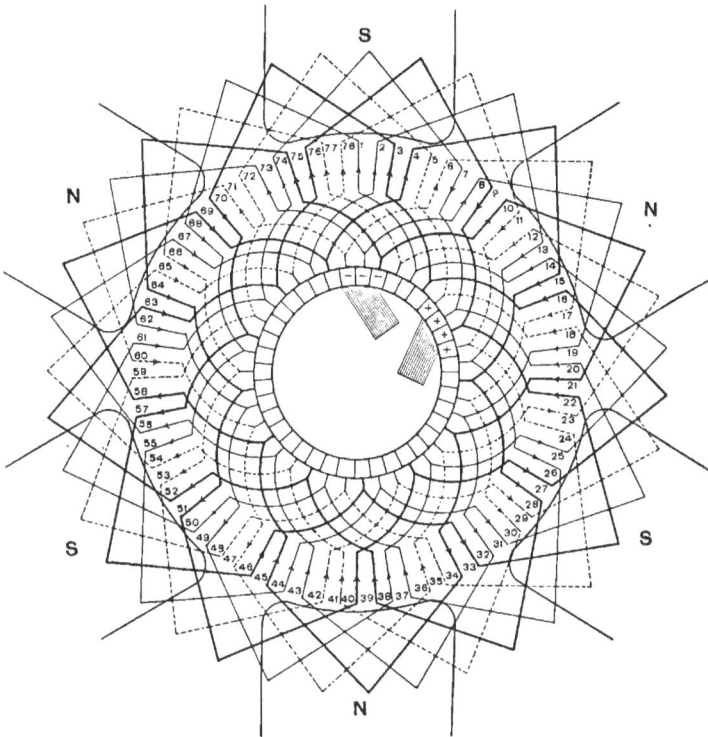

Fig. 62
TWO CIRCUIT, TRIPLE WINDING.

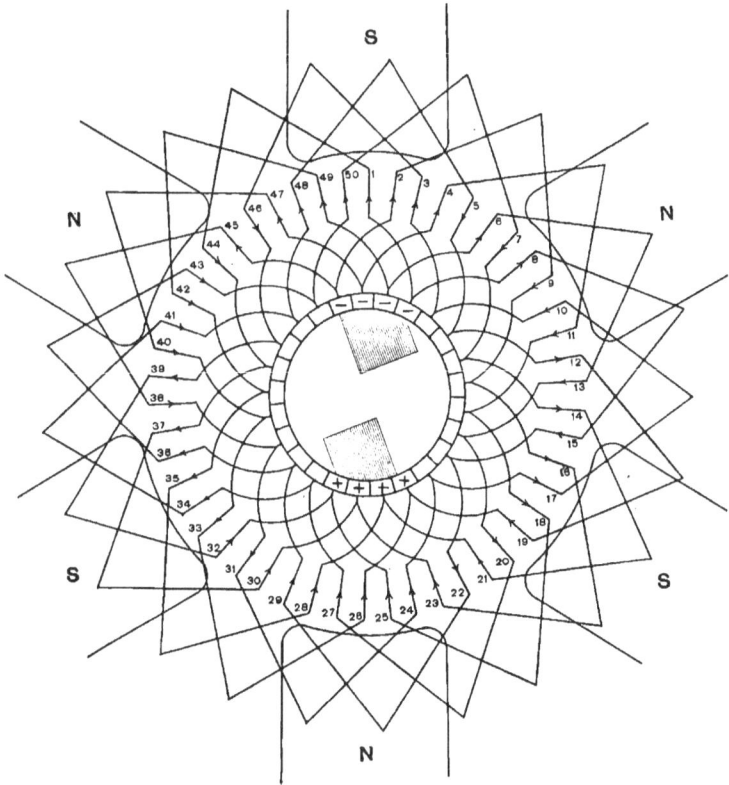

Fig. 63

TWO CIRCUIT, QUADRUPLE WINDING

Figure 63 is a six-pole, two-circuit, singly re-entrant, quadruple winding. Symbolically = (QQQ) $n = 6$, and $m = 4$. In order that it should be singly re-entrant, it was necessary for the greatest common factor of "y" and "m" to be 1. Therefore "y" was taken equal to 7.

$$C = ny \pm 2 \; m = 6 \times 7 \pm 2 \times 4 = 34 \text{ or } 50.$$

Fifty conductors have been taken.

In the position shown, the circuits through the armature are : —

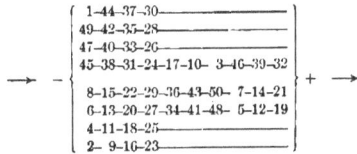

$$
\rightarrow \; - \;
\left\{
\begin{array}{l}
1\text{--}44\text{--}37\text{--}30\text{------------} \\
49\text{--}42\text{--}35\text{--}28\text{------------} \\
47\text{--}40\text{--}33\text{--}26\text{------------} \\
45\text{--}38\text{--}31\text{--}24\text{--}17\text{--}10\text{--} \; 3\text{--}46\text{--}39\text{--}32 \\
\\
8\text{--}15\text{--}22\text{--}29\text{--}36\text{--}43\text{--}50\text{--} \; 7\text{--}14\text{--}21 \\
6\text{--}13\text{--}20\text{--}27\text{--}34\text{--}41\text{--}48\text{--} \; 5\text{--}12\text{--}19 \\
4\text{--}11\text{--}18\text{--}25\text{------------} \\
2\text{--} \; 9\text{--}16\text{--}23\text{------------}
\end{array}
\right\}
\; + \; \rightarrow
$$

Figure 64 is a six-pole, two-circuit, quadruply re-entrant, quadruple winding. It would be represented symbolically as $\bigcirc\bigcirc\bigcirc\bigcirc$. $n=6$, and $m=4$. In order that it should be quadruply re-entrant, it was necessary for the greatest common factor of "y" and "m" to be 4. Therefore "y" was taken equal to 8.

$$C = ny \pm 2m = 6 \times 8 \pm 2 \times 4 = 40 \text{ or } 56.$$

Fifty-six conductors have been taken. "y" is alternately 7 and 9, as it is obviously impossible to let $y = 8$.

In the position shown, the circuits through the armature are : —

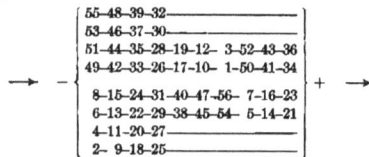

$$\longrightarrow \quad - \left\{ \begin{array}{l} 55\text{--}48\text{--}39\text{--}32\text{------------} \\ 53\text{--}46\text{--}37\text{--}30\text{------------} \\ 51\text{--}44\text{--}35\text{--}28\text{--}19\text{--}12\text{--}\ 3\text{--}52\text{--}43\text{--}36 \\ 49\text{--}42\text{--}33\text{--}26\text{--}17\text{--}10\text{--}\ 1\text{--}50\text{--}41\text{--}34 \\ \\ 8\text{--}15\text{--}24\text{--}31\text{--}40\text{--}47\text{--}56\text{--}\ 7\text{--}16\text{--}23 \\ 6\text{--}13\text{--}22\text{--}29\text{--}38\text{--}45\text{--}54\text{--}\ 5\text{--}14\text{--}21 \\ 4\text{--}11\text{--}20\text{--}27\text{------------} \\ 2\text{--}\ 9\text{--}18\text{--}25\text{------------} \end{array} \right\} \quad + \quad \longrightarrow$$

Fig. 64.

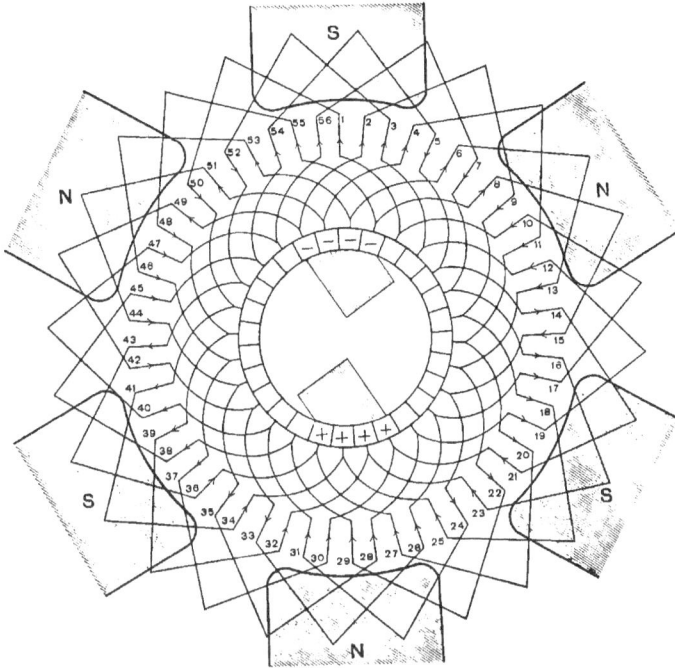

TWO CIRCUIT QUADRUPLE WINDING.

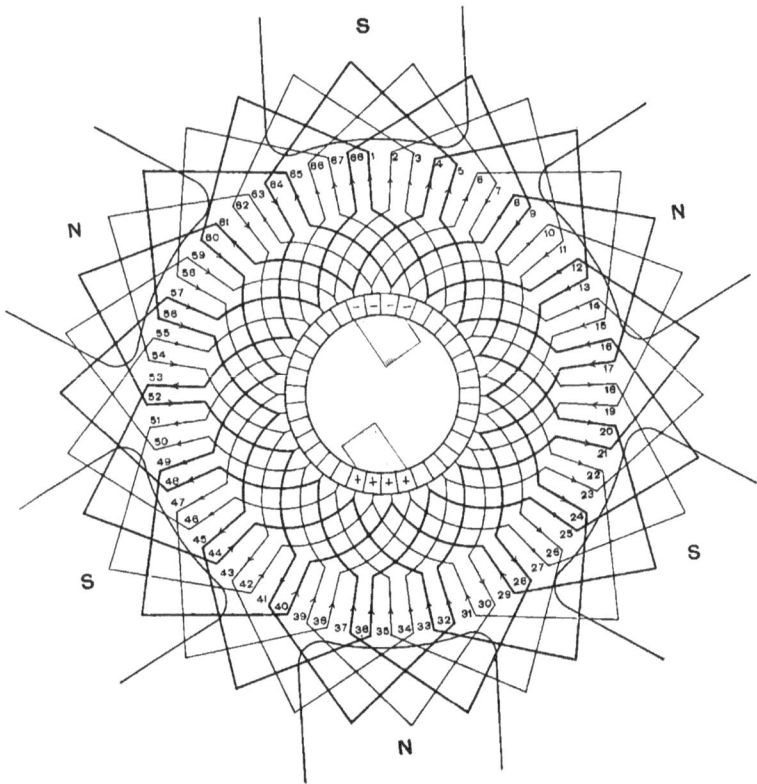

Fig. 65

TWO CIRCUIT, QUADRUPLE WINDING.

Figure 65 is a six-pole, two-circuit, doubly re-entrant, quadruple winding. It would be represented symbolically as ⓪⓪. $n = 6$, and $m = 4$. In order that it should be doubly re-entrant, it was necessary for the greatest common factor of "y" and "m" to be 2. Therefore "y" was taken equal to 10.

$$C = ny \pm 2\,m = 6 \times 10 \pm 2 \times 4 = 52 \text{ or } 68.$$

Sixty-eight conductors have been chosen. "y" is alternately 9 and 11, because its average value, being even, could not be used.

The two independently re-entrant windings have been represented respectively by light and by heavy lines.

In the position shown, the circuits through the armature are : —

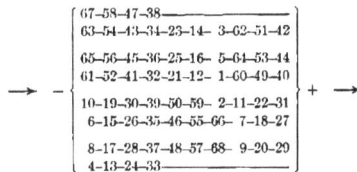

$$\rightarrow - \left\{ \begin{array}{l} 67\text{--}58\text{--}47\text{--}38\text{------} \\ 63\text{--}54\text{--}43\text{--}34\text{--}23\text{--}14\text{--}\ 3\text{--}62\text{--}51\text{--}42 \\[4pt] 65\text{--}56\text{--}45\text{--}36\text{--}25\text{--}16\text{--}\ 5\text{--}64\text{--}53\text{--}44 \\ 61\text{--}52\text{--}41\text{--}32\text{--}21\text{--}12\text{--}\ 1\text{--}60\text{--}49\text{--}40 \\[4pt] 10\text{--}19\text{--}30\text{--}39\text{--}50\text{--}59\text{--}\ 2\text{--}11\text{--}22\text{--}31 \\ \ 6\text{--}15\text{--}26\text{--}35\text{--}46\text{--}55\text{--}66\text{--}\ 7\text{--}18\text{--}27 \\[4pt] \ 8\text{--}17\text{--}28\text{--}37\text{--}48\text{--}57\text{--}68\text{--}\ 9\text{--}20\text{--}29 \\ \ 4\text{--}13\text{--}24\text{--}33\text{------} \end{array} \right\} + \rightarrow$$

Figure 66 is a six-pole, two-circuit, quadruply re-entrant, quadruple winding [○○○○]. $n=6$, and $m=4$. In order that it should be quadruply re-entrant, it was necessary that the greatest common factor of "y" and "m" should be 4. Therefore "y" was taken equal to 12.

$$C = ny \pm 2\,m = 6 \times 12 \pm 2 \times 4 = 64 \text{ or } 80.$$

Eighty conductors have been taken. "y" is alternately 11 and 13, its average value being even.

The four independently re-entrant windings have been represented by four varieties of lines.

In the position shown, the circuits through the armature are : —

$$\longrightarrow \quad - \left\{ \begin{array}{l} 77\text{--}66\text{--}53\text{--}42\text{--}29\text{--}18\text{--}\ 5\text{--}74\text{--}61\text{--}50 \\ 75\text{--}64\text{--}51\text{--}40\text{--}27\text{--}16\text{--}\ 3\text{--}72\text{--}59\text{--}48 \\ 73\text{--}62\text{--}49\text{--}38\text{--}25\text{--}14\text{--}\ 1\text{--}70\text{--}57\text{--}46 \\ 71\text{--}60\text{--}47\text{--}36\text{--}23\text{--}12\text{--}79\text{--}68\text{--}55\text{--}44 \\ 10\text{--}21\text{--}34\text{--}45\text{--}58\text{--}69\text{--}\ 2\text{--}13\text{--}26\text{--}37 \\ 8\text{--}19\text{--}32\text{--}43\text{--}56\text{--}67\text{--}80\text{--}11\text{--}24\text{--}35 \\ 6\text{--}17\text{--}30\text{--}41\text{--}54\text{--}65\text{--}78\text{--}\ 9\text{--}22\text{--}33 \\ 4\text{--}15\text{--}28\text{--}39\text{--}52\text{--}63\text{--}76\text{--}\ 7\text{--}20\text{--}31 \end{array} \right\} \quad + \quad \longrightarrow$$

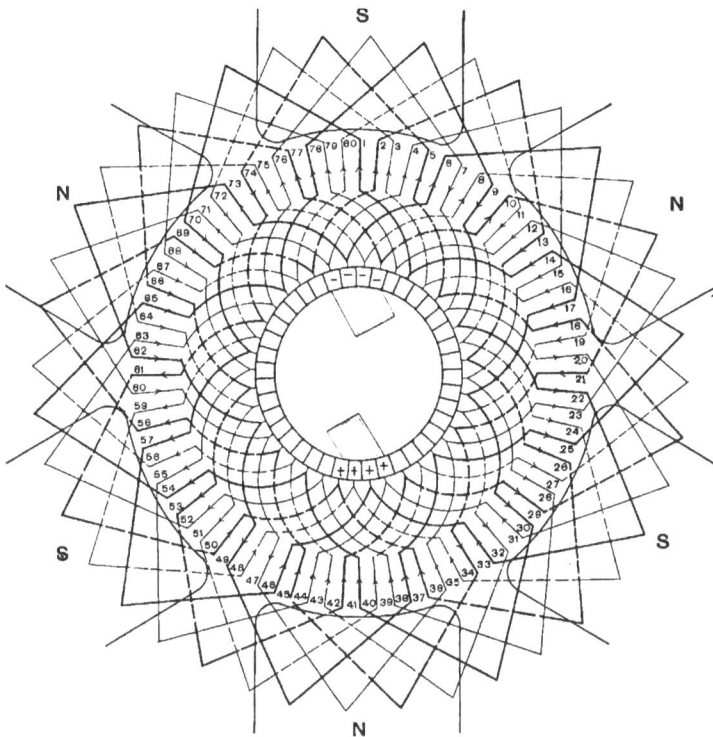

Fig. 66

TWO CIRCUIT, QUADRUPLE WINDING.

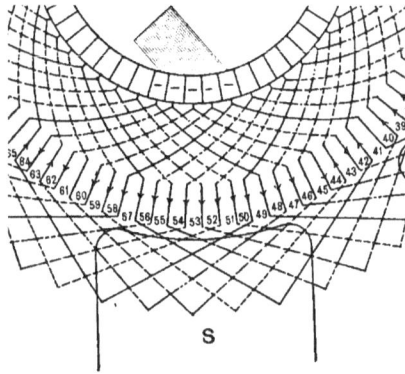

Fig. 67

TWO CIRCUIT, QUADRUPLE WINDING.

Figure 67 is a six-pole, two-circuit, quadruply re-entrant, quadruple winding. It would be represented symbolically as $\bigcirc\bigcirc\bigcirc\bigcirc$. $n = 6$, and $m = 4$. In order that it should be quadruply re-entrant, it was necessary that the greatest common factor of "y" and "m" should be 4. Therefore "y" was taken equal to 16.

$$C = ny \pm 2\,m = 6 \times 16 \pm 2 \times 4 = 88 \text{ or } 104.$$

One hundred and four conductors have been taken. "y" is 17 at the front end, and 15 at the back end, thus averaging 16.

The four independently re-entrant windings have been represented by four different styles of lines.

In the position shown, the circuits through the armature are : —

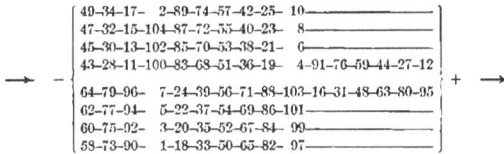

$$- \left\{ \begin{array}{l} 49\text{--}34\text{--}17\text{--}\ \ 2\text{--}89\text{--}74\text{--}57\text{--}42\text{--}25\text{--}\ 10\text{------}\\ 47\text{--}32\text{--}15\text{--}104\text{--}87\text{--}72\text{--}55\text{--}40\text{--}23\text{--}\ \ 8\text{------}\\ 45\text{--}30\text{--}13\text{--}102\text{--}85\text{--}70\text{--}53\text{--}38\text{--}21\text{--}\ \ 6\text{------}\\ 43\text{--}28\text{--}11\text{--}100\text{--}83\text{--}68\text{--}51\text{--}36\text{--}19\text{--}\ \ 4\text{--}91\text{--}76\text{--}59\text{--}44\text{--}27\text{--}12\\ 64\text{--}79\text{--}96\text{--}\ \ 7\text{--}24\text{--}39\text{--}56\text{--}71\text{--}88\text{--}103\text{--}16\text{--}31\text{--}48\text{--}63\text{--}80\text{--}95\\ 62\text{--}77\text{--}94\text{--}\ \ 5\text{--}22\text{--}37\text{--}54\text{--}69\text{--}86\text{--}101\text{------}\\ 60\text{--}75\text{--}92\text{--}\ \ 3\text{--}20\text{--}35\text{--}52\text{--}67\text{--}84\text{--}\ 99\text{------}\\ 58\text{--}73\text{--}90\text{--}\ \ 1\text{--}18\text{--}33\text{--}50\text{--}65\text{--}82\text{--}\ 97\text{------} \end{array} \right\} +$$

Figure 68 differs from Fig. 67 in the use of the negative instead of the positive sign in the formula. It is given to emphasize the fact that this has no influence on the type of winding. It requires, however, a greater length of copper for a given number of conductors. Like Fig. 67, it is a six-pole, two-circuit, quadruply re-entrant, quadruple winding.   It would be represented symbolically as $\bigcirc \bigcirc \bigcirc \bigcirc$. $n=6$, and $m=4$. In order that it should be quadruply re-entrant, it was necessary for the greatest common factor of "y" and "m" to be 4. Therefore "y" was taken equal to 16.

$$C = ny \pm 2\,m = 6 \times 16 \pm 2 \times 4 = 88 \text{ or } 104.$$

Eighty-eight conductors have been taken. "y" is 17 at the front, and 15 at the back end.

The four independently re-entrant windings have been represented by different kinds of lines.

In the position shown, the circuits through the armature are: —

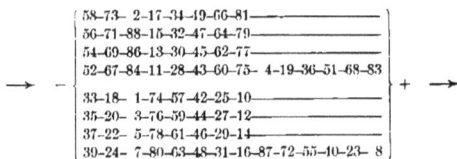

$$\rightarrow \quad - \left\{ \begin{array}{l} 58\text{-}73\text{-} \ 2\text{-}17\text{-}34\text{-}49\text{-}66\text{-}81\text{---} \\ 56\text{-}71\text{-}88\text{-}15\text{-}32\text{-}47\text{-}64\text{-}79\text{---} \\ 54\text{-}69\text{-}86\text{-}13\text{-}30\text{-}45\text{-}62\text{-}77\text{---} \\ 52\text{-}67\text{-}84\text{-}11\text{-}28\text{-}43\text{-}60\text{-}75\text{-} \ 4\text{-}19\text{-}36\text{-}51\text{-}68\text{-}83 \\ \\ 33\text{-}18\text{-} \ 1\text{-}74\text{-}57\text{-}42\text{-}25\text{-}10\text{---} \\ 35\text{-}20\text{-} \ 3\text{-}76\text{-}59\text{-}44\text{-}27\text{-}12\text{---} \\ 37\text{-}22\text{-} \ 5\text{-}78\text{-}61\text{-}46\text{-}29\text{-}14\text{---} \\ 39\text{-}24\text{-} \ 7\text{-}80\text{-}63\text{-}48\text{-}31\text{-}16\text{-}87\text{-}72\text{-}55\text{-}40\text{-}23\text{-} \ 8 \end{array} \right\} \quad + \quad \rightarrow$$

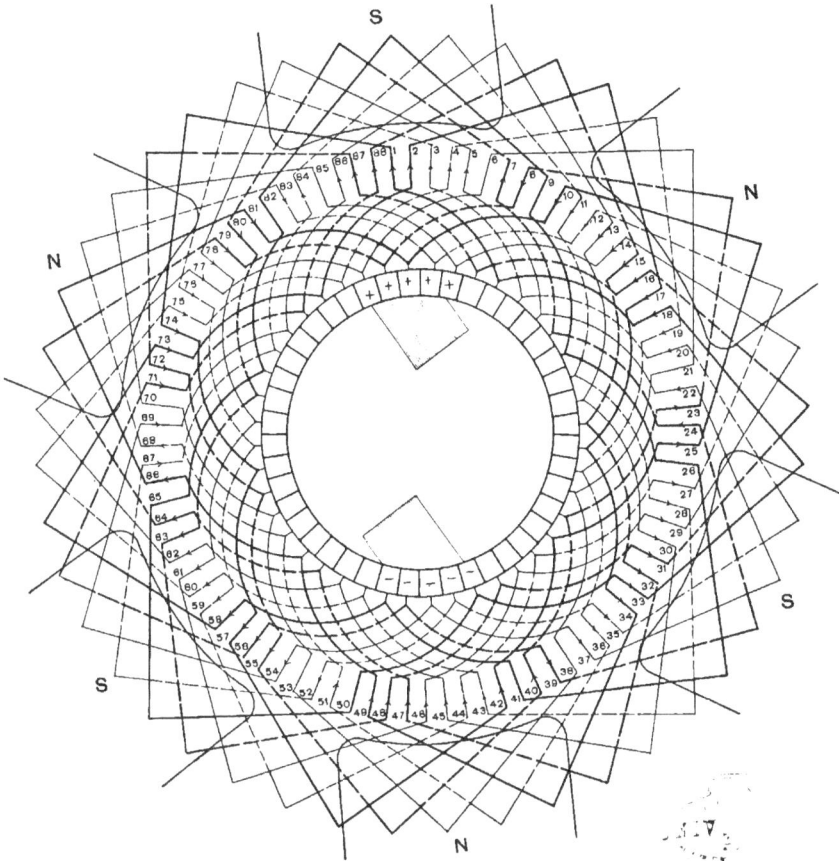

Fig. 68

TWO CIRCUIT, QUADRUPLE WINDING.

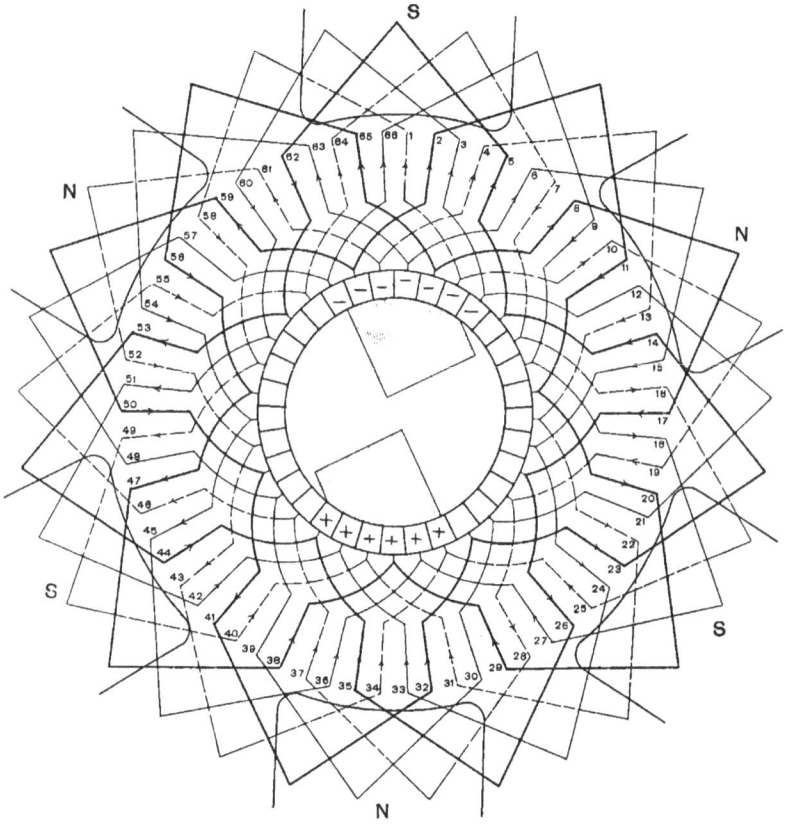

Fig. 69

TWO CIRCUIT, SEXTUPLE WINDING.

The next four diagrams (Figs. 69, 70, 71, 72) form a group of sextuple windings. It is thought that an examination of this group will bring out very clearly the method of applying and the interpretation of the rules concerning two-circuit, multiple windings. The following table will be of assistance in studying them : —

Figure.	n	y	m	C	G.C.F. of m and y.	Name of Winding.	Symbol.
69	6	9	6	66	3	Two-circuit, triply re-entrant, sextuple winding.	⊚⊚⊚
70	6	10	6	72	2	Two-circuit, doubly re-entrant, sextuple winding.	⊚⊚ ⊚⊚
71	6	11	6	78	1	Two-circuit, singly re-entrant, sextuple winding.	⊚⊚⊚⊚⊚
72	6	12	6	84	6	Two-circuit, sextuply re-entrant, sextuple winding.	○○○○○○

Figure 69 is a six-pole, two-circuit, triply re-entrant, sextuple winding. It would be symbolically represented as ⊚⊚⊚. $n = 6$, and $m = 6$. In order that it should be triply re-entrant, it was necessary that the greatest common factor of "m" and "y" should be 3. Therefore "y" was taken equal to 9.

$$C = ny \pm 2m = 6 \times 9 + 2 \times 6 = 42 \text{ or } 66.$$

Sixty-six conductors were taken. The three independently re-entrant windings have been represented respectively by light, heavy, and broken lines.

In the position shown, the circuits through the armature are : —

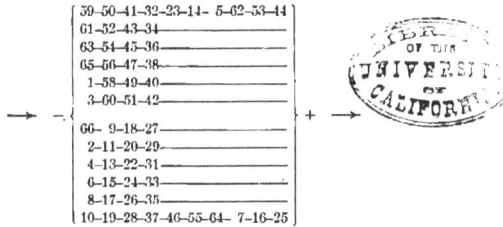

$$\longrightarrow \; -\left\{ \begin{array}{l} 59\text{-}50\text{-}41\text{-}32\text{-}23\text{-}14\text{-} \ 5\text{-}62\text{-}53\text{-}44 \\ 61\text{-}52\text{-}43\text{-}34 \text{------} \\ 63\text{-}54\text{-}45\text{-}36 \text{------} \\ 65\text{-}56\text{-}47\text{-}38 \text{------} \\ 1\text{-}58\text{-}49\text{-}40 \text{------} \\ 3\text{-}60\text{-}51\text{-}42 \text{------} \\ 66\text{-} \ 9\text{-}18\text{-}27 \text{------} \\ 2\text{-}11\text{-}20\text{-}29 \text{------} \\ 4\text{-}13\text{-}22\text{-}31 \text{------} \\ 6\text{-}15\text{-}24\text{-}33 \text{------} \\ 8\text{-}17\text{-}26\text{-}35 \text{------} \\ 10\text{-}19\text{-}28\text{-}37\text{-}46\text{-}55\text{-}64\text{-} \ 7\text{-}16\text{-}25 \end{array} \right\} \; + \; \longrightarrow$$

Figure 70 is a six-pole, two-circuit, doubly re-entrant, sextuple winding.   It would be represented symbolically as ⓖⓖ ⓖⓖ. $n = 6$, and $m = 6$.   In order that it should be doubly re-entrant, it was necessary that the greatest common factor of "m" and "y" should be 2.   Therefore "y" was taken equal to 10.

$$C = ny \pm 2\,m = 6 \times 10 \pm 2 \times 6 = 48 \text{ or } 72.$$

Seventy-two conductors have been taken.   The two independently re-entrant windings have been represented respectively by full and dotted lines.

In the given position, the circuits through the armature are: —

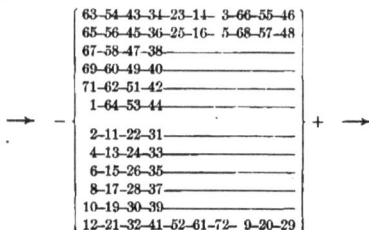

$$\longrightarrow \quad - \left\{ \begin{array}{l} 63\text{--}54\text{--}43\text{--}34\text{--}23\text{--}14\text{--}\ 3\text{--}66\text{--}55\text{--}46 \\ 65\text{--}56\text{--}45\text{--}36\text{--}25\text{--}16\text{--}\ 5\text{--}68\text{--}57\text{--}48 \\ 67\text{--}58\text{--}47\text{--}38 \\ 69\text{--}60\text{--}49\text{--}40 \\ 71\text{--}62\text{--}51\text{--}42 \\ \ 1\text{--}64\text{--}53\text{--}44 \\ \ 2\text{--}11\text{--}22\text{--}31 \\ \ 4\text{--}13\text{--}24\text{--}33 \\ \ 6\text{--}15\text{--}26\text{--}35 \\ \ 8\text{--}17\text{--}28\text{--}37 \\ 10\text{--}19\text{--}30\text{--}39 \\ 12\text{--}21\text{--}32\text{--}41\text{--}52\text{--}61\text{--}72\text{--}\ 9\text{--}20\text{--}29 \end{array} \right\} \quad + \longrightarrow$$

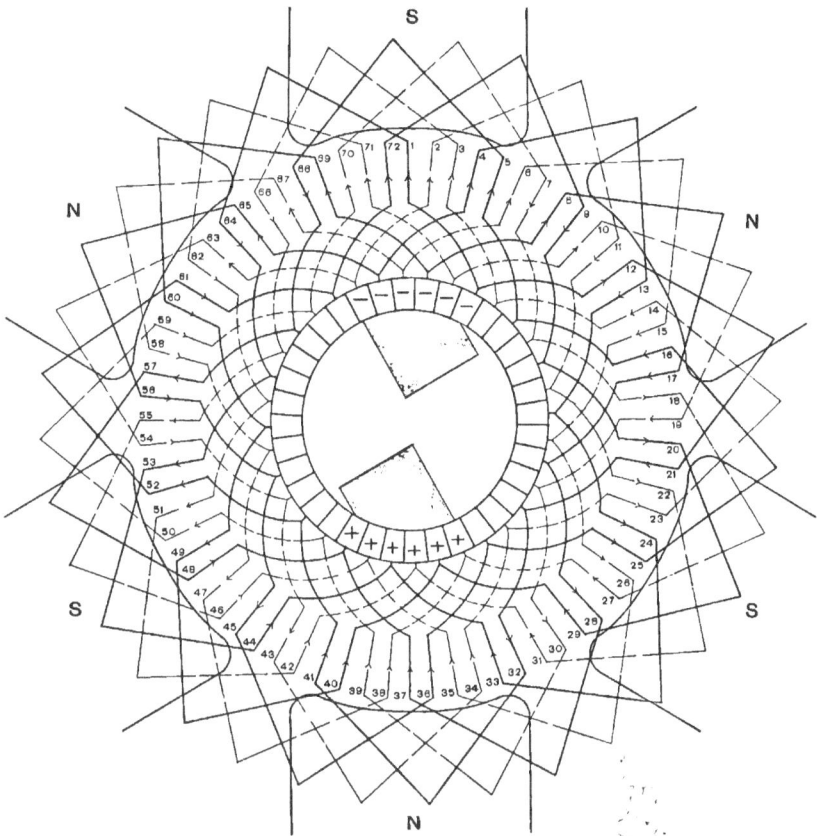

Fig. 70

TWO CIRCUIT, SEXTUPLE WINDING.

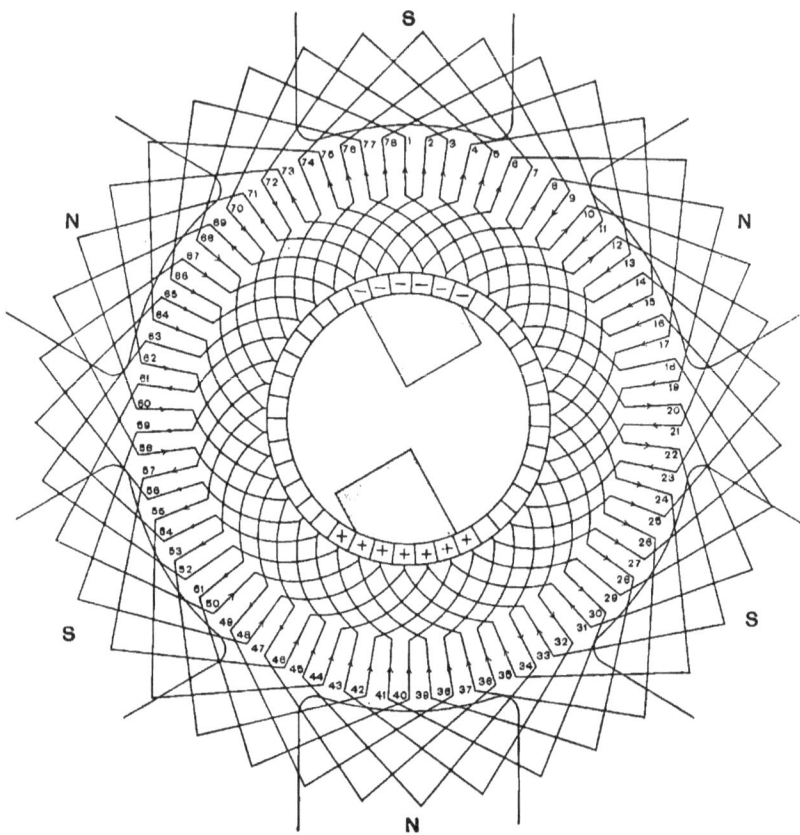

Fig. 71

TWO CIRCUIT, SEXTUPLE WINDING.

Figure 71 is a six-pole, two-circuit, singly re-entrant, sextuple winding. It would be represented symbolically as (○○○○○). $n = 6$, and $m = 6$. In order that it should be singly re-entrant, it was necessary that the greatest common factor of "m" and "y" should be 1. Therefore "y" was taken equal to 11.

$$C = ny \pm 2 \, m = 6 \times 11 \pm 2 \times 6 = 54 \text{ or } 78.$$

Seventy-eight conductors have been chosen.

In the given position, the circuits through the armature are : —

$$\rightarrow \quad - \quad \left\{ \begin{array}{l} 69\text{–}58\text{–}47\text{–}36\text{–}25\text{–}14\text{–} \ 3\text{–}70\text{–}59\text{–}48 \\ 71\text{–}60\text{–}49\text{–}38\text{–}27\text{–}16\text{–} \ 5\text{–}72\text{–}61\text{–}50 \\ 73\text{–}62\text{–}51\text{–}40\text{——————} \\ 75\text{–}64\text{–}53\text{–}42\text{——————} \\ 77\text{–}66\text{–}55\text{–}44\text{——————} \\ 1\text{–}68\text{–}57\text{–}46\text{——————} \\ \\ 12\text{–}23\text{–}34\text{–}45\text{–}56\text{–}67\text{–}78\text{–}11\text{–}22\text{–}33 \\ 10\text{–}21\text{–}32\text{–}43\text{–}54\text{–}65\text{–}76\text{–} \ 9\text{–}20\text{–}31 \\ 8\text{–}19\text{–}30\text{–}41\text{——————} \\ 6\text{–}17\text{–}28\text{–}39\text{——————} \\ 4\text{–}15\text{–}26\text{–}37\text{——————} \\ 2\text{–}13\text{–}24\text{–}35\text{——————} \end{array} \right\} \quad + \quad \rightarrow$$

Figure 72 is a six-pole, two-circuit, sextuply re-entrant, sextuple winding. It would be represented symbolically as ○○○○○○. $n=6$, and $m=6$. In order that it should be sextuply re-entrant, it was necessary that the greatest common factor of "m" and "y" should be 6. Therefore "y" was taken equal to 12.

$$C = ny \pm 2\,m = 6 \times 12 \pm 2 \times 6 = 60 \text{ or } 84.$$

Eighty-four conductors have been taken.

The six independently re-entrant windings are represented respectively by different styles of lines. "y," of course, is taken alternately 11 and 13.

In the given position, the circuits through the armature are : —

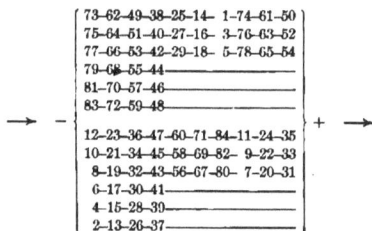

$$\longrightarrow \;-\; \left\{ \begin{array}{l} 73\text{--}62\text{--}49\text{--}38\text{--}25\text{--}14\text{--} \ 1\text{--}74\text{--}61\text{--}50 \\ 75\text{--}64\text{--}51\text{--}40\text{--}27\text{--}16\text{--} \ 3\text{--}76\text{--}63\text{--}52 \\ 77\text{--}66\text{--}53\text{--}42\text{--}29\text{--}18\text{--} \ 5\text{--}78\text{--}65\text{--}54 \\ 79\text{--}68\text{--}55\text{--}44\text{------} \\ 81\text{--}70\text{--}57\text{--}46\text{------} \\ 83\text{--}72\text{--}59\text{--}48\text{------} \\[4pt] 12\text{--}23\text{--}36\text{--}47\text{--}60\text{--}71\text{--}84\text{--}11\text{--}24\text{--}35 \\ 10\text{--}21\text{--}34\text{--}45\text{--}58\text{--}69\text{--}82\text{--} \ 9\text{--}22\text{--}33 \\ \ 8\text{--}19\text{--}32\text{--}43\text{--}56\text{--}67\text{--}80\text{--} \ 7\text{--}20\text{--}31 \\ \ 6\text{--}17\text{--}30\text{--}41\text{------} \\ \ 4\text{--}15\text{--}28\text{--}39\text{------} \\ \ 2\text{--}13\text{--}26\text{--}37\text{------} \end{array} \right\} \;+\; \longrightarrow$$

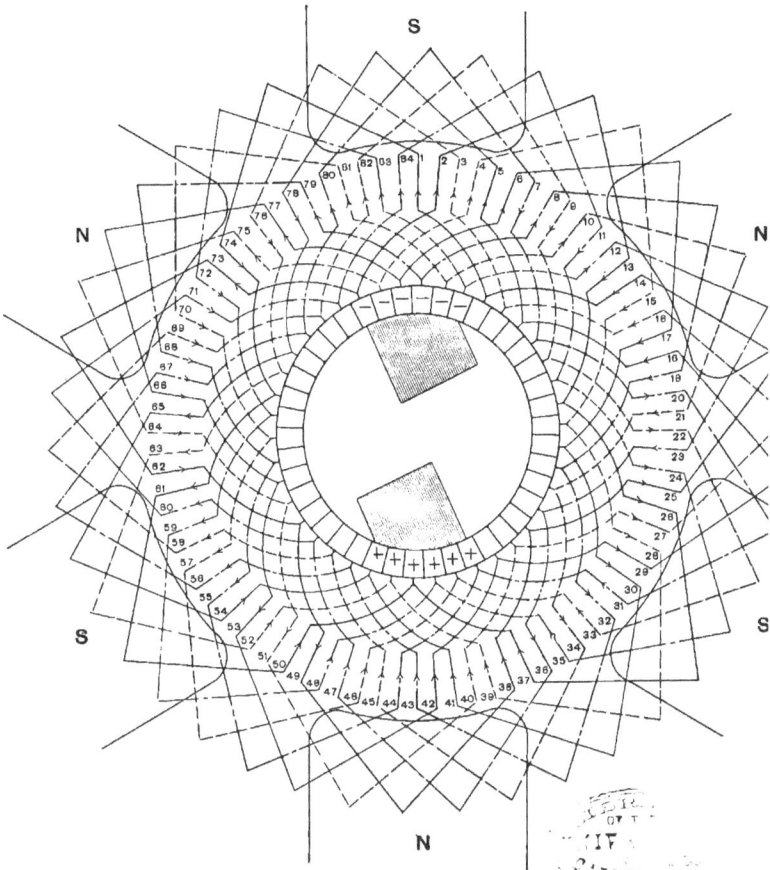

Fig. 72

TWO CIRCUIT, SEXTUPLE WINDING.

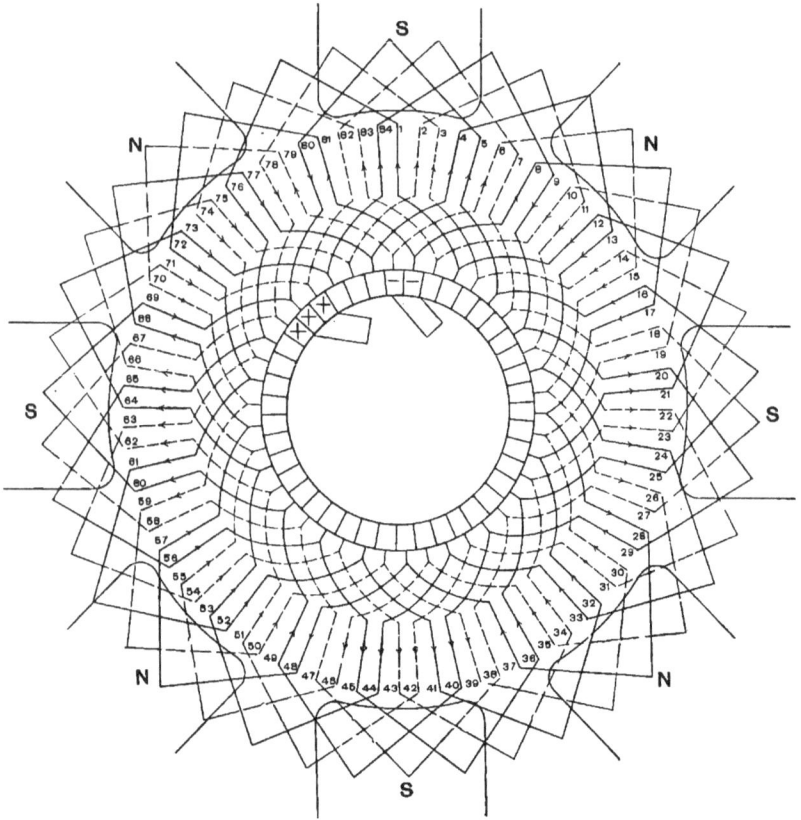

Fig. 73
TWO CIRCUIT, DOUBLE WINDING.

Figure 73 is an eight-pole, two-circuit, doubly re-entrant, double winding. It would be represented symbolically as $\bigcirc\,\bigcirc$. $n = 8$, and $m = 2$. In order that it should be doubly re-entrant, it was necessary that the greatest common factor of "m" and "y" should be 2. Therefore "y" was taken equal to 10.

$$C = ny \pm 2\,m = 8 \times 10 \pm 2 \times 2 = 76 \text{ or } 84.$$

Eighty-four conductors have been taken.

The two independently re-entrant windings are represented respectively by full and dotted lines. "y" is taken alternately 11 and 9, the average pitch being 10.

In the given position, the circuits through the armature are: —

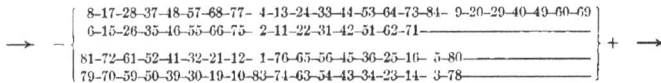

$$\longrightarrow \quad - \left\{ \begin{array}{l} 8\text{-}17\text{-}28\text{-}37\text{-}48\text{-}57\text{-}68\text{-}77\text{-} \ 4\text{-}13\text{-}24\text{-}33\text{-}44\text{-}53\text{-}64\text{-}73\text{-}84\text{-} \ 9\text{-}20\text{-}29\text{-}40\text{-}49\text{-}60\text{-}69 \\ 6\text{-}15\text{-}26\text{-}35\text{-}46\text{-}55\text{-}66\text{-}75\text{-} \ 2\text{-}11\text{-}22\text{-}31\text{-}42\text{-}51\text{-}62\text{-}71\text{-}\!\!-\!\!-\!\!-\!\!-\!\!-\!\!- \\ 81\text{-}72\text{-}61\text{-}52\text{-}41\text{-}32\text{-}21\text{-}12\text{-} \ 1\text{-}76\text{-}65\text{-}56\text{-}45\text{-}36\text{-}25\text{-}16\text{-} \ 5\text{-}80\!\!-\!\!-\!\!-\!\!-\!\!-\!\!-\!\!- \\ 79\text{-}70\text{-}59\text{-}50\text{-}39\text{-}30\text{-}19\text{-}10\text{-}83\text{-}74\text{-}63\text{-}54\text{-}43\text{-}34\text{-}23\text{-}14\text{-} \ 3\text{-}78\!\!-\!\!-\!\!-\!\!-\!\!- \end{array} \right\} \quad + \quad \longrightarrow$$

Figure 74 is an eight-pole, two-circuit, singly re-entrant, double winding. It would be represented symbolically as ⓖ. $n = 3$, and $m = 2$. In order that it should be singly re-entrant, it was necessary that the greatest common factor of "y" and "m" should be 1. Therefore "y" was taken equal to 11.

$$C = ny \pm 2m = 8 \times 11 \pm 2 \times 2 = 84 \text{ or } 92.$$

Eighty-four conductors have been taken *just as in the preceding figure.* In the given position, the circuits through the armature are : —

$$\longrightarrow \quad - \left\{ \begin{array}{l} 8\text{–}19\text{–}30\text{–}41\text{–}52\text{–}63\text{–}74\text{–}\ 1\text{–}12\text{–}23\text{–}34\text{–}45\text{–}56\text{–}67\text{——————}\\ 6\text{–}17\text{–}28\text{–}39\text{–}50\text{–}61\text{–}72\text{–}83\text{–}10\text{–}21\text{–}32\text{–}43\text{–}54\text{–}65\text{–}76\text{–}\ 3\text{–}14\text{–}25\text{–}36\text{–}47\text{–}58\text{–}69\text{——}\\ 81\text{–}70\text{–}59\text{–}48\text{–}37\text{–}26\text{–}15\text{–}\ 4\text{–}77\text{–}66\text{–}55\text{–}44\text{–}33\text{–}22\text{–}11\text{–}84\text{–}73\text{–}62\text{–}51\text{–}40\text{–}29\text{–}18\text{–}\ 7\text{–}80\\ 79\text{–}68\text{–}57\text{–}46\text{–}35\text{–}24\text{–}13\text{–}\ 2\text{–}75\text{–}64\text{–}53\text{–}42\text{–}31\text{–}20\text{–}\ 9\text{–}82\text{——————} \end{array} \right\} \quad + \quad \longrightarrow$$

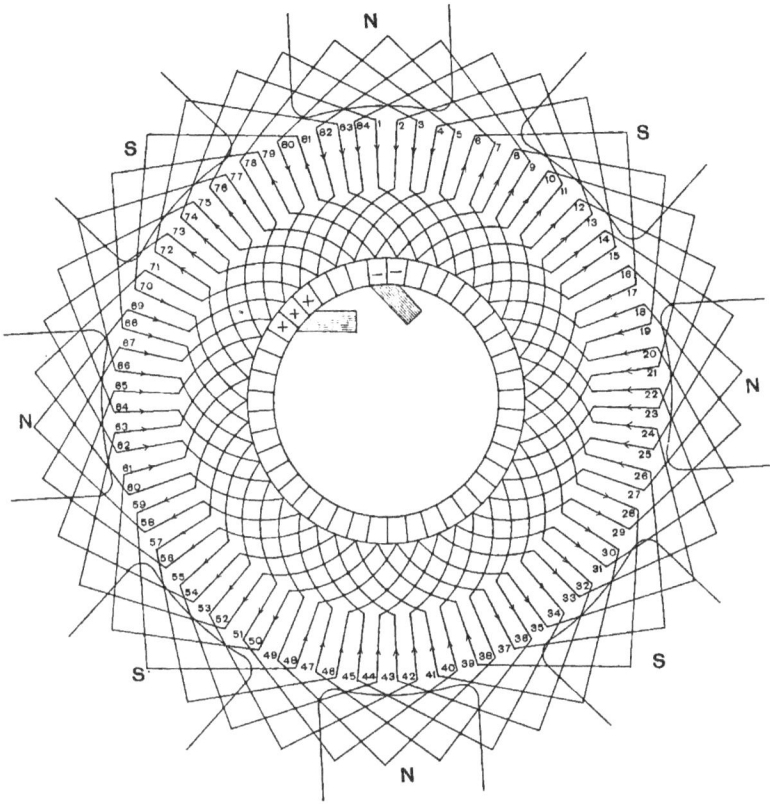

Fig. 74

TWO CIRCUIT, DOUBLE WINDING.

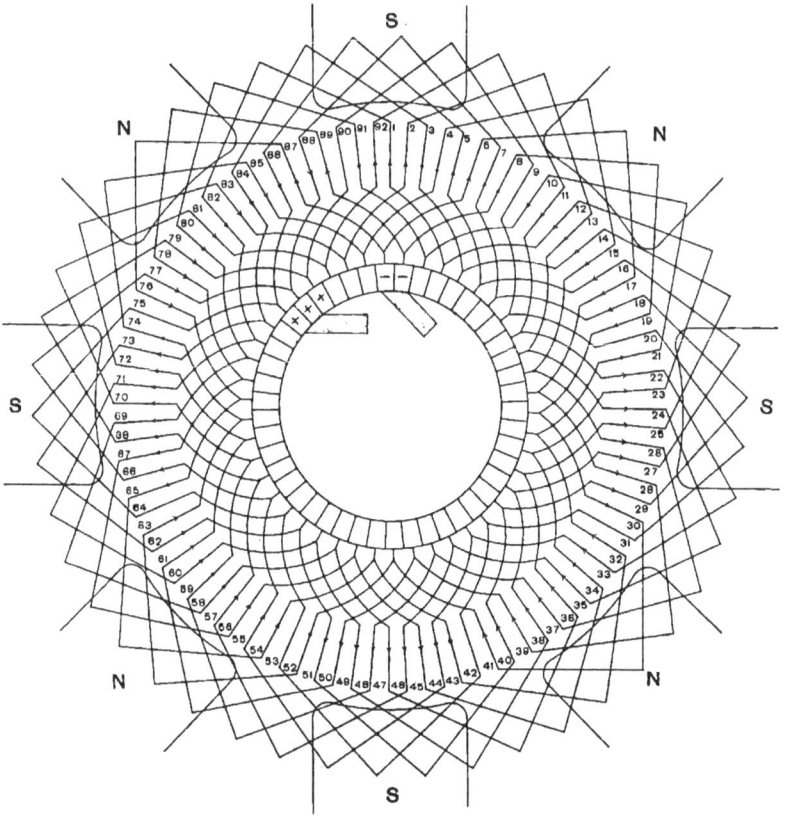

Fig. 75

TWO CIRCUIT, DOUBLE WINDING.

Figure 74 was obtained by using the negative sign in the formula—

$$C = ny \pm 2\,m.$$

This is, as has been pointed out, rather wasteful of copper, and was only done to demonstrate the fact that in certain cases with a given number of conductors, either a singly or a doubly re-entrant, double winding may be used.

In Fig. 75, the positive sign was used. It will, however, not be necessary to analyze it, it not being materially different from Fig. 74.

Numerous interesting deductions concerning two-circuit, multiple-wound, drum armatures may be made from the data contained in the tables in Chapter XVIII.

CHAPTER XI.

THE armature coils of dynamos have, in addition to their function of establishing the electromotive force required external to the armature, the function of setting up in the arc of commutation an electromotive force to reverse the current in them as they successively pass the collecting brushes (by arc of commutation is meant the arc in which the current in the armature coils is reversed, the extent of this are being determined by the length of the arc of contact of the collecting brushes). In the ordinary methods of armature winding the electromotive force for reversing the current in the coils is obtained by giving the collecting brushes an angular lead, the amount of which depends upon the distribution of the magnetic flux in the air gap, the coefficient of self-induction of the armature coils when in the arc of commutation, and the rate of change of the current in the coils, while the current is being reversed. In generators this angular lead is in such direction that the magnetomotive force of the armature is opposed to the magnetomotive force of the field magnets to an extent proportional to the angle of load, in consequence of which the reversing field becomes of diminished intensity for an increase of current in the armature, when it needs to be increased.

Mr. Sayers, of Glasgow, has patented a winding in which the commutation of the current in the main armature coils is effected by an additional set of coils which may be termed commutating coils. These coils are applicable to any form of armature winding suitable for commutating machines. One of these coils is connected between each commutator bar and the connections joining the main armature coils in series with each other. These commutating coils are located on the periphery of the armature in such a position with respect to the main coils that the magnetomotive force of the main coils tends with increasing current to increase the flux through them, and further so that the magnetomotive force of the armature acts *with* the magnetomotive force of the field magnets instead of *against* it as in ordinary dynamos. It is possible, therefore, through a certain range of output to sparklessly operate a generator at constant voltage without changing the load of the brushes or the excitation of the field magnets. It may be noted that when one of the main coils is short-circuited by the collecting brushes it is through two of these commutating coils, and the electromotive force from these coils effective for reversing the current in the main coil is the excess of the electromotive force generated in the leading coil over that in the following coil. The position, then, of the reversing field, if effective, is fixed as to angular extent between very narrow limits. It does not appear to the writer that the reversing field can be so localized for great changes of current in the armature as one might infer from reading the discussion of Mr. Sayers' paper at the Institution of Electrical Engineers. (See Vol. XXII., pages 377–441, Journal Ins. Elect. Engrs., London). Within certain limits, however, it appears that the magnetomotive force of the armature may be utilized in creating proper strength of reversing field.

This method, as applied to a bi-polar drum winding, is illustrated in Fig. 76. It will be seen to consist of a regular drum winding, with the difference that the connections from the winding to the commutator segments, instead of consisting of short leads, consist of auxiliary force conductors which pass from the winding, backward, a short distance against the direction of rotation, and then parallel to the regular face conductors to the back of the armature. The conductor then passes forward in the direction of rotation, and again crossing the armature, is carried to the commutator segment.

In the diagram, the current in the coil A^2 has just been reversed. The coil A^1 is, by the two adjacent commutator segments under the brush, short-circuited while its main conductors are still moving through intense fields, tending to maintain the current in its original direction. But this short circuit contains, in series with the main coil, the two connections to the commutator segments, both of which are so linked with the magnetic flux from the pole piece, that electromotive forces are induced. Of the electromotive forces induced in the two commutator loops, that in the loop drawn in the figure is added to that of the short-circuited main coil, but this loop is farther out of the magnetic field than the remaining loop (not drawn) of the short-circuited section. This latter loop, leading from the segment next adjacent on the left of that shown at

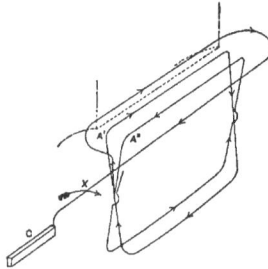

Fig. 76.

C, being well under the pole pieces, has induced in it a strong electromotive force, which opposes that in the rest of the short-circuited section, and enables a current to be generated in the direction of that in the half of the armature circuit of which it is soon to become a part.

In such a drum winding, Mr. Sayers refers to these commutator connections as "reverser bars." As they carry the current only during the short time that their corresponding sections are passing under the brushes, they may be of much smaller cross-section than the main conductors.

It will be seen from the above description that the winding is particularly adapted for use with ironclad armatures with very small air gaps, for the effectiveness of the arrangement is largely dependent upon the differential inductive action upon two successive reverser bars, and the more abrupt the demarcation of the magnetic flux, the greater will be this differential effect.

It should be clearly understood that this winding is equally applicable to rings, discs, and other types of armature.

Part II.

WINDINGS FOR ALTERNATING–CURRENT DYNAMOS AND MOTORS.

CHAPTER XII.

ALTERNATING-CURRENT WINDINGS.

In general, any of the continuous-current armature windings may be employed for alternating-current work, but the special considerations leading to the use of alternating currents generally make it necessary to abandon the styles of winding best suited to continuous-current work, and to use windings specially adapted to the conditions of alternating-current practice.

Attention should be called to the fact that all the re-entrant (or closed circuit) continuous-current windings must necessarily be two-circuit or multiple-circuit windings, while alternating-current armatures may, and almost always do from practical considerations, have one-circuit windings, i.e. one circuit per phase. From this it follows that any continuous-current winding may be used for alternating-current work, but an alternating-current winding cannot generally be used for continuous-current work. In other words, the windings of alternating-current armatures are essentially non-re-entrant (or open circuit) windings, with the exception of the ring-connected polyphase windings, which are re-entrant (or closed circuit) windings. These latter are, therefore, the only windings which are applicable to alternating-continuous current, commutating machines.

Usually, high voltages are desired, and in such cases windings are generally adopted in which heavily insulated coils are imbedded in slots in the armature surface. Often, for single-phase alternators, one slot or coil per pole piece is used, as this permits of the most effective disposition of the armature conductors as regards generation of electromotive force. If more slots or coils are used, or, in the case of face windings, if the conductors are more evenly distributed over the face of the armature, the electromotive forces generated in the various conductors are in different phases, and the total electromotive force is less than the algebraic sum of the effective electromotive forces induced in each conductor. But, on the other hand, the subdivision of the conductors in several slots or angular positions per pole, in the case of face windings, their more uniform distribution over the peripheral surface, decreases the self-induction of the windings with its attendant disadvantages. It also utilizes more completely the available space and tends to bring about a better distribution of the necessary heating of core and conductors. Therefore, in cases where the voltage and the corresponding necessary insulation permit, the conductors are sometimes spread out to a greater or less extent from the elementary groups necessary in cases where very high potentials are used.

Windings in which such a subdivision is adopted, will be referred to as having a multi-coil construction, as distinguished from the form in which the conductors are assembled in one group per pole piece, which latter will be called uni-coil windings.

The terms uni- and multi-*slot* have been applied to alternating-current ironclad armatures, but the modified nomenclature described in the preceding paragraph will be preferable, in that it does not distinguish between armatures where the groups are arranged on the periphery, and those in which the groups are imbedded in slots. A little consideration will show the advisability of this nomenclature, as it will often permit one description to suffice for a winding which may be used either for ironclad or smooth-core construction.

It will be seen later, that in most *multiphase windings*, multi-coil construction involves only very little sacrifice of electromotive force for a given total length of armature conductor, and in good designs is generally adopted to as great an extent as proper space allowance for the insulation will permit.

Often in alternating current installations, step-up or step-down transformers, or both, are used, and in such cases the other extreme is approached, and the apparatus is built for very low voltages. This permits the use of very small space for insulation ; and conductors of large cross-section, often arranged with only one conductor per group, are used. Here the multi-coil construction is less difficult, although still attended to some extent with the disadvantage of obtaining less than the maximum possible voltage per unit length of armature conductor.

Examples of windings adapted respectively to both of the above extremes will be given in the following chapters.

It will now be readily understood that the ordinary continuous-current windings are not, in the great majority of cases, adaptable to the work to be done. They should, however, always be kept in mind, and will often be found to work in nicely in special cases.

A class of apparatus, best termed alternating continuous-current, commutating machines, is now being found of much value in various ways. They are in a general way used for feeding continuous-current circuits, from single-phase or multiphase circuits (or *vice versa*), and also sometimes for feeding alternating circuits of one class (for example, single- or quarter-phase) from those of another (say three-phase). This type of armature may usually be best laid out by employing regular continuous-current windings and tapping them off in the proper manner. Examples will be given.

A wide variety of styles of armature construction have been employed in alternating-current machinery. Rings, drums (both ironclad and smooth-core), discs, and very many other types have been successfully built. Iron cores are used by some makers, and carefully avoided by others. Internal and external rotating parts have each found advocates. This great variety renders detailed treatment difficult, and in the following discussion it has been generally assumed that the windings are laid on the periphery of a drum, either on the surface, or imbedded in slots, and that the necessary connections are made at the ends of the armature. These peripheral conductors are represented diagrammatically by radial lines, and the end connections by crooked lines. Thus, re-entrant polygons drawn with heavy lines may be taken to represent coils of the desired number of turns, the lighter lines representing the connections of these coils to each other.

In the case of bar windings, no difficulty will be found in understanding the diagrams, as they correspond quite nearly to the continuous-current windings. Small, heavy circles in the middle of the diagram represent collector rings. If a winding is desired, for a disc or some other type, the diagrams will generally be found amply suggestive. Pancake coils and other types of windings, not specifically described, may be readily planned by slight modifications of the diagrams.

No examples have been given of gramme-ring alternating-current windings, as these may be found in text books, and are so easily understood as to require no discussion.

Before concluding these general considerations, it is desirable to emphasize the following points regarding the relative merits of uni- and multi-coil construction : —

With a given number of conductors arranged in a multi-coil winding, less terminal voltage will be obtained at no load than would be the case if they had been arranged in a uni-coil winding, and the discrepancy will be greater in proportion to the number of coils into which the conductors per pole piece are subdivided, assuming that the spacing of the groups of conductors is uniform over the entire periphery.

Thus, if the terminal voltage at no load be taken as 1 for a uni-coil construction, it will, for the same total number of conductors, be .707 for a two-coil, .667 for a three-coil, .654 for a four-coil, etc.

But when the machine *is loaded*, the current in the armature *causes reactions which play an important part* in *determining the voltage at the generator terminals*, and this may only be maintained constant as the load comes on, by increasing the field excitation, often by a very considerable amount. Now, with a given number of armature conductors, carrying a given current, these reactions are greatest when the armature conductors are concentrated in *one group per pole piece*, that is, when the uni-coil construction is adopted, and they decrease to a considerable degree as the conductors are subdivided into small groups distributed over the entire armature surface, that is, they decrease when the multi-coil construction is used. The ratios given above for the relative voltages at *no load*, for uni- and multi-coil construction, *do not*, therefore, represent the relative values of the windings under working conditions, and it is believed that careful consideration should in many cases be given to both styles of winding, before deciding upon the one best suited for the purpose.

Multi-coil design also results in a much more equitable distribution of the conductors, and, in the case of ironclad construction, permits of coils of small depth and width which cannot fail to be much more readily maintained at a low temperature for a given cross-section of conductor, or, if desirable to take advantage of this point in another way, it should be practicable to use a somewhat smaller cross-section of conductor for a given temperature limit. And similarly, when we consider smooth-core construction, we find that the distribution of conductors over the entire surface carries with it great advantages from a mechanical standpoint.

CHAPTER XIII.

SINGLE–PHASE WINDINGS.

FIGURE 77 is a diagram of a winding for single-phase alternating-current generators and synchronous motors, which has been very extensively used. It has one group per pole piece, consisting of adjacent halves of two coils of the proper number of turns. These are interconnected as shown by the light lines. The adjacent halves of the two coils are usually arranged side by side, but it might sometimes be of advantage to place them one over the other. The arrangement of two coils side by side has been satisfactorily applied in various types of ironclad armatures. In Figs. 102 and 119 are given examples of this style of winding connected respectively for quarter-phase and for three-phase work. It should be noted, however, that the same armature can be used for three-phase purposes only by having fields with different numbers of pole pieces.

The avoidance of crossings at the ends, and the extreme simplicity of this style of winding, are its chief advantages.

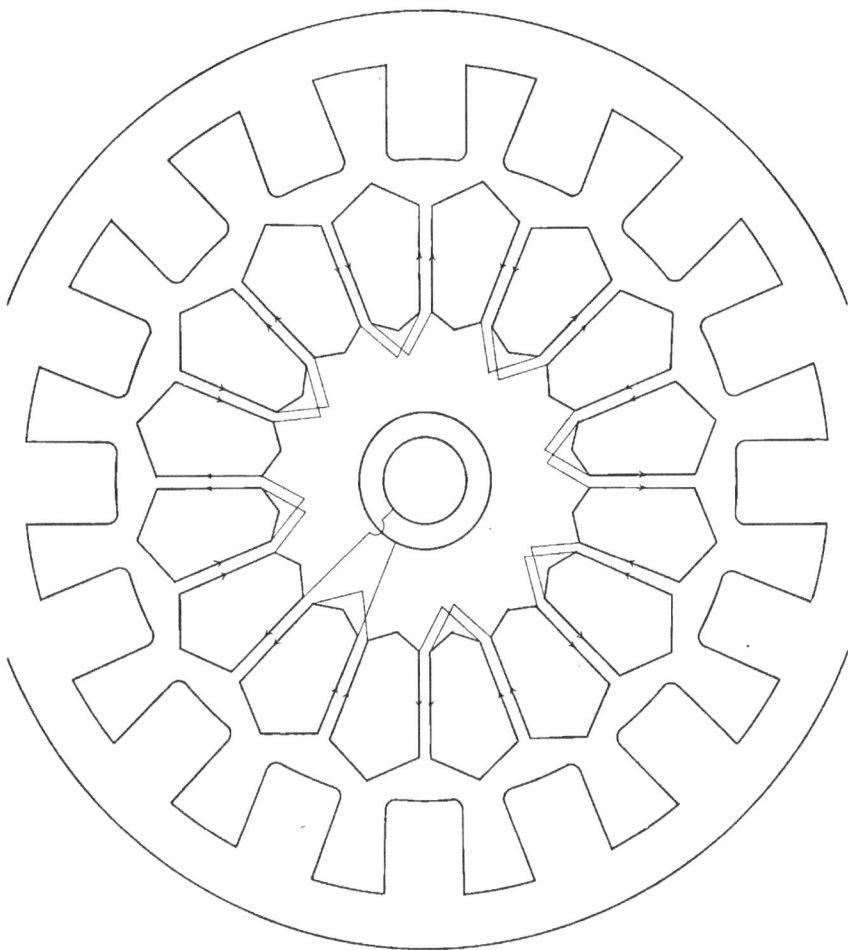

Fig. 77

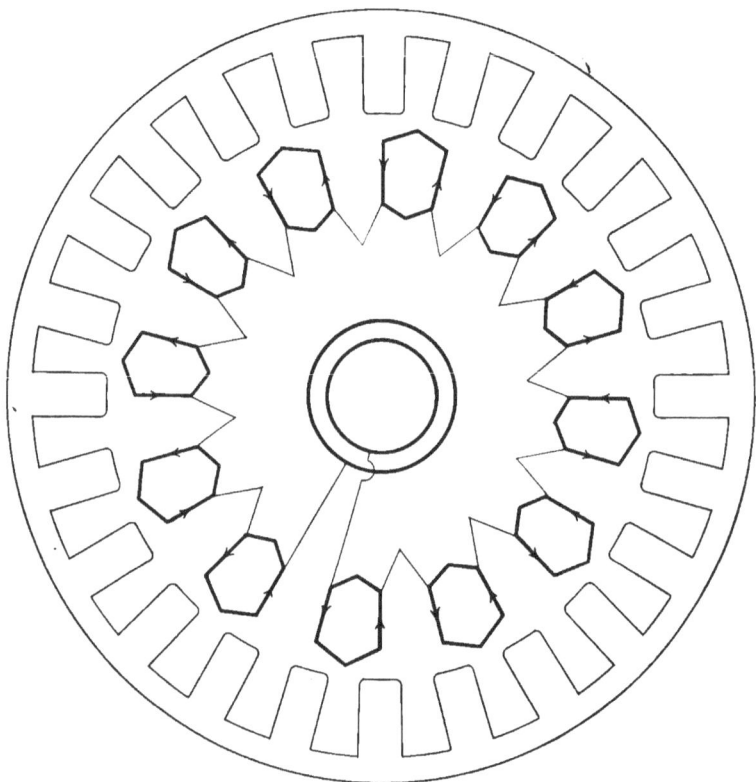

Fig. 78

In Fig. 78 is given another uni-coil winding, but here only one coil is placed in each slot. In many cases this might be preferable to the arrangement shown in Fig. 77, but the ends of the armatures are not so completely occupied by the ends of the coils, which wastes room and tends to bring about a less even distribution of the loss by heating. The use of only half as many coils is, of course, generally an advantage, on account of simplicity, but it is usually necessary for each coil to be wound deeper, which is objectionable from a thermal standpoint, as well as from the fact that a greater depth of space has to be allowed for the winding at the ends of the armature.

It should not be overlooked that if half the number of pole pieces is odd, the armature coils could not be connected up in two parallels, which would in practice be a very considerable objection, as it would limit the use of the armature for other purposes than that contemplated in laying out the original design.

One feature of this winding worthy of consideration is the great ease of insulation, it being, in this respect, superior to Fig. 77, one of the groups of which consists of adjacent halves of two coils, having between them the entire voltage of the armature.

Figure 79 is a bar winding, with one bar per pole piece, corresponding to the coil winding of Fig. 78. This would be used for low voltages, and in the case of generators of large capacity, such windings are practicable for high voltages. It is typical of the simplest form of a multipolar, single-phase alternator, and has been used in some very large machines.

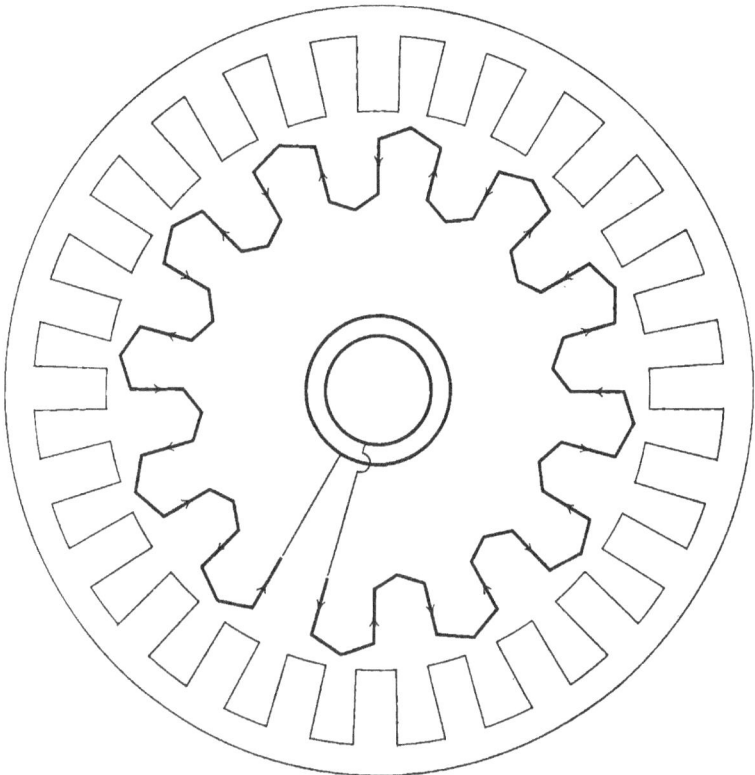

Fig. 79

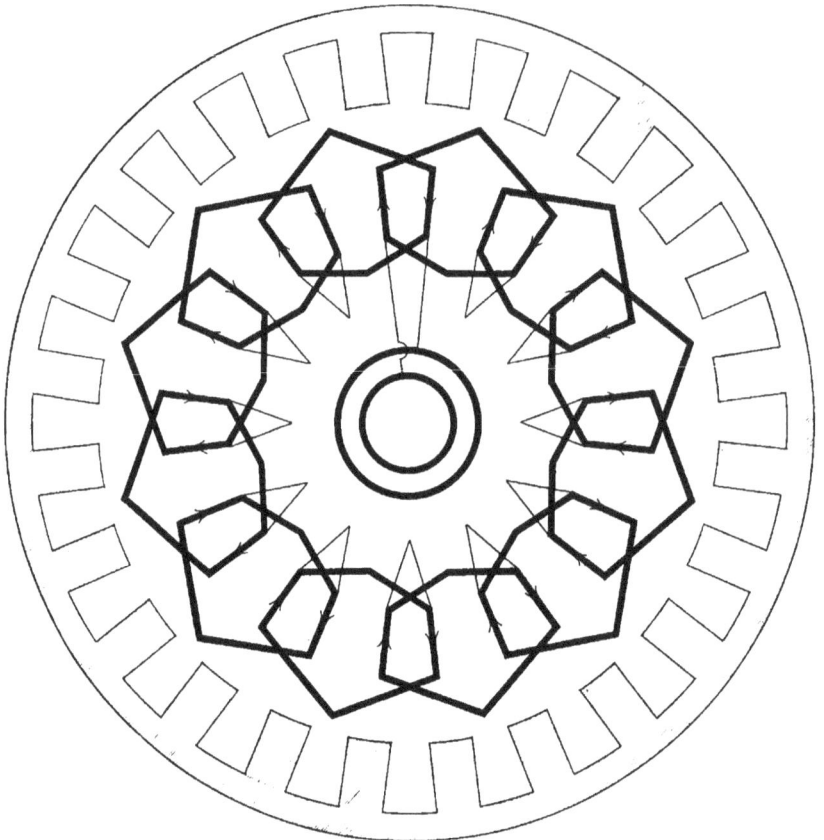

Fig. 80

Figure 80 is another uni-coil winding. It is given largely as a matter of interest; for, as will be seen, it has undesirable crossings and very long end connections, which would be very wasteful of copper unless the length of the magnet cores parallel to the shaft is great compared with the length of the pole arc. Even in such a case there would be no advantage over Fig. 78, unless for the fact that Fig. 80 is a very good winding for a three-phase alternator of one-third the number of poles, and the case might occur where it would be of advantage to use the same armature and winding for both cases. This would make an excellent three-phase winding for one-third as many poles, and would then be similar to the three-phase winding given in Fig. 116.

The corresponding diagram for a bar winding, with one bar per pole piece, is sufficiently evident from Fig. 80, and, in view of its unimportance, will not be given.

The following diagrams are multi-coil, single-phase alternators.   As a class they have been very thoroughly discussed in the general remarks of the preceding chapter.

Figure 81 represents a very simple two-coil winding.   It is to be noted that this winding is mechanically identical, with the exception of the interconnection of the coils, with the winding of Fig. 78, but it is put in a frame with *half as many* poles as there are groups of conductors, instead of, as was the case in Fig. 78, being laid out for a frame with a number of poles *equal* to the number of groups of conductors.

As already pointed out, such multi-coil windings do not at no load generate so great an electromotive force per unit of length of face conductor, as uni-coil windings.   It has, however, been also shown on page 164 that this objection does not have such great weight as would at first sight appear to be the case.

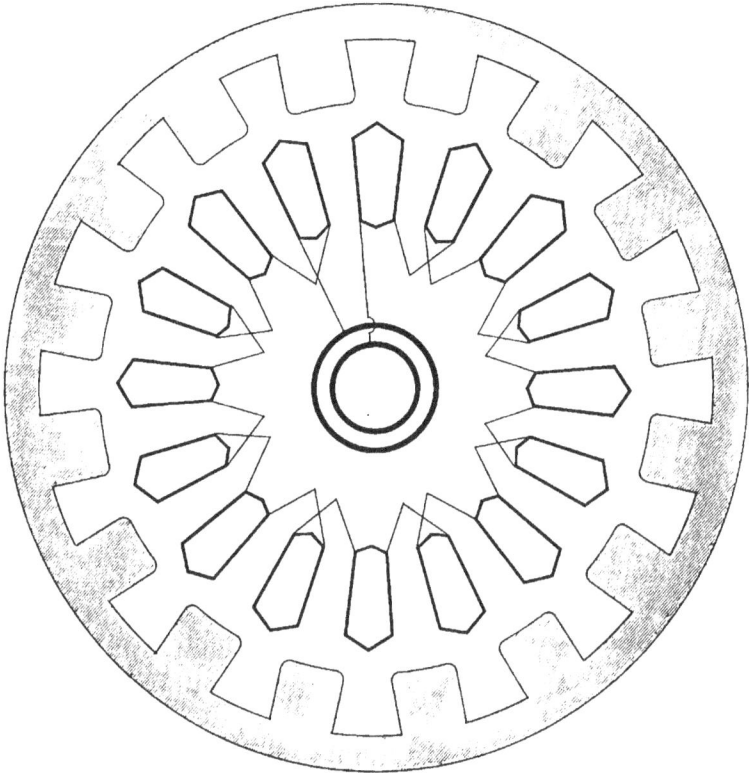

Fig. 81

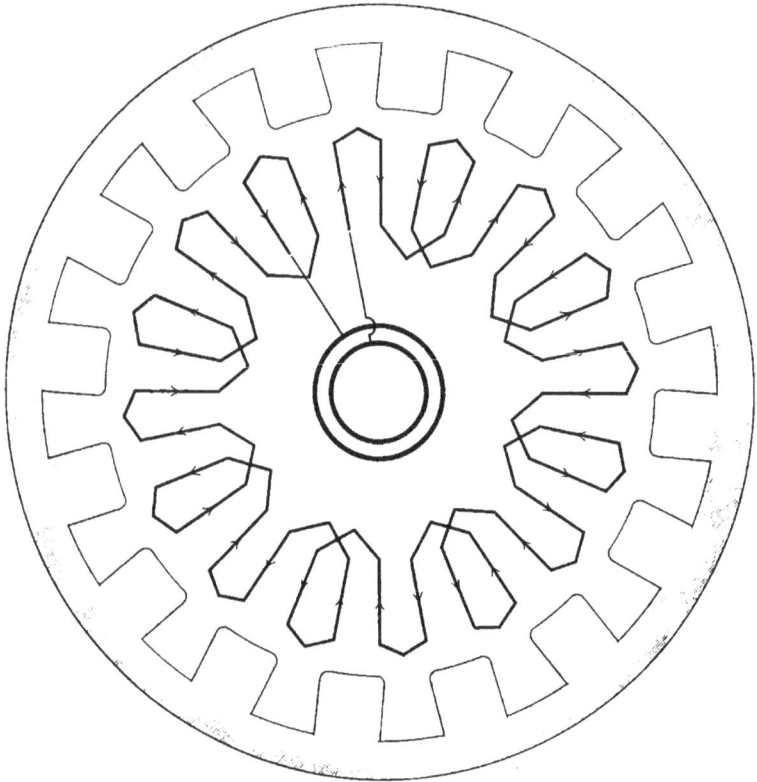

Fig. 82

Figure 82 gives a bar winding with two bars per pole piece. It corresponds to the coil winding of Fig. 81. These two windings (Figs. 81 and 82) could probably be used to advantage in many cases, but, of course, their disadvantages should be carefully considered.

Figure 83 represents another two-coil winding. It would seldom be used, as it has the faults and lacks the merit of the winding given in Fig. 81.

If, however, the coils, instead of being evenly spaced, were brought into groups of two, not very far apart, it would, to some extent, have part of the advantages of the uni-coil construction, and would partly overcome some of the faults of the latter. If modified in this way, it would partake of the nature of the windings given in Figs. 97, 98, and 99, and the remarks made in connection with these figures should be referred to.

If Figs. 81 and 82 should be similarly treated (that is, if the coils should be brought into groups of two coils each, not very far apart), the result would be a winding comparable to those given in Figs. 97 and 99.

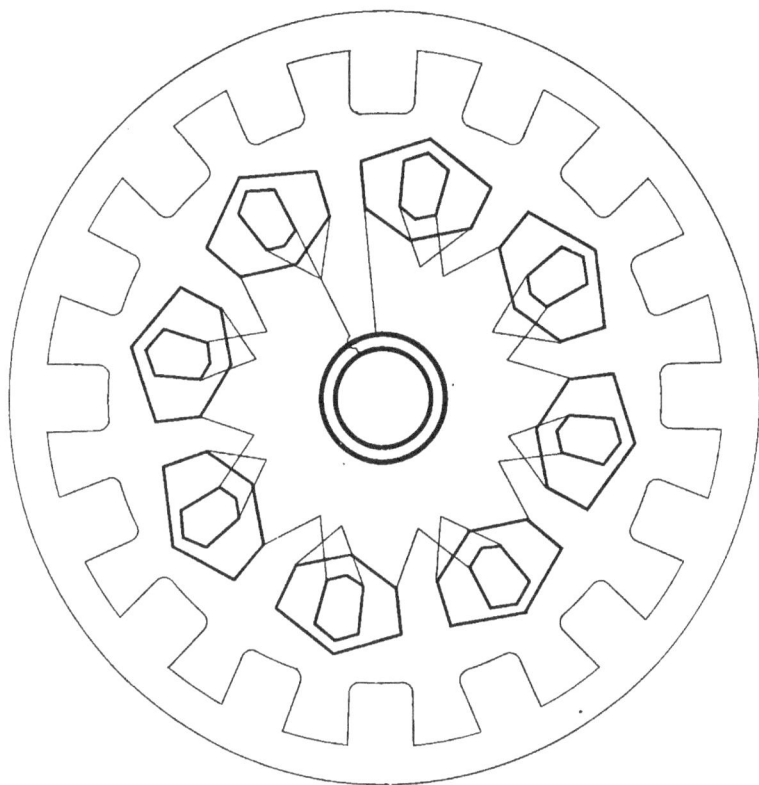

Fig. 83

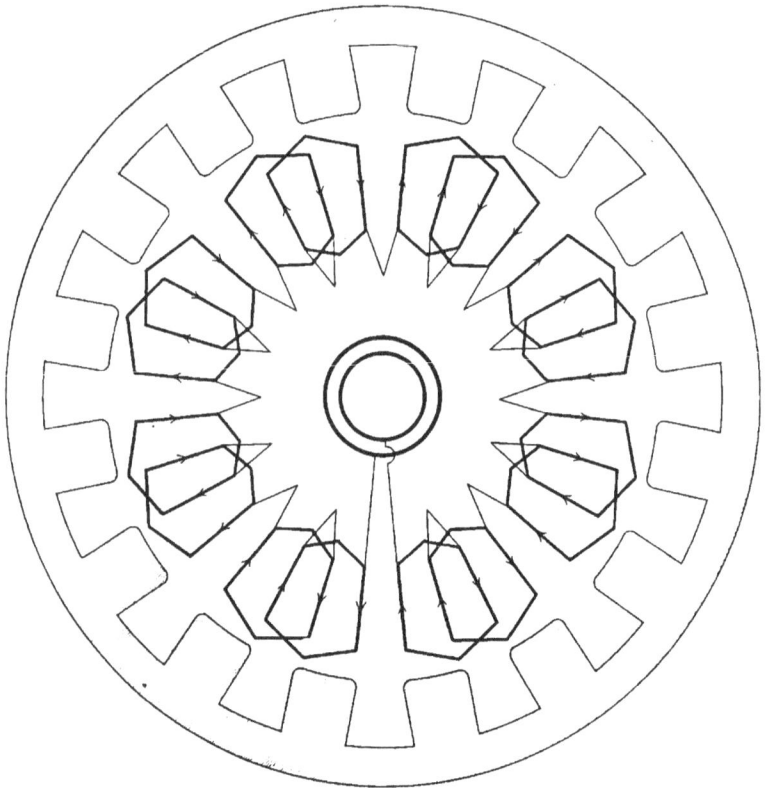

Fig. 84

Figure 84 is a diagram of another two-coil winding. It is connected as a single-phase alternator, but except for the manner of interconnection of the coils it is identical with the quarter-phase winding given in Fig. 100. It will give the same voltage as would Fig. 100, if the two components of the quarter-phase winding should be connected in series.

For this reason (that is, because when reconnected, it makes a good quarter-phase winding), it might sometimes be used, but of course, would, as stated in connection with previous windings, require a greater length of wire to generate the same voltage than a uni-coil winding, and would naturally have a greater armature self-induction. But, of course, the decrease in self-induction due to the multi-coil construction would somewhat compensate for this increase.

Figure 85 gives a diagram for a single-phase bar winding, corresponding to Fig. 84. It is only of interest as showing that it is identical with Fig. 82, except that the long-end connections which were at the collector ring end in Fig. 82 are now at the other end.

It should be noted that all these multi-coil windings now under consideration would, for a given terminal voltage, require much more field excitation at no load than corresponding uni-coil windings. But at full load they would, in some cases, require little if any more field excitation than would be the case with uni-coil windings. As a result of these considerations it will be seen to be necessary in any particular case to observe the requirements for the field excitation as regards permissible regulation, heating, etc., when deciding upon the type of armature winding to adopt.

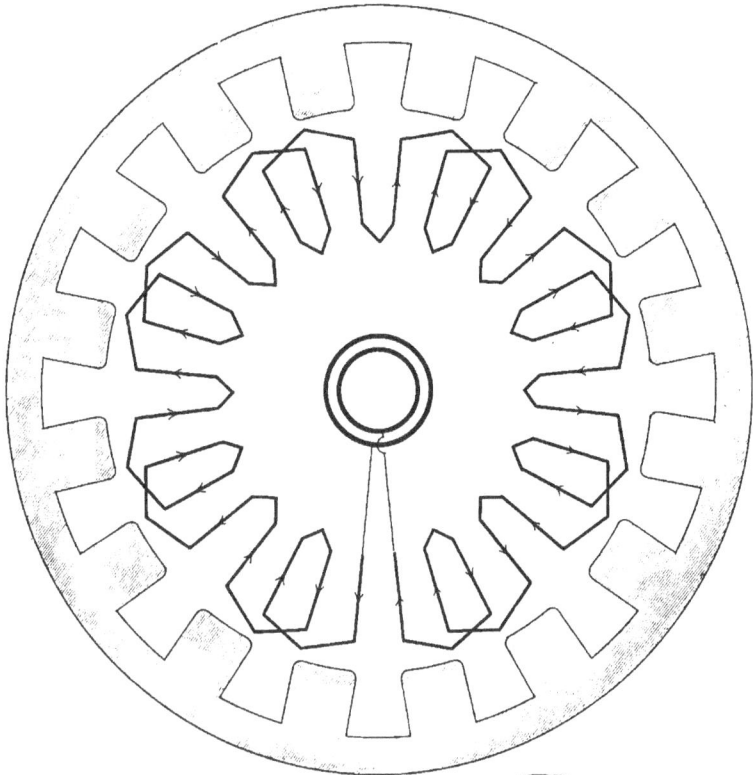

Fig. 85

Fig. 86

Figure 86 should be compared with Fig. 80. It is quite like the latter, except that it has two coils per pole piece instead of one. It would, of course, not be used, as it has such long end connections.

The number of poles is sixteen. Such a winding with twelve, eighteen, or twenty-four poles could be used in a three-phase armature of one-third the number of poles by merely changing the interconnections of the coils. Figure 123 gives such a diagram for a three-phase alternator in an eight-pole frame.

The mechanical arrangement of such windings as those given in Figs. 80, 86, and 123 is exceptionally good, although in the case of Figs. 80 and 86, they are much less simple, as single-phase windings, than those that do not cross.

Figure 87 represents a winding with two groups of coils per pole, and two coils per group. It will be seen to be identical with the two-phase winding of Fig. 108, except that it is connected up as a single-phase winding. With the exception of the sequence of interconnection of the coils, it may be considered to be two windings like Fig. 77, one of which is displaced 90°, so that its conductors lie half way between those of the other.

Its end connections permit of good mechanical arrangement; very much, in fact, like that of Figs. 80, 86, and 128.

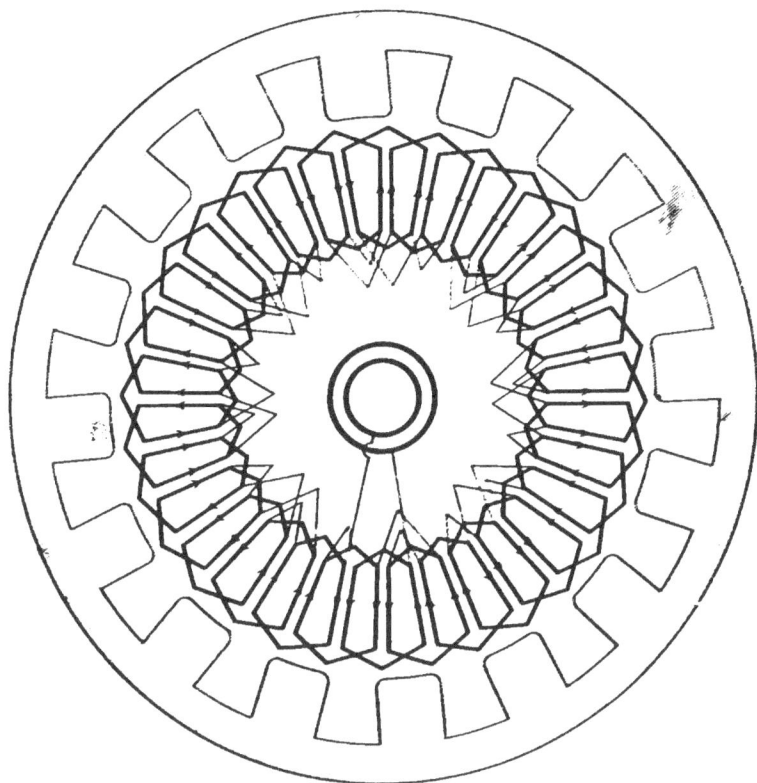

Fig. 87

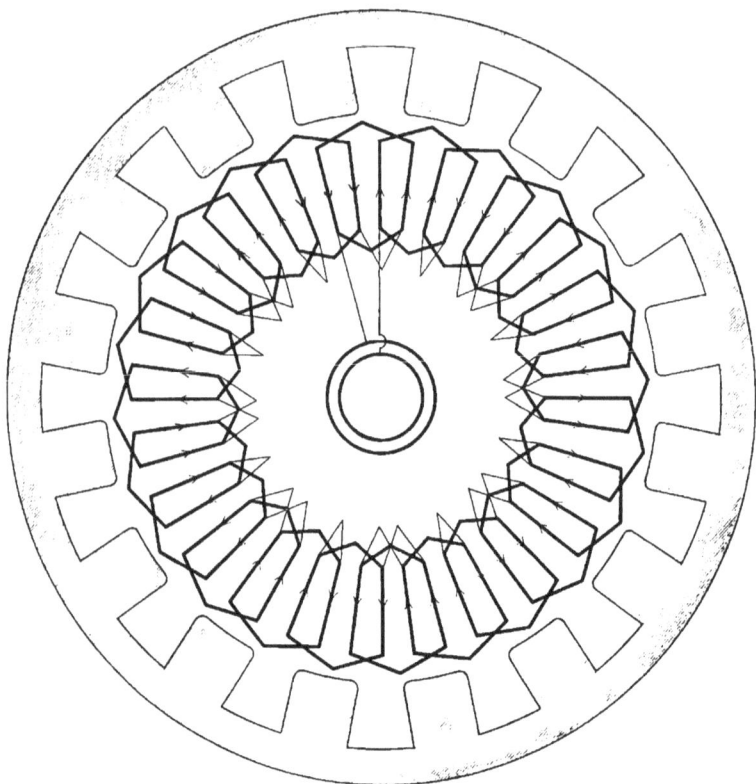

Fig. 88'

Figure 88 shows a useful three-coil winding. It has all the advantages and disadvantages already noted of multi-coil armatures.

The end connections can be very nicely arranged, so as to permit of winding on forms and slipping them into slots. Only two different shapes of forms are necessary ; one-half of the coils would be wound in one of them, and the rest in the other.

It will be seen that it is really the three-phase winding of Fig. 116 connected up as a single-phase winding. For this reason, among others, it might be expected to be of service where it would be of advantage to have armatures which could be used interchangeably for single- or three-phase work. Most three-phase windings could, of course, be similarly used.

As a single-phase winding *per se*, Fig. 88 is excelled by the windings of Figs. 92 and 94, which require a smaller length of end conductors.

Figure 89 is the bar winding corresponding to the coil winding of Fig. 88. It is not a generally useful winding. Among other faults it has three different lengths of end connections, half of them being very long. In this respect it is excelled by the winding given in Fig. 93. The end connections at one end are perfectly regular, but this would seldom be considered to compensate for the needlessly great length of copper employed.

This winding is an example of the importance of thoroughly examining many diagrams before adopting a winding for a certain case ; for it is not at once apparent that this winding could be improved upon, and if thought of first, might be chosen without further investigation.

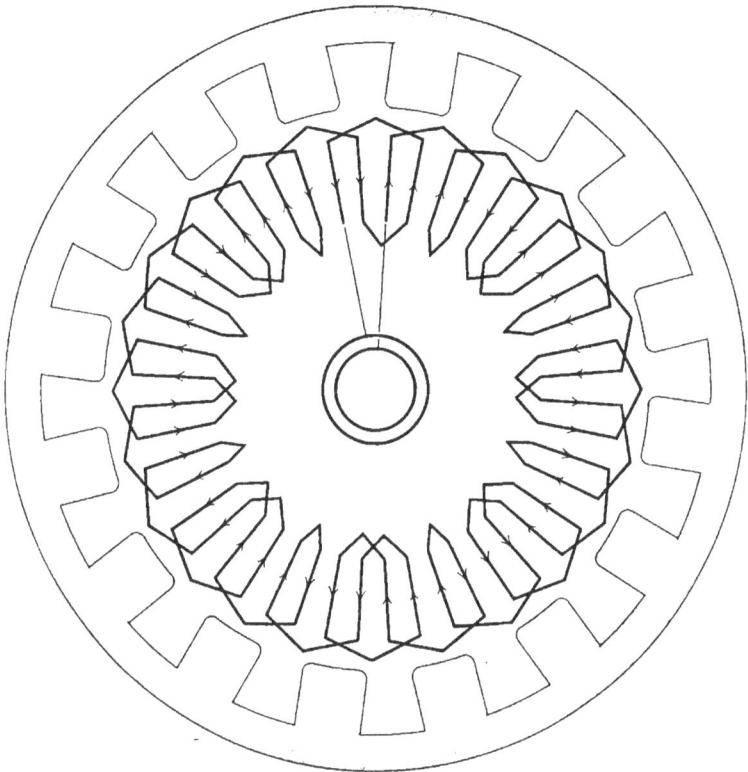

Fig. 89

Fig. 90

Figure 90 gives a coil winding very similar to that of Fig. 88.   But the end crossings would render it very inconvenient, and the space at the ends of the armature is not so well utilized as it was in Fig. 88.   This would tend to an undesirable concentration of the heating.

Unlike Fig. 88, the winding would not interfere with the armature, being made in segments for convenience of shipment.   But Figs. 92 and 94, which require less copper in the end connections, also possess this advantage, Fig. 94 to the greatest extent of all.

Figure 91 has all the faults of Figs. 89 and 90.   It is the bar winding corresponding to Fig. 90.   It is inferior to the winding shown in Fig. 93.

It has the advantage that the winding is more symmetrical as a whole than many better windings, and it is for this reason readily constructed and connected up, with little liability of error.   It is a great help for the winder to be able to intelligently perform his work, and windings that are, electrically and mechanically, to a small extent inferior, might in some cases consistently be adopted because of the simplicity of winding.   They also permit of the more ready locating and correcting of faults that are liable to develop during the practical operation of the machinery.

Fig. 91

Fig. 92

Figure 92 is another three-coil winding. It gives the same results as Figs. 88 and 90, but with less copper, as it has shorter end connections. It is also simpler, as there is much less overlapping at the ends. Only two sizes of coils are necessary.

The chief point of inferiority to Figs. 88 and 90 is that it cannot be connected up as a three-phase armature.

Even Fig. 92 is not so good as Fig. 94 (to be described later), which latter has still shorter end connections and less crossings.

There is no good bar winding corresponding to Fig. 92.

Figure 92 possesses the advantage noted in the discussion of Fig. 90, that the armature may be built and shipped in sections without interfering with the winding.

Figure 93 is the best bar winding for three bars per pole piece. It is distinctly superior to Figs. 89 and 91, as it has much shorter end connections. It requires, moreover, only two different lengths of end connections, whereas Figs. 89 and 91 each require three.

The following diagram is a section of a bar winding with five bars per pole piece : —

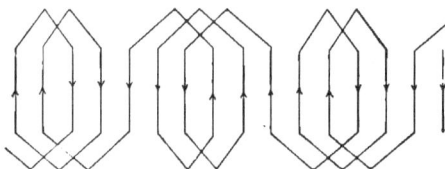

Fig. 93

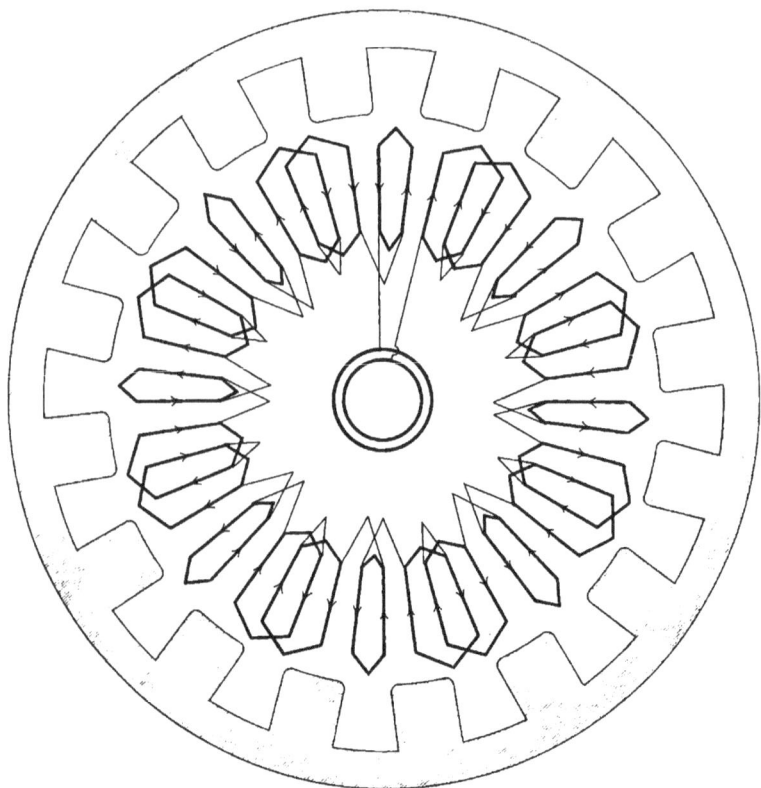

Fig. 94

Figure 94 is the coil winding corresponding to the bar winding of Fig. 93.

This coil winding is superior to that of Figs. 88, 90, and 92, in that it gives the same result with much shorter end connections and with fewer crossings of the end connections. Like Fig. 92, it cannot be connected up as a three-phase alternator, it being in this respect inferior to Figs. 88 and 90.

The winding of Fig. 94 could readily be built in sections in cases where it would be necessary to ship the armature in segments.

Figure 95 is a coil winding electrically equivalent to Figs. 88, 90, 92, and 94.

Windings of this class may readily be derived from the example given in Fig. 95, for any desired number of coils per pole piece. It often works out well from a mechanical standpoint, and although the end connections are necessarily longer than in the preceding windings, it will frequently be found useful.

The various coils might with advantage be grouped to a greater or less extent, in accordance with the principles exemplified in Figs. 97, 98, and 99, which, together with the accompanying text, should be consulted in this connection.

Fig. 95

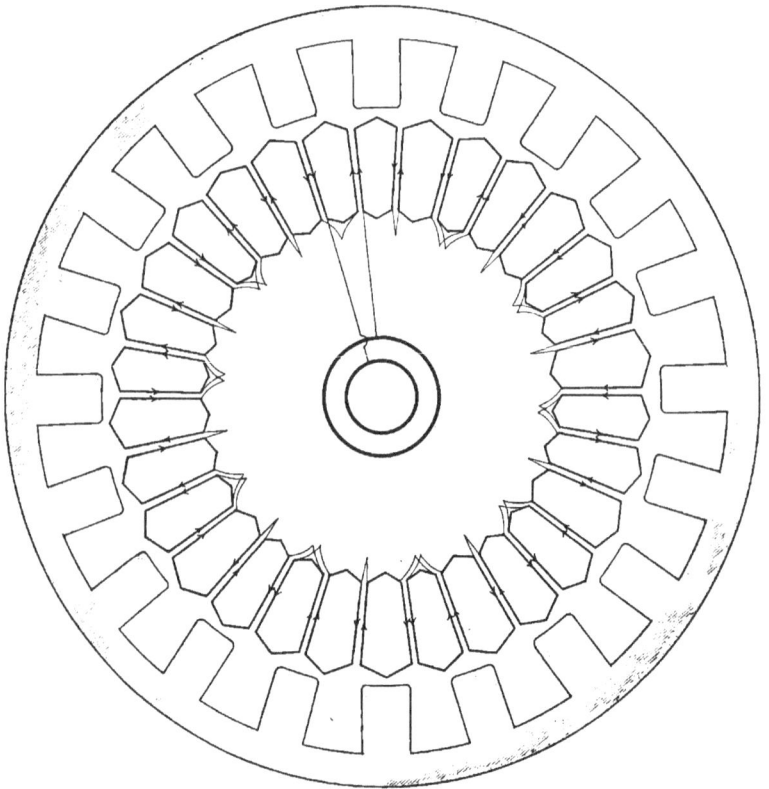

Fig. 96

Figure 96 gives a coil winding with one and one-half coils per pole piece. It has two coils per group. It is really a winding such as Fig. 77, put in a field with two-thirds as many poles as the armature has coils. Thus in Fig. 96 there are thirty armature coils and twenty field poles. There is disadvantageous counter-induction which makes the use of more armature copper necessary than would be used in a uni-coil winding. The armature could, however, be used interchangeably in fields with n and with $\frac{2}{3}n$ poles, which property permits of the use of the armature in cases where different speeds or periodicities may be called for.

Also by changing the interconnections of the coils, an excellent three-phase armature is obtained. The three-phase connections of such a winding are given in Fig. 119.

Moreover, owing to the fact that when one side of a coil is under a field pole, the other is between two poles, the self-induction of such a winding is low, and is fairly uniform for all positions of the armature.

Many of the multi-coil windings given heretofore have been somewhat undesirable by reason of the counter-induction, which made it necessary to have a greater length of conductor for a given voltage than would have been necessary if the conductors had been concentrated in one coil per pole piece.

Figure 97 is a winding which, while retaining to a great extent many of the advantages of multi-coil windings, is usually as good with regard to its freedom from counter-induction as a uni-coil winding with evenly spread coils.

It is in fact one of the two windings of the quarter-phase diagram of Fig. 104.

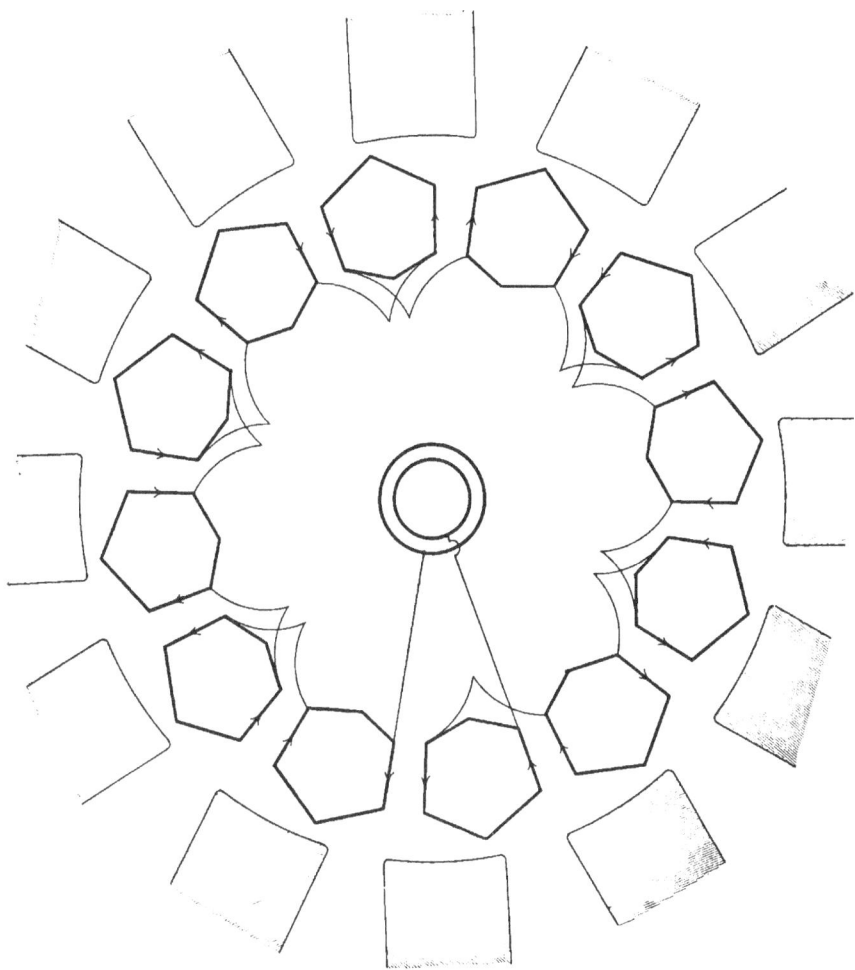

Fig. 97

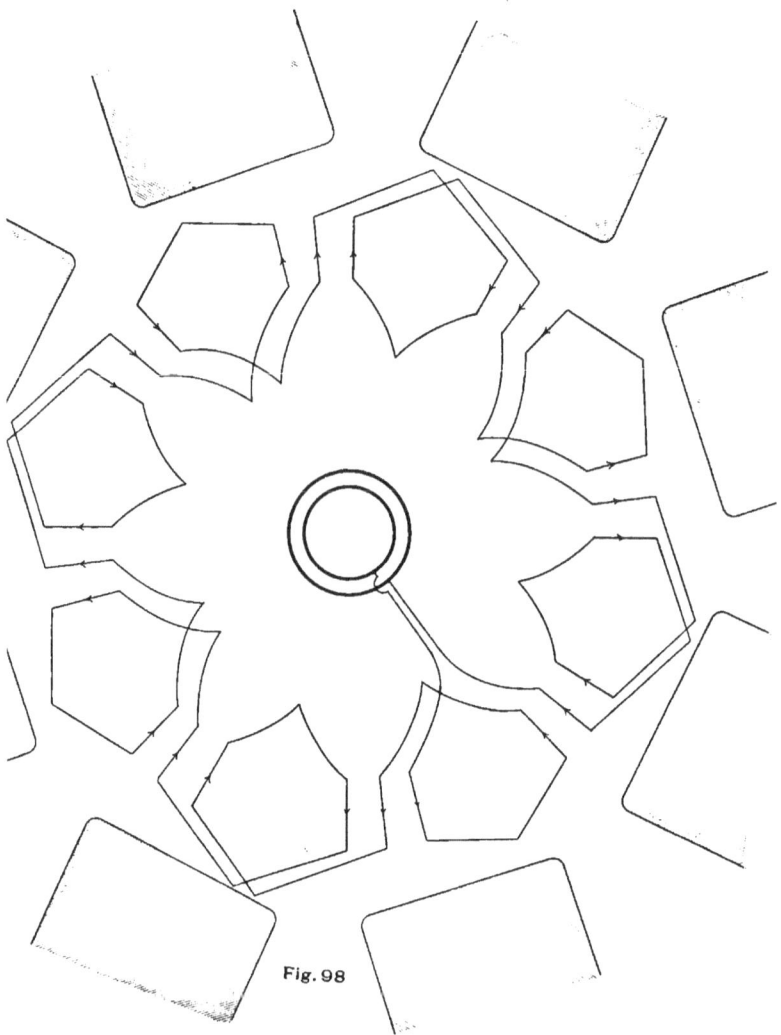

Fig. 98

Figure 98 does not differ essentially from Fig. 97 as far as regards the point that it is intended to illustrate. It, also, is one of the two windings of a quarter-phase armature, being in fact derived from the quarter-phase diagram of Fig. 112.

Other excellent diagrams of this type may be derived by considering one of the two windings of the quarter-phase armatures shown in Figs. 105, 106, 107, and 111.

Figure 99, like its predecessors, Figs. 97 and 98, has its coils arranged in groups in the periphery of the armature. It has to some extent their advantages and disadvantages. It differs from them in utilizing two-thirds of the available space, instead of one-half, and is more of a compromise with the uniformly distributed windings.

It is obvious that windings such as the three just given may readily be derived from any of the evenly distributed multiphase windings by simply discarding one or more of the windings belonging to the respective phases of such diagrams. They may also be derived from many of the single-phase windings by shifting the coils laterally from the normal position into the desired groups.

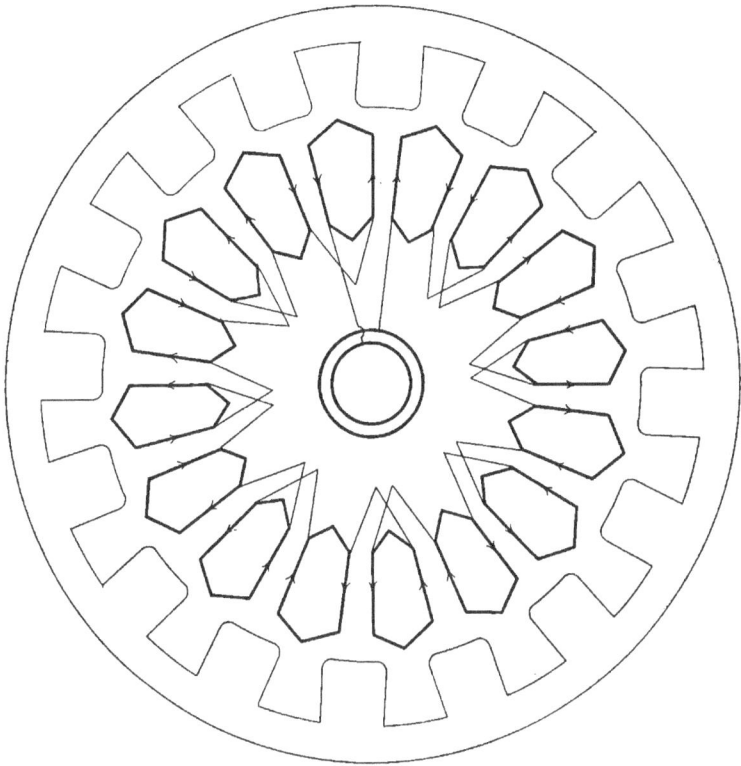

Fig. 99

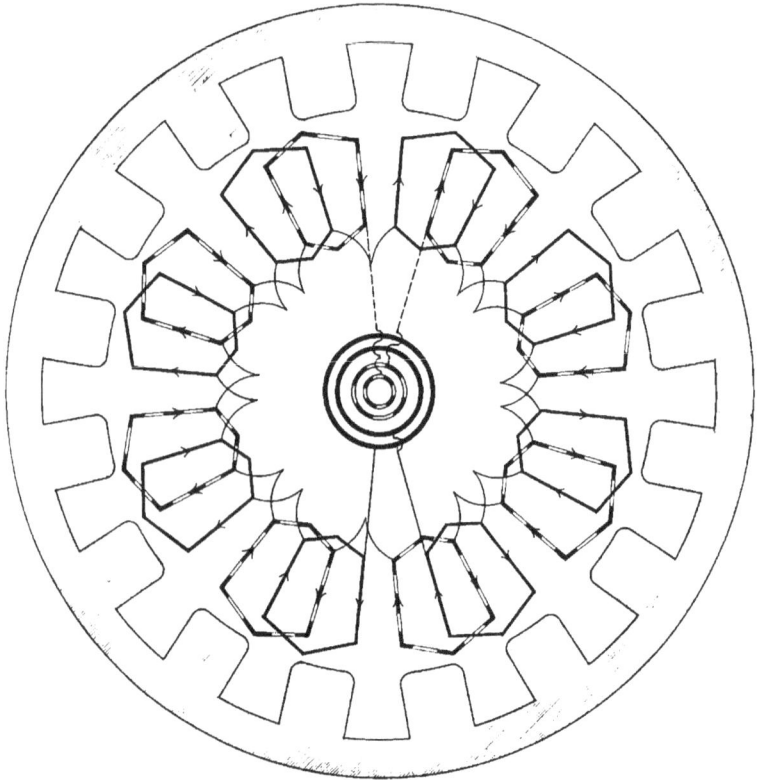

Fig. 100

CHAPTER XIV.

QUARTER-PHASE WINDINGS.

FIGURE 100 represents a quarter-phase coil winding with one group of conductors per pole piece per phase. In accordance with the nomenclature already adopted, this would be known as a uni-coil winding; although it has but *one* coil per pole piece per *phase*, it has *two* coils per *pole piece*.

The two windings are represented, respectively, by full and broken lines. The winding is quite simple, but has the objection of crossings at the ends. In this respect it is inferior to the style of winding represented by the diagram of Fig. 102.

Three collector rings could be used, one of them being common to each winding. In the diagrams, however, four collector rings will be shown, this being the method now generally used. In connection with a system employing three collector rings, the standard quarter-phase commutating machines (to be described later) could not be used.

213

Figure 101 is the bar winding corresponding to Fig. 100. It does not well utilize all of the available space on the armature ends. This is generally not a great objection in the case of uni-coil windings, as there is in such cases plenty of room on the ends, but, other things being equal, it is of course preferable to have windings uniformly distributed at the ends as well as on the surface. In this connection Fig. 109 should be studied, and it will be seen that by placing two conductors in a group a perfectly symmetrical design is obtained with one group per pole piece.

A decided objection to this arrangement would be that adjacent conductors would have between them large differences of potential, whereas in Fig. 101 there are but few points in which neighboring conductors have between them any considerable percentage of the total terminal voltage.

Fig. 101

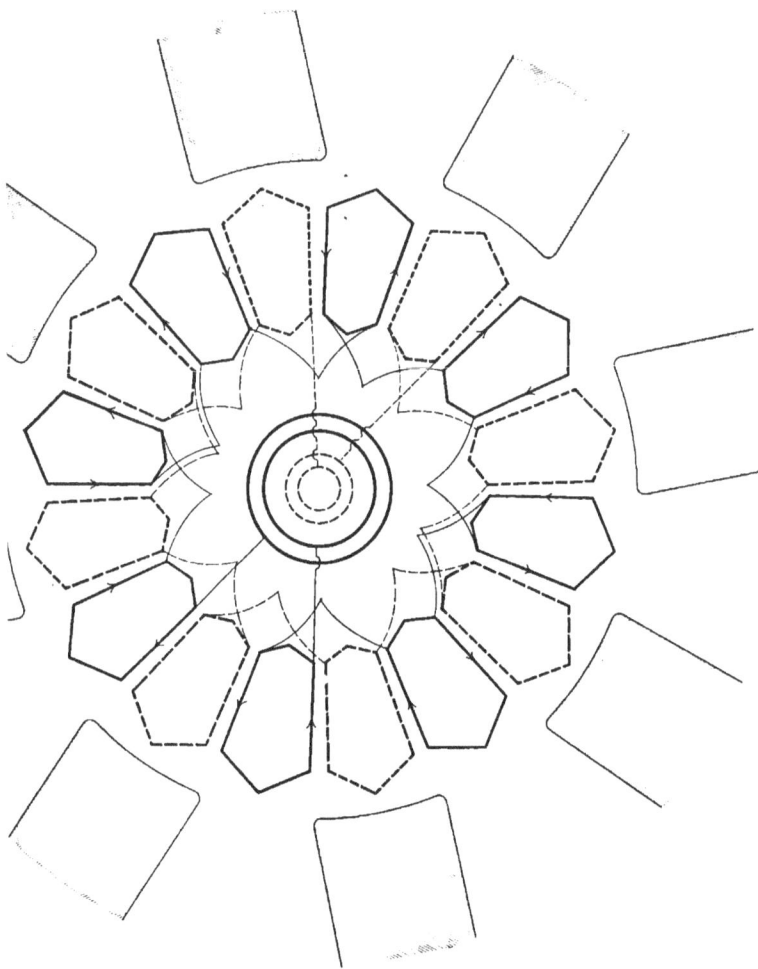

Fig 102

Figure 102 is a non-overlapping quarter-phase winding
with one group of conductors per pole piece per phase. It has
the advantage over Fig. 100 that there are no crossings
at the ends of the armature, and that it utilizes the end space
more completely, thus bringing about a better distribution of
the necessary heating losses in the copper. Its chief fault is
that if the width of the pole face is over one-half of the
distance between pole centers, the coils never embrace the
total flux from one pole piece. However, at full load, the
area occupied by the flux is narrower, and a greater portion
would be included than at no load, so that this objection
would not be so serious as would appear at first sight.
Moreover, the necessary space allowance for the field
winding will in many cases not permit the width of the pole
piece to be sufficiently great to cause any trouble in this
respect. Mechanically, this is an excellent winding, being,
in fact, the single-phase winding given in Fig. 77, for double
the number of poles.

The remarks made in connection with Fig. 96 (single-
phase alternating winding with one and one-half slots per
pole piece) should also be considered in studying this winding. Consult also Fig. 119 and corresponding text.

Figure 103, which like Fig. 102 has two coils per group, is not open to the objection discussed on the preceding page. It has, however, crossings at the ends. It is to be preferred to Fig. 100 for the reason that the end space is more effectively utilized, but the additional crossings would require a somewhat greater length of wire than would be necessary in Fig. 100.

Bar windings could be built corresponding to the coil windings of Figs. 102 and 103. They would not be symmetrical at both ends, but might advantageously prove applicable for certain cases. The two bars of a group could be placed either over each other, or side by side. With smooth-core construction the latter arrangement would be adopted, and often also in ironclad armatures with bar windings.

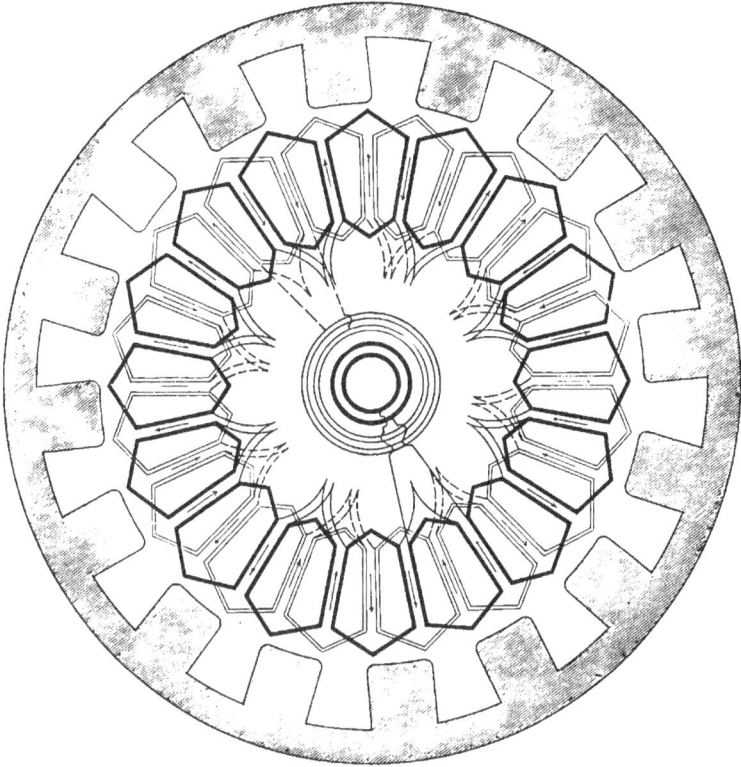

Fig. 103

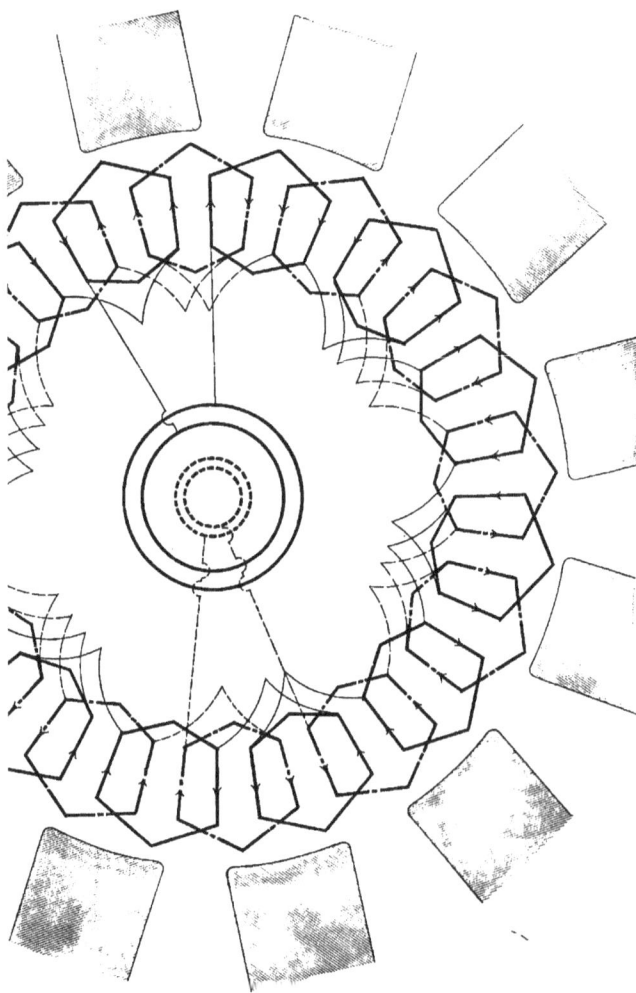

Fig. 104

Figure 104 is a quarter-phase coil winding with two conductors per pole piece per phase.   It is entirely symmetrical,
and utilizes all the winding space to the best advantage.   The
crossings at the ends are unavoidable, but may be made
thoroughly satisfactory from a mechanical standpoint by
proceding in the manner shown most clearly in the diagram
of Fig. 123.

Such windings are applicable to quarter-phase armatures
with any even number of coils per pole piece per phase.

In studying Fig. 104 it will be instructive to examine
Fig. 97, which is one of the two windings of Fig. 104.

Figure 105 is electrically equivalent to Fig. 104. The winding might sometimes be used, although it would for most purposes be excelled by Fig. 104.

It will be noted that the end connections are longer, and that they occupy a greater depth. Much of the end space is wasted. This winding is superior to that of Fig. 104, in that the coils are so located as to make it very plain how the connections should run. This would be of great assistance to the winder, and would, moreover, facilitate the detection and correction of faults that might develop in practical working.

An armature with such a winding could be built and shipped in segments.

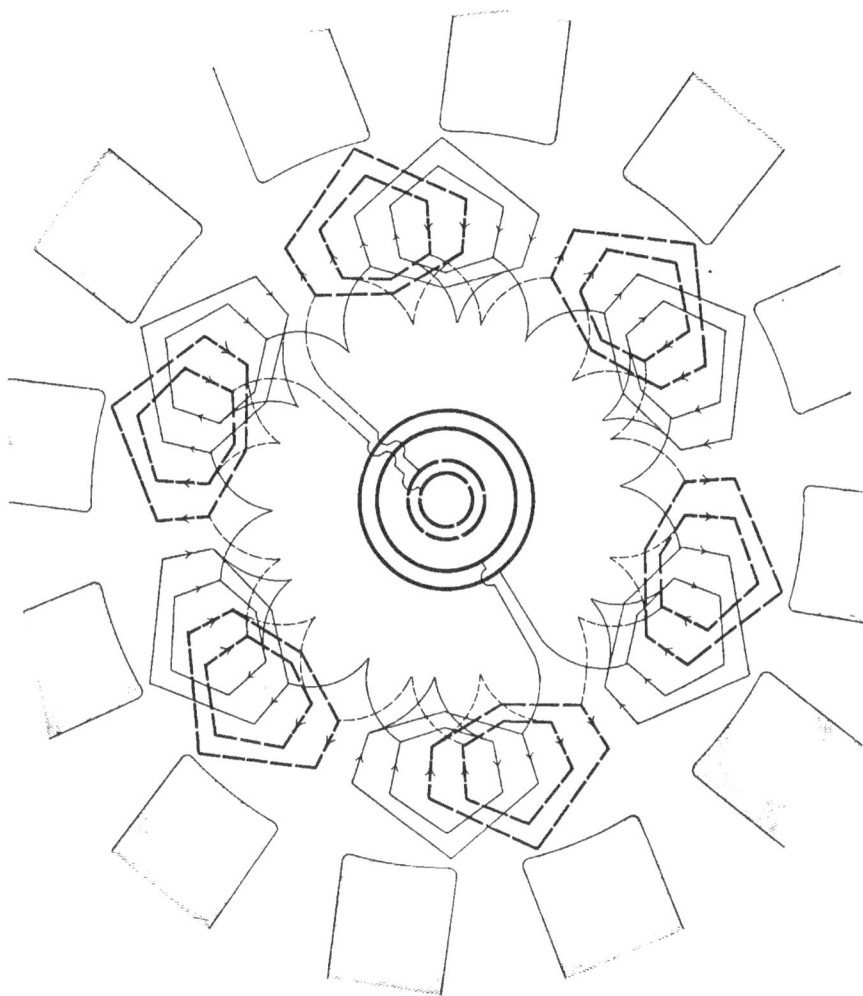

Fig. 105

Fig. 106

Figure 106 is a bar winding differing but little in principle from the coil winding of Fig. 105. The space is uniformly occupied at the collector ring end, but is not at the other end.

This lack of uniformity in end connections is not of very great moment in bar windings with few bars per pole piece. Other things being equal, however, it would on the whole seem best to avoid it, although in special cases such disposition of the end-connections allows room much needed for mechanical arrangements.

Figure 107 is a bar winding corresponding to Fig. 104. It is a good example of the fact that very symmetrical coil windings often correspond to very unsymmetrical bar* windings, and *vice versa*. But, as noted on the preceding page, this lack of symmetry is in such cases not a great objection, and has, incidentally, some redeeming features.

One of the two windings of this diagram would, as mentioned on page 209, work out very well for a single-phase armature.

Fig. 107

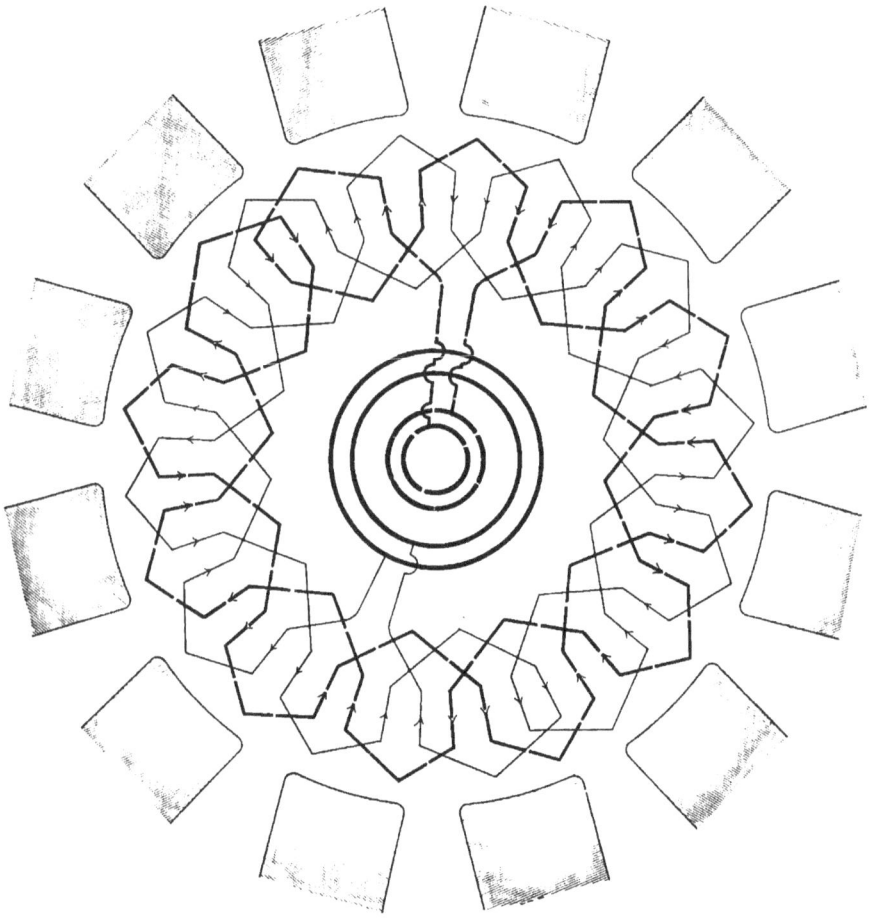

Fig. 108

Figure 108 is a much better bar winding than Fig. 107, though electrically equivalent.

It will be seen to be unsymmetrical at two points at the end distant from the collector   This irregularity consists in the end connections of the two adjacent bars starting off in the same direction, instead of, as in all other parts of the winding except these two, going in opposite directions. Four of the end connections have to be longer than the rest.

This winding is practically the same as the following one, Fig. 109, except that the above-described irregularity is introduced instead of making use of the cross-connections shown in Fig. 109.

Figure 109 is a symmetrical quarter-phase bar winding with two conductors per pole piece per phase. If used for an ironclad or projection armature, it may have four slots per pole piece with one conductor per slot, or two slots per pole piece with two conductors per slot.

Examination will show that it is essentially a twelve-pole armature with four separate series of windings of twelve bars each. These four windings are connected up into two windings of twenty-four conductors each.

At the front end $y=5$, and at the back end $y=3$, therefore average $y=4$.

As pointed out in the discussion of Fig. 101, Figs. 108 and 109 have the fault that neighboring conductors have between them large percentages of the total potential of the armature, and this would sometimes be objectionable in cases of high potential windings.

It will doubtless have been observed that in the case of quarter-phase windings, multi-coil construction does not have to so great an extent the fault pointed out in the case of corresponding single-phase windings, of useless counter-electromotive forces.

The coils of one phase usually embrace practically the entire flux, because the two groups of conductors, forming respectively the two sides of a coil, are usually separated by a group forming one side of a coil belonging to the winding of the other phase.

This advantage is possessed in a still greater degree by the three-phase windings, which will be discussed later.

Exceptions to the above statement often occur in cases where single and multi-phase alternating windings are obtained from ordinary direct-current windings.

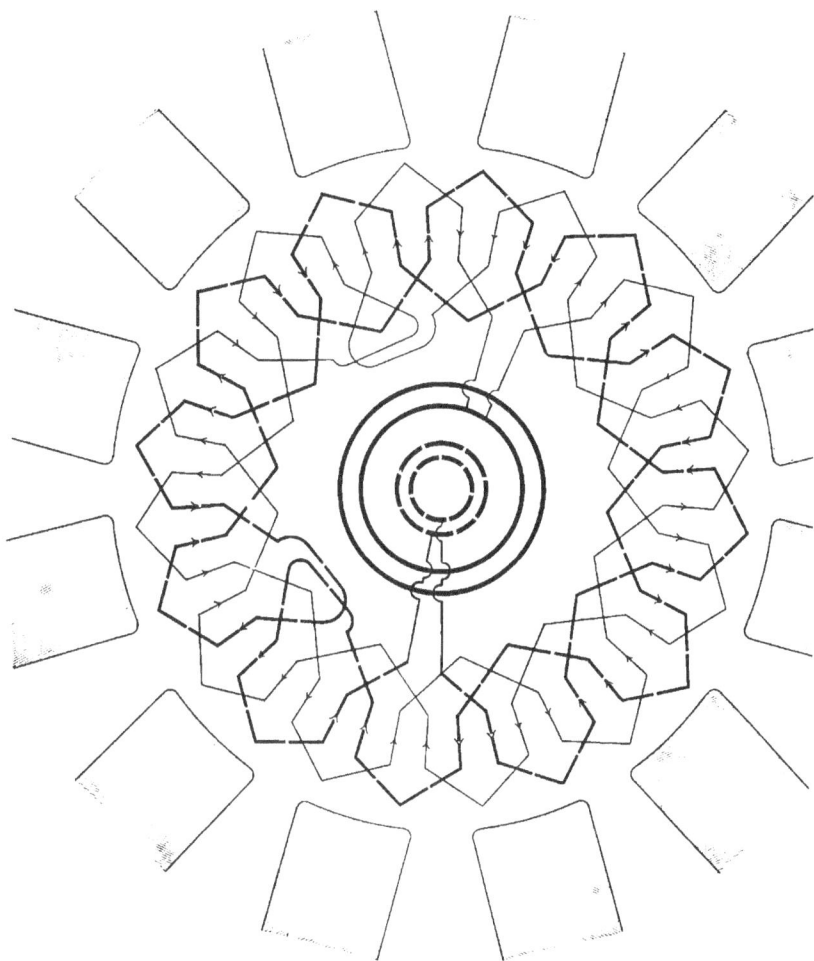

Fig. 109

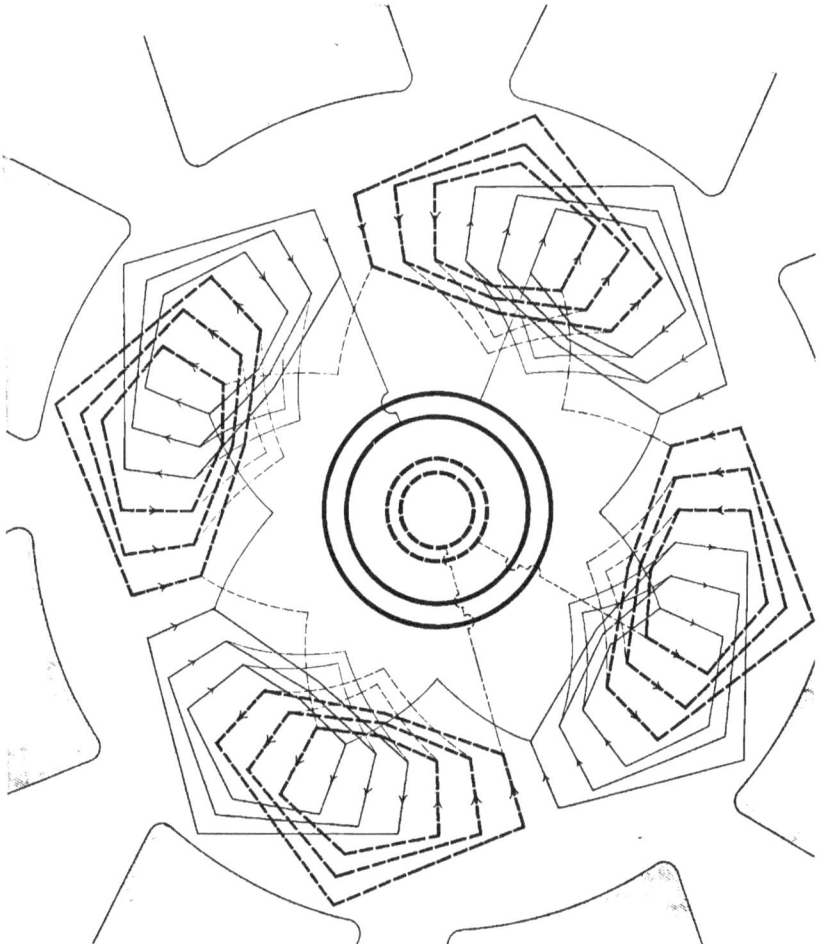

Fig. 110

Figure 110 represents a quarter-phase coil winding with three slots per pole piece per phase. It does not utilize very uniformly the end space on the armature, the end connections being three layers deep at some points and much less at others.

An advantage of this winding is the well-defined nature of the coils, rendering it easy to see just how they should be connected. The winding might also be necessary, if it should be required that the armature should be built so that it could be shipped in segments.

Figure 111 is electrically equivalent to Fig. 110, but the end connections are only two layers deep, are shorter, and are better distributed over the ends of the armature. Where the number of coils per pole piece per phase must be odd, windings such as those given in Figs. 110 and 111 must for quarter-phase armatures often be chosen. It is quite apparent that, except in special cases, the style of diagram shown in Fig. 111 will give the best result.

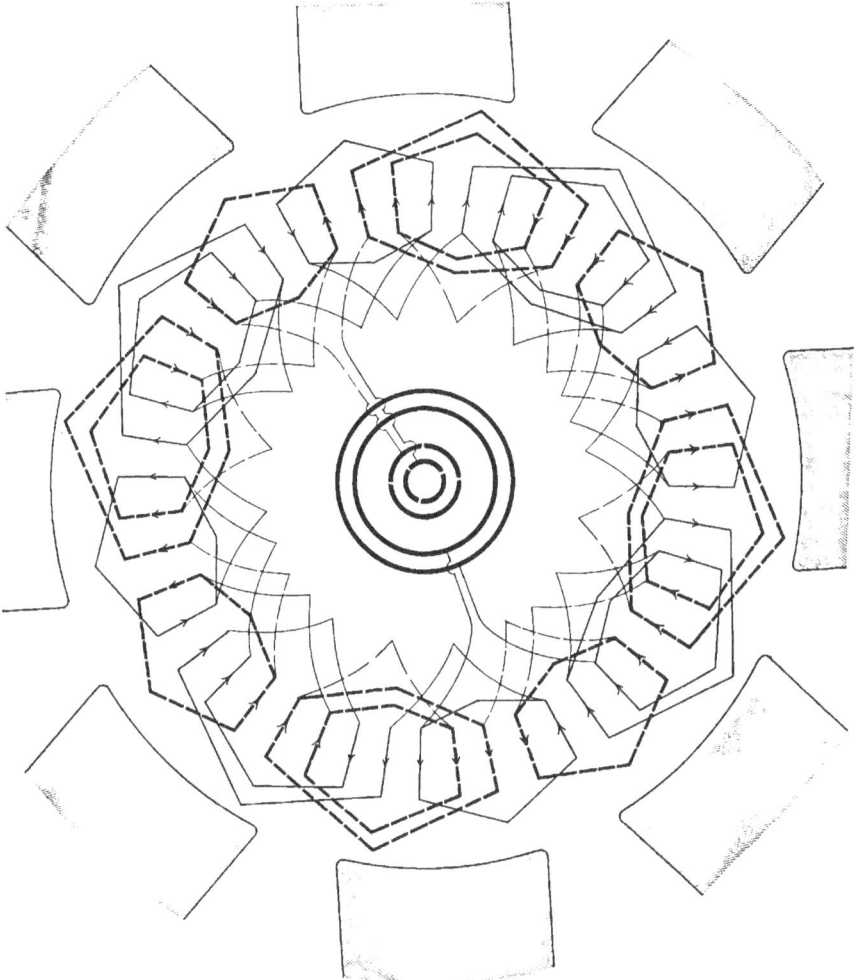

Fig. 111

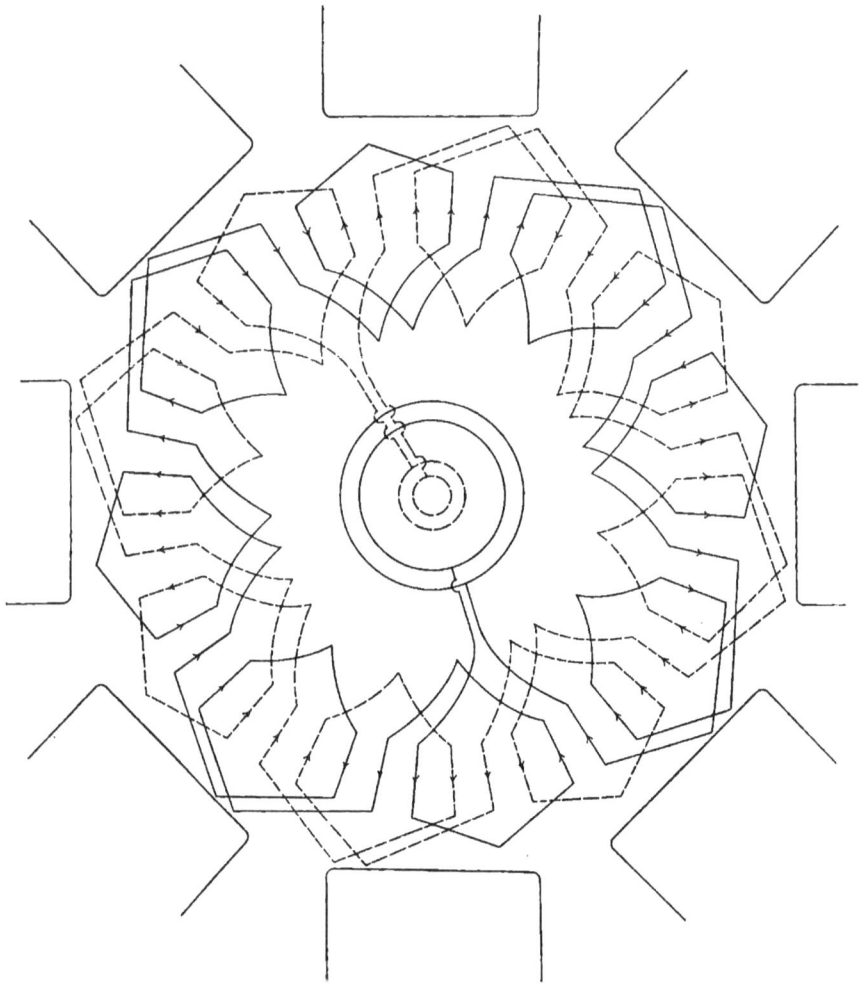

Fig. 112

Figure 112 is a bar winding corresponding to the coil winding of Fig. 111. Although not symmetrical, the end connections are fairly well distributed, and there would be in but very few places any great percentage of the total difference of potential between adjacent conductors. Several different lengths of end connections would necessarily have to be employed.

One of the two windings of this diagram has already been given in Fig. 98 in Chapter XIII. on Single-Phase Windings.

Figure 118 represents a quarter-phase bar winding with
four conductors per pole piece per phase.   It is perfectly
symmetrical, and may have one, two, or four conductors per
slot, as desired.

This winding is like that of Fig. 109, except that four
sets of elementary windings are connected in series to form
one of the two phases, instead of two sets, as was the case
in Fig. 109.

If one-half or one-quarter as great a terminal electromotive force should be desired, two, or all four, of these
elementary windings could be connected in parallel between
the collector rings, instead of joining them in series as
shown.

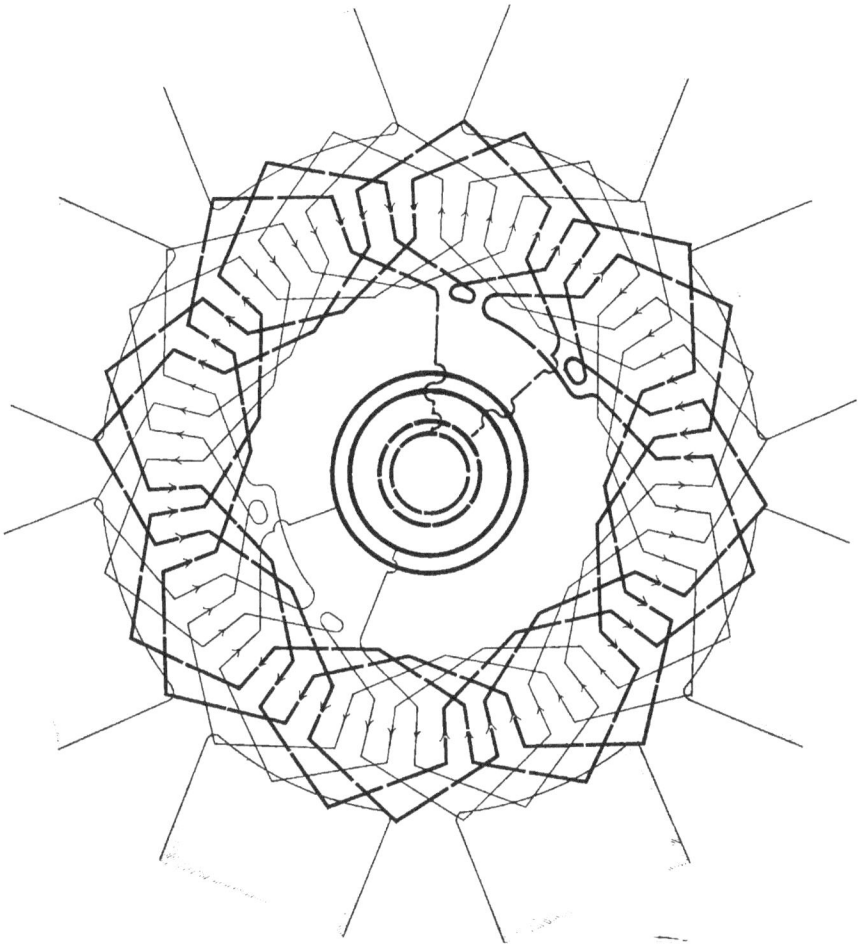

Fig. 113

N

Fig. 114

TWO-CIRCUIT WINDING FOR QUARTER-PHASE CONTINUOUS CURRENT COMMUTATING MACHINE.

Figure 114 is the diagram for the winding for a commutating machine for deriving a continuous current from a quarter-phase alternating supply, or *vice versa*, or for a generator for supplying both continuous and quarter-phase systems.

Examination will show that it is the two-circuit single winding of Fig. 43 (Chap. VIII.), tapped off from four approximately equidistant points to four collector rings. As the winding consists of sixty-eight conductors, there should be seventeen conductors in each section, but for the convenience of having all the connections to the collector rings made at one end, the divisions are 16, 16, 18, and 18. With the large numbers of conductors used in practice, the irregularity produced by one conductor more or less would be of less importance, though always undesirable. In such a winding four points only of the armature are tapped independently of the number of poles.

TWELVE-CIRCUIT WINDING FOR QUARTER-PHASE CONTINUOUS-CURRENT COMMUTATING MACHINE.

Figure 115 is another winding for a quarter-phase continuous-current commutating machine. It is fundamentally a multiple-circuit, continuous-current winding, and requires four leads (one to each collector ring) for each *pair* of poles.

It is to be remembered that in quarter-phase continuous-current commutating machines, the effective voltage between collector rings 180° apart equals the continuous-current voltage multiplied by .707 (or divided by 1.414).

Fig. 115

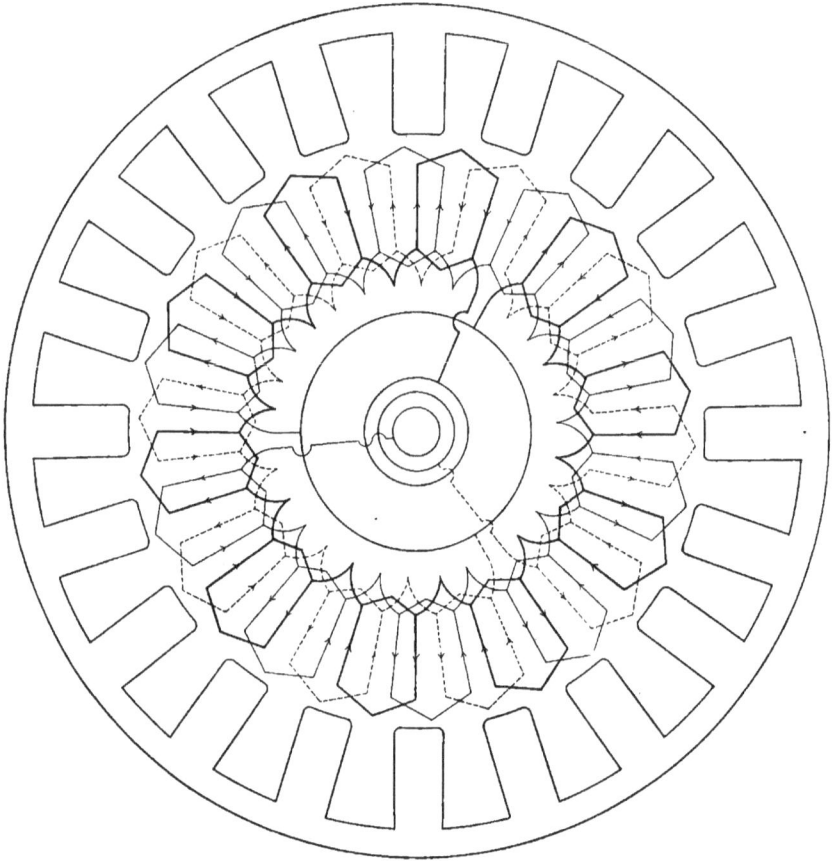

Fig. 116

CHAPTER XV.

THREE-PHASE WINDINGS.

FIGURE 116 is a three-phase coil winding with one set of conductors per pole piece per phase. The coils belonging to the three windings may be distinguished from each other by the three different styles of lines. The armature is connected in a manner technically known as the "Y" connection. The characteristic of this style of connecting three-phase windings is that one end of each of the three windings is brought to a common connection, the other three ends being carried to three collector rings.

Inasmuch as three-phase alternators have but recently been used to any considerable extent in practice, it may not be out of place to give as concisely as possible a few of the leading considerations involved in their practical construction and operation, as far as relates to the armature windings.

One complete cycle is passed through by any armature conductor while passing from a certain point opposite one pole piece, say the middle of the north pole, to the corresponding point opposite the next pole piece of the *same polarity*. This angular distance is usually spoken of as 360°, independently of the number of poles of the machine. Now, a three-phase armature winding is merely three single-phase windings, laid on the same armature, the conductors of the three windings, however, being located 120° (one-third of a cycle) behind each other. Any conductor of one winding is, therefore, at any instant, in a different phase from that of the conductors of the other windings. Thus, in the position represented in Fig. 116, the conductors represented by heavy lines are directly opposite the middle of the pole pieces, the light line conductors are located 120° behind them, and the dotted conductors are 120° behind the light conductors and 240° behind the heavy conductors.

Now it follows from the relative positions of the conductors of the three phases, that the electromotive forces generated in the three windings are 120° behind each other, and if they are sine waves, they may be represented, as in the following figure, by three sine curves displaced 120° behind each other.

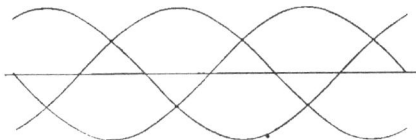

If the three circuits are equally loaded, these curves may also be considered to represent the corresponding instantaneous values of the current.

245

It will be noted that at every instant, the algebraic sum of the three currents is zero. Now instead of having three *pairs* of lines and brushes and collector rings, one end of each of the three windings is brought to a common connection, and a conductor from this common connection could be used as a common return for each of the three circuits. But, since the resultant current at every instant is zero, this conductor becomes superfluous and is omitted.

If the voltage between any ring and the common connection, that is, the voltage per phase, is equal to v, then the volts V between any pair of collector rings will be, —

$$V = \sqrt{3}\, v \text{ or } 1.732\, v.$$

The effective current will be equal in each of the three lines, and may be represented by C. With a non-inductive load, the watts output, W, will be, —

$$W = 3\; Cv = \frac{3\;CV}{\sqrt{3}} = 1.732\; CV.$$

If the load is inductive, the current C, for a given output W, will be greater than with a non-inductive load.

A safe and easily understood way of connecting the three windings correctly to the three collector rings and the common connection, is to consider that the winding whose conductors occupy the position in the middle of the pole piece, is carrying the maximum current, and to indicate its direction on the winding diagram by an arrow. The currents at the same instant in the conductors immediately next to it on the right and left are in the same direction, and should be so marked by arrow-heads. Now, from the sine curves given above, it will be seen that where one curve has a maximum value, the other two have a value half as great, and in the opposite direction. Therefore consider that the current in the winding occupying the position at the middle of the pole face is flowing away from the common connection. Then the currents in the other two windings, which are each of half the magnitude of the former, must both be flowing into the common connection; therefore join those ends of the three windings to the common connection, which will bring about this condition at this instant. Carry the other three ends to the three rings. This has been done in the upper diagram of Fig. 117, which represents a " Y " connected three-phase winding.

Another way of connecting up three-phase armatures is to connect the three windings in series in a closed circuit, and at every third of the total way through the circuit thus formed, to carry off a lead to one of the collector rings.

In the case of this, technically called the "delta" (Δ) connection, the current C in the line (*i.e.* beyond the collector rings) is $C = \sqrt{3}\, c$, or $C = 1.732\, c$, where $c =$ current in the winding. The volts per winding are in this case equal to the volts between each pair of collector rings ; that is, to the volts per phase. The watts output of a machine are, —

$$W = 3\; cV = \frac{3\;CV}{\sqrt{3}} = 1.732\; CV.$$

Examples of each of these two connections are given in Fig. 117.

The upper diagram represents a " Y " connected three-phase armature, and the lower diagram represents the very same armature, but with a "delta" (Δ) connection.

In connecting up the separate windings for a "delta" (Δ) connection, it is most convenient to choose the instant when the conductors of one phase are opposite the middle of a pole piece. Then assume those conductors to be carrying the maximum current, which is illustrated in the figure by the larger arrow-head.

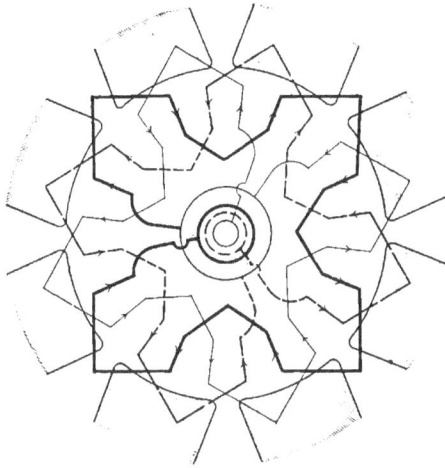

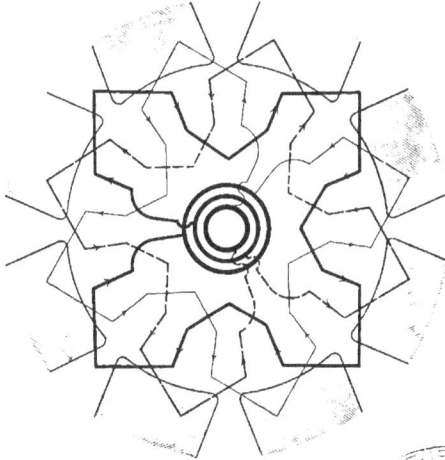

Fig. 117

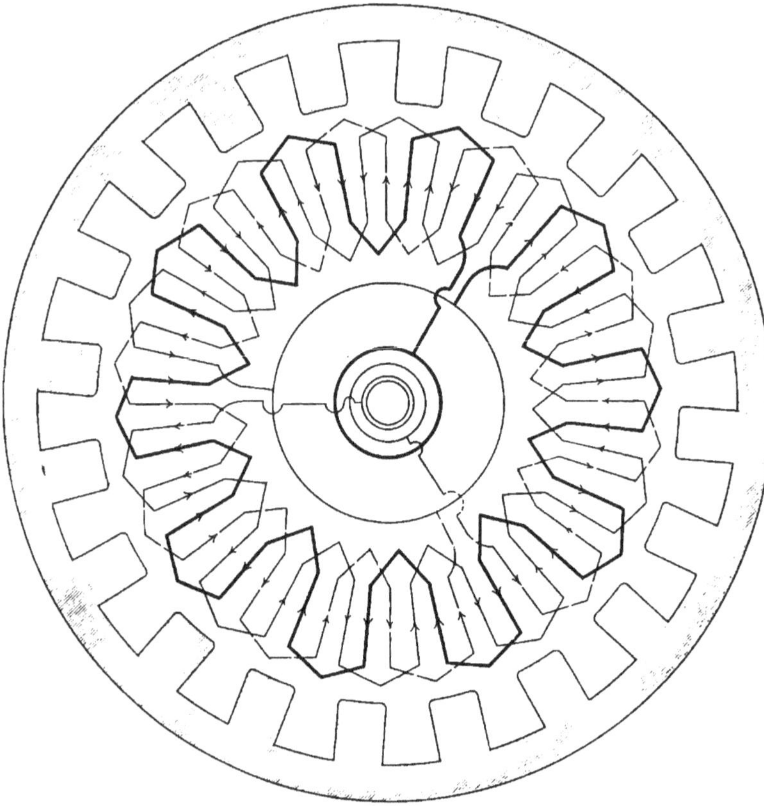

Fig. 118

The other two windings are at the same instant having induced in them currents of only one-half this magnitude. The condition of affairs in line and in winding is, for the instant, as represented in the following diagram.

From this it is seen, that, starting from the middle collector ring (corresponding to point a in the diagram), and following the direction of the current, we must pass through the heavy winding, carrying the large current to the outer ring (corresponding to point b of diagram). In the other direction, we must pass from the middle ring (i.e. point a), through the dotted winding, which carries one-half as great a current, to the inner collector ring (corresponding to point c of diagram). Then we must continue through the light winding, still in the direction of the current, until we again reach the outer collector ring, or point b of diagram.

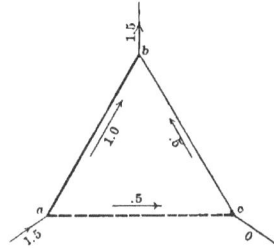

Any of the following three-phase diagrams may be connected either "delta" or "Y," but they will usually be shown with the "Y" connection.

It is well to keep in mind that if a "Y" connected armature is changed over to the "delta" connection, it may with the same regulation and heating give 1.732 times as much current, but only $\frac{1}{1.732}$ times the voltage. The reverse holds true in changing from "delta" (Δ) to "Y."

Figure 118 is the bar winding corresponding to Fig. 116. It has one bar per pole piece per phase. This winding, while partaking of all the advantages and disadvantages of multi-coil construction, would be particularly unsatisfactory for a three-phase *motor* on account of the dead points that it would develop at starting. These dead points are much less marked with multi-coil windings and with windings like those in Figs. 119 and 120.

In the case of induction motors, it is customary to make use of such windings as those given in Figs. 126 and 127, where smoother action is obtained partly by virtue of the choice of a number of conductors, prime, or nearly so, to the number of poles.

Figure 119 is a non-overlapping, three-phase, coil winding, with only one and one-half coils per pole piece per
phase.   It is the winding which was given with its single-
phase connection, in Fig. 96.   This should make a very
excellent three-phase winding, as there is no crossing of
the coils.   It is a regular thirty-pole, single-phase winding,
connected up as a three-phase armature for twenty poles.
This diagram should be compared with Fig. 77, Fig. 96, and
Fig. 102.   It should be particularly suitable for use in three-
phase motor work, as it should have very weakly defined
dead points.   In a projection armature, when a slot is
opposite a certain pole piece, spaces between two slots will
be opposite the adjacent pole pieces, thus giving a more
equitable distribution of the magnetic flux.

The inductance of such a winding is low and fairly
uniform, for the reason that when one side of a coil occupies
a position under a pole piece, the other side of the coil is
between two pole pieces.

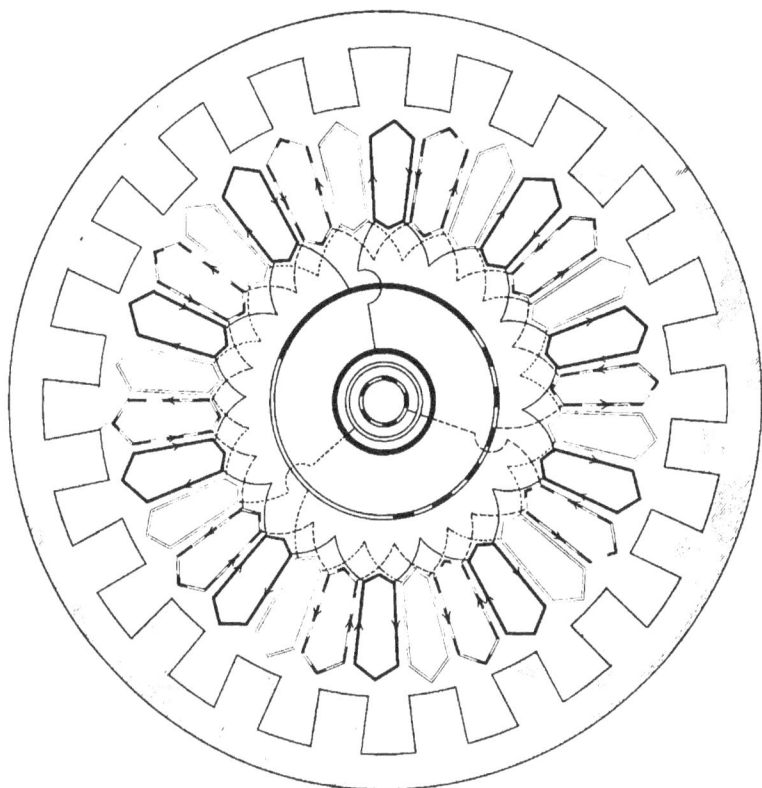

Fig. 119

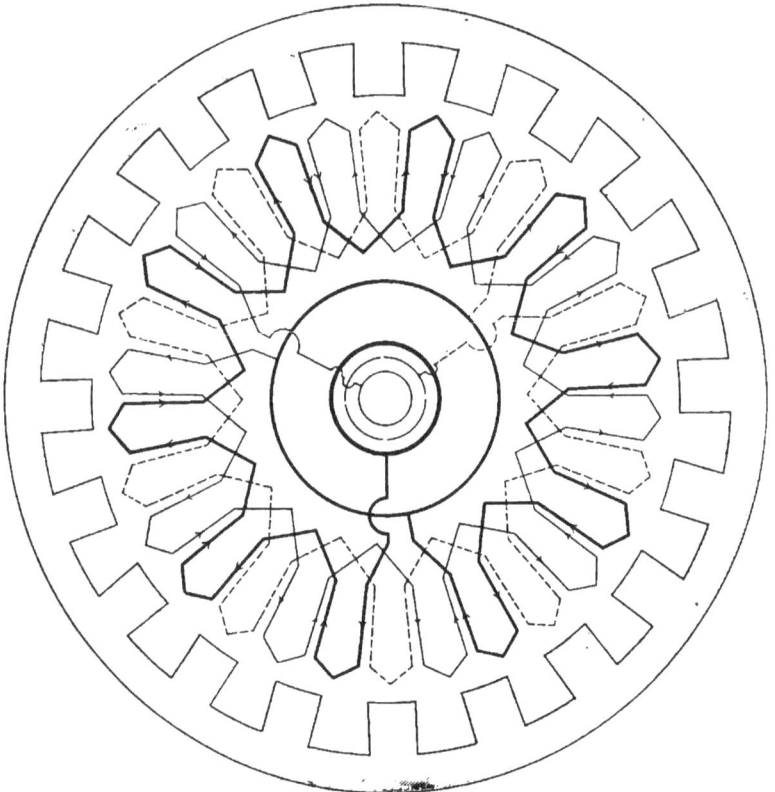

Fig. 120

Figure 120 represents the corresponding bar winding. In the case of projection or ironclad armatures, it would have two bars per slot, which might be arranged one over the other or side by side. It is interesting to note that each slot would contain one bar of each of two windings, two bars of the same winding never occupying the same slot.

All the remarks regarding the winding of Fig. 119 apply equally well to Fig. 120.

Figure 121 is a three-phase coil winding, with two groups
of conductors per pole piece per phase. The mechanical
arrangement of the coils at the ends of the armature could
not be designed nearly so satisfactorily from a mechanical
point of view, as in the style of winding given in Fig. 123.
It is believed that in most instances the style of winding
shown in Fig. 123 will be found to give the best results.

Fig. 121

Fig. 122

Figure 122 is the bar winding corresponding to Fig. 121. The end connections are perfectly symmetrical and well distributed at one end, but are far from it at the other. Its point of superiority over Fig. 124 is that it has, as a rule, no great differences of potential between adjacent conductors.

As already stated, the irregular distribution of the end conductors is not, at least in the case of bar windings, so great an objection in cases where there are comparatively few bars per pole piece. And in this instance there is a sort of a regularity about their grouping, that might be found of advantage on account of the large spaces that it makes available for mechanical arrangements.

Figure 123, which was devised by Mr. Thorburn Reid, who has devised a number of useful windings, is superior in the mechanical arrangement of the coils, to the winding of Fig. 121. The corresponding bar winding is not drawn, but it may be readily seen that it would have no very obvious advantages.

Coil windings of the same style as that of Fig. 123 may be constructed with any number of coils per pole piece per phase, and are frequently superior to other arrangements.

It is thought that the style of lining adopted in the diagram will indicate fairly well the arrangement of the end connections, if care is taken to note that the conductors of some groups of coils are carried directly over in the same plane as the face wires, to the conductors forming the other side of the group. The end conductors of the other coils have to be bent down out of the plane of the face conductors and then back again into their plane. The coils are usually wound in forms and then laid in place on the armature.

Fig. 123

Fig. 124

Figure 124 is a three-phase bar winding, with two bars per pole piece per phase. It is perfectly symmetrical, and may have either one or two conductors per group. It is inferior to Fig. 122, in that, from the nature of the winding, there are much greater differences of potential between adjacent conductors than in Fig. 122.

In Fig. 124, the pitch is 5 at one end and 7 at the other. Two sets of conductors, each set having as many conductors as there are pole pieces, are joined in series to form each one of the three windings. If an armature for half the voltage had been wished, the two sets of conductors forming each winding would have been connected in parallel.

This winding, as well as the next (Fig. 125), is of the same general character as those shown in Figs. 109 and 113.

Figure 125 is similar in all respects to Fig. 124, except that it has three conductors per pole piece per phase. The pitch is 9 at both ends.   It could be connected so as to give one-third as great a terminal electromotive force by joining the three elementary groups of which each winding is formed, in parallel, instead of in series.

In connection with Figs. 124 and 125, emphasis should be laid on the fact that in virtue of the nature of these windings, whereby adjacent conductors have between them large differences of potential, valuable space has to be sacrificed to make room for the proper thickness of insulation, which, with types of winding not possessing this character, could be usefully employed.

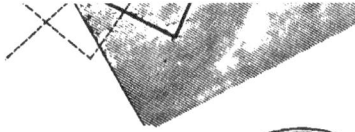

Fig. 125

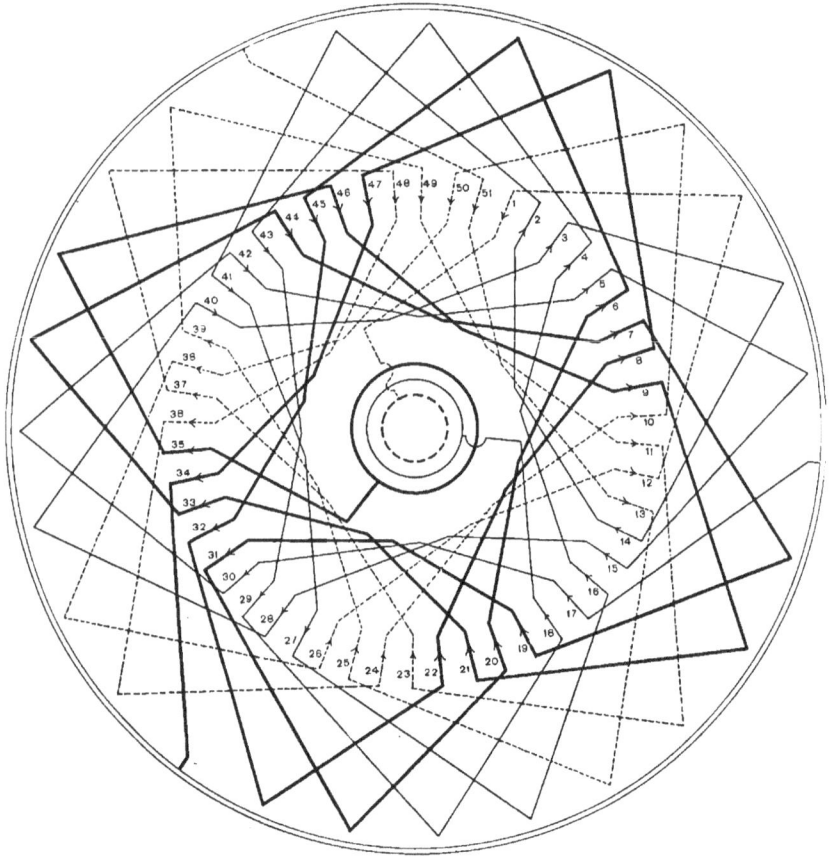

Fig. 126

Figure 126 is a four-pole, three-phase bar winding of a very irregular character. It has fifty-one conductors, seventeen per phase. There are, therefore, unequal numbers of conductors, both per phase and per pole, opposite the different pole pieces.

This style of winding has been used with success in induction motors, where it is important to choose a number of slots on the armature, which is prime, or nearly so, to the number of slots on the field. It may be well to state that, in the case of induction motors, the *field*, in the most successful types, consists merely of an assembly of annular punchings with radial slots within which the cylindrical drum *armature* revolves. It is practically a transformer, one of the elements, usually the secondary, being movable. It has become customary to call the moving element, the *armature*, and the stationary, the *field*. In the types, and for the voltages generally employed, it has been found best to use a coil winding for the field, the coils often being wound on forms and slipped into the slots. In the armature, which is practically a short-circuited secondary, the number of conductors and slots is determined by the permissible inductance, the actual voltage of the armature being to a great extent immaterial. In certain types the ratio of field to armature conductors has been something like 6 : 1. It is in connection with such motors as these, that the winding diagram of Fig. 126 will be found of greatest service. There cannot well be more than one bar per slot, because of the irregularity of the end connections.

Figure 127 is another three-phase bar winding with fifty-one conductors. It has six poles, and is even more irregular than the winding of Fig. 126. It, like Fig. 126, will find its chief use in the design of induction apparatus. Windings, almost as irregular, might be used in large polyphase generators, where it is desired to have but one conductor per slot.

Fig. 127

Fig. 128

TWO-CIRCUIT WINDING FOR THREE-PHASE CONTINUOUS-CURRENT, COMMUTATING MACHINE.

Figure 128 represents the same winding as Fig. 114, except that here it is tapped off at three nearly equidistant points instead of at four, as was the case in Fig. 114.

The result is a winding for a three-phase, continuous-current, commutating machine.

The total sixty-eight bars are divided up into sets of twenty-two, twenty-two, and twenty-four conductors, respectively, which are represented on the diagram by heavy, light, and dotted lines.

If the conductors are arranged in groups of two each, as would frequently be the case in projection armatures, where two conductors would often be placed together in each slot, it is of interest to note that these two conductors never belong to the same phase.

SIX-CIRCUIT WINDING FOR THREE-PHASE, CONTINUOUS-CURRENT, COMMUTATING MACHINE.

Figure 129 is still another three-phase, continuous-current, commutating machine, but with a six-circuit winding. It requires three leads per pair of poles ; therefore, in this case, nine leads. It is quite analogous to the quarter-phase, continuous-current, commutating machine of Fig. 115.

It is of interest to notice the relation of the voltage between collector rings to the continuous-current voltage at the commutator, in the case of three-phase, continuous-current, commutating machines. It will have been observed that they have "delta" connected windings.

Let $V =$ continuous-current voltage at the commutator ; then, taking the point of zero potential to be at the middle of the winding, the electromotive force of each half of the winding is $\dfrac{V}{2}$. But the corresponding *effective* alternating electromotive force will be $\dfrac{V}{2\sqrt{2}}$. This, therefore, will correspond to the voltage between common connection (point of zero potential), and collector ring, for an *equivalent* "Y" connected three-phase armature winding. Now the voltage between the collector rings of the "delta" connected armature winding will be $\sqrt{3}$ times as great as the voltage to the common connection of this *equivalent* "Y" winding, therefore the voltage between the collector rings will be, —

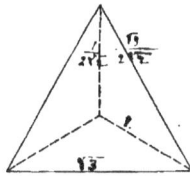

$$\frac{\sqrt{3}\,V}{2\sqrt{2}} = .612\,V,$$

where $V =$ continuous-current voltage at commutator.

Inasmuch as a "delta" connected winding cannot be readily conceived to have a point of zero potential, the above subterfuge of substituting for it, the *equivalent* "Y" connected winding, will often be found to facilitate the handling of three-phase winding problems. When doing so, the *equivalent* "Y" potential and the *equivalent* "Y" current may be spoken of as attributes of a "delta" connected armature. In the accompanying figure, an *equivalent* "Y" connected winding is diagrammatically shown dotted within a "delta" connected winding.

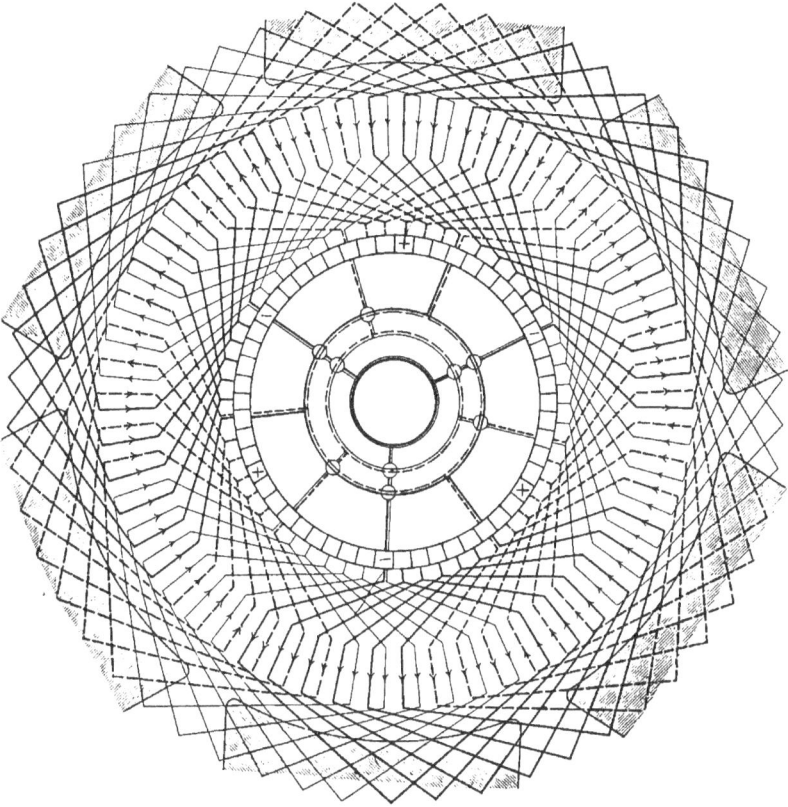

Fig. 129

Part III.

WINDING FORMULÆ AND TABLES.

CHAPTER XVI.

FORMULÆ FOR ELECTROMOTIVE FORCE.

COMPREHENSIVE formulæ for the calculation of the electromotive force set up in armatures may be derived from the formula for the voltage in a circuit, in which the variation of magnetic flux is a simple harmonic function of the time. These formulæ are :—

1. $V = 6.28\ TNM\ 10^{-8}$, the maximum voltage set up in a cycle ; 1.000
2. $V = 4.44\ TNM\ 10^{-8}$, the effective voltage set up in a cycle ; $.707$
3. $V = 4.00\ TNM\ 10^{-8}$, the mean or average voltage set up in a cycle, $.637$

where V is the voltage generated, in volts ; T the number of turns in series, M the number of *cgs* lines included or excluded by each of the T turns in a magnetic cycle, and N the number of magnetic cycles per second.

In armatures of alternators, the effective, or square root of the mean square of the electromotive forces is required, since this is proportional to the effective voltage, *i.e.* the voltage to maintain current C (square root of the mean square of the current), in a non-inductive resistance. In this case it is supposed that the T turns are so situated as to be simultaneously affected by any change of the magnetic flux, otherwise the voltage for each of the turns differently situated must be calculated separately and properly combined to obtain the resultant voltage.

In the case of multi-phase alternating-current machines, the voltage in each circuit should be calculated, and the resultant voltage derived according to the method of connection, and addition of vectors according to the angle by which the several phases differ from each other.

In quarter-phase machines with common connection, the resultant voltage is $\sqrt{2}$, or 1.414 times the voltage generated in one circuit.

In three-phase apparatus, the resultant voltage is the same as the voltage generated in one circuit when the circuits are connected "delta"; and $\sqrt{3}$, or 1.732 times the voltage generated in one circuit when the circuits are connected "Y."

In alternating-current commutating machines, the ratio of the voltage between the continuous and the alternating current circuits is 1 : .707 in the case of single-phase and quarter-phase commutating machines, and 1 : .612 in the case of three-phase commutating machines. In other words, if the voltage at the continuous current side is known, the voltage between collector rings will be .707 times as great in the case of single and quarter phase commutating machines, and will be .612 times as great in the case of three-phase commutating machines.

In armatures of continuous-current dynamos, the voltage at the terminals is constant during any period considered, and is the integral of all the voltages successively set up in the different armature coils according to their position in the magnetic field; and since in this case only average voltages are considered, the resultant voltage is independent of any manner in which the magnetic flux may vary through the coils.

Formula 3 is applicable to all continuous-current armatures, whether ring, drum or disc, two-circuit or multiple circuit, and whether the winding be single or multiple.

The simplicity and wide applicability of these formulæ make them preferable to many others that are difficult to interpret, because of the many accessory conditions that must be kept in mind.

Although, by the constants given above, the voltages may be obtained at the alternating current, as well as at the continuous current terminals of commutating machines, the former, i.e. the voltages at the alternating current terminals, may be obtained from the following formulæ, in which V is the required voltage between collector rings, T is the number of turns in series between collector rings, M is the magnetic flux from one pole piece into the armature, and N is the number of cycles per second: —

For single and quarter phase commutating machines, $V = 2.83\ TNM\ 10^{-8}$.

For three-phase commutating machines, $V = 3.69\ TNM^{-8}$.

CHAPTER XVII.

METHOD OF APPLYING THE ARMATURE WINDING TABLES.

THE nature and use of the tables may be most easily understood by applying them to the solution of a few examples.

EXAMPLE 1. — If we wish a two-circuit, triple winding for a drum armature, with about 670 conductors and six poles, what is the exact number of conductors that must be employed to give us a singly re-entrant winding?

Turning to page 312, we find that a two-circuit, triple winding with 670 conductors, is impossible for six poles, but that 672 conductors may be used ; and to have the winding singly re-entrant, the front and back pitches must each equal 113. If the front and back pitches should be taken equal to 111, a triply re-entrant winding would result.

EXAMPLE 2. — We next wish to ascertain how many volts this machine will give when the armature is driven at 440 r.p.m., if the flux from each pole piece into the armature equals 2.25 megalines.

The table of Drum Winding Constants on page 280 tells us that with 100 conductors, 100 r.p.m., and a flux equal to one megaline, the terminal volts will, for a six-pole machine, be equal to 1.667. Therefore, in the case before us, we have

$$V = 1.667 \times 6.72 \times 4.40 \times 2.25 = 111 \text{ volts.}$$

From the same table we find that for a two-circuit, triple winding with six poles, we have .200 average volts between commutator segments per megaline and per 100 r.p.m. So, in this case, we shall have .200 × 2.25 × 4.40 = 1.98 average volts between commutator segments.

EXAMPLE 3. — Certain conditions fix the flux of a dynamo from one pole piece into the armature at 8.30 megalines, and the speed at 100 r.p.m. If we wish to employ an eight-pole, two-circuit, double winding, how many conductors do we need, to obtain 150 volts?

Consulting the table of Drum Winding Constants, on page 280, we find that for eight-pole, two-circuit, double windings, we have 3.33 volts per 100 conductors with 100 r.p.m., and one megaline of flux. Therefore, we shall require $\dfrac{150}{3.33} \times \dfrac{100}{8.30} = 544$ conductors.

By reference to page 301, it will be seen that for eight poles, the nearest number of conductors that we can use in order to have a two-circuit, double winding, is 540 or 548. Suppose we use 540 conductors. If we wish a doubly re-entrant winding, we shall take the pitch at one end equal to 67, and that at the other end equal to 69.

EXAMPLE 4. — A slotted armature is to have ten poles, and a two-circuit, triple winding, with eight conductors per slot.

By reference to the table of Summarized Conditions for Two-Circuit, Triple Windings, on page 283, we find that it may be either singly or triply re-entrant, according to the number of conductors used.

The winding is to have 424 conductors. Turning to page 310, it is seen that the pitch must be 43 at both ends, and that for 424 conductors the winding must be singly re-entrant.

If the flux is 20.0 megalines, and the speed 105 r.p.m., we find from page 280 that the voltage will be

$$2.78 \times 4.24 \times 1.05 \times 20.0 = 247 \text{ volts.}$$

The average volts per bar are

$$.556 \times 20.0 \times 1.05 = 11.7 \text{ volts.}$$

EXAMPLE 5. — An eight-pole armature has a multiple-circuit, double winding, with 1258 conductors. By consulting page 343, we find that it is singly re-entrant, and that the pitch should be 155 at one end, and 159 at the other. It is, of course, understood that these pitches are taken in opposite directions. One of them might have been indicated as positive, and the other as negative. It may be well to point out here that the letters F and B at the head of the tables, meaning respectively, "front" and "back," are interchangeable, meaning merely that the one figure represents the pitch at one end, and the other figure, that at the other end. This is true in regard to all the tables, both two-circuit and multiple-circuit.

Returning to Example 5, the voltage of the machine, assuming the flux equals 7.85 megalines, and a speed of 300 r.p.m., is found by the table of Drum Winding Constants on page 280, to be

$$.833 \times 12.58 \times 3.00 \times 7.85 = 247 \text{ volts.}$$

The average volts per bar are

$$.1333 \times 7.85 \times 3.00 = 3.14 \text{ volts.}$$

EXAMPLE 6. — A two-circuit, single winding is wanted, with four conductors per slot.

From the table of Summarized Conditions for Two-Circuit, Single Windings, on page 281, it may be seen that this is only possible with 6, 10, 14, etc., poles; being impossible with 4, 8, 12, 16, etc., poles. The winding is designed for fourteen poles, and 660 conductors. We find from page 329, that the pitch is 47 at both ends. The machine gives 160 volts, and the speed is 75 r.p.m. By the aid of the table on page 280, we find that the flux is equal to

$$\frac{160}{11.67 \times 6.60 \times .75} = 2.77 \text{ megalines.}$$

Average volts per commutator segment $= 3.27 \times 2.77 \times .75 = 6.80$ volts.

The above examples have all been chosen merely to illustrate the use of the tables, and the relative magnitudes employed in any one example are *not* such as would occur in practice.

The tables on pages 280, 281, 282, and 283 are constructed on the assumption that no interpolated commutator segments are employed, and that no portion of the normal number of commutator segments is omitted, and when this is not the case, the results should be properly modified, as may readily be done.

In all the tables, a proper interpretation of the term "conductors" should be made. As stated in the introductory chapter, "groups of conductors" may often be substituted therefor.

It is believed that after becoming familiar with the arrangement of the tables, their use will be found to be of value in a great variety of problems connected with armature windings. Any single result can, however, be obtained by an application of the rules and formulæ given in the text, but after these rules and formulæ are once understood, it will be found that subsequent problems will generally be most conveniently solved by means of the tables.

CHAPTER XVIII.

ARMATURE WINDING TABLES.

DRUM WINDING CONSTANTS.

					NUMBER OF POLES						
			CLASS OF WINDING.	4	6	8	10	12	14	16
DRUM ARMATURES.	VOLTS PER 100 CONDUCTORS PER 100 R. P. M. AND FLUX=ONE MEGALINE.	MULTIPLE CIRCUIT	Single	1.667	1.667	1.667	1.667	1.667	1.667	1.667
			Double	.833	.833	.833	.833	.833	.833	.833
			Triple	.556	.556	.556	.556	.556	.556	.556
		TWO CIRCUIT	Single	3.33	5.00	6.67	8.33	10.00	11.67	13.33
			Double	1.667	2.50	3.33	4.17	5.00	5.83	6.67
			Triple	1.111	1.667	2.22	2.78	3.33	3.89	4.44
	AVERAGE VOLTS BETWEEN COMMUTATOR SEGMENTS PER MEGA LINE & PER 100 R. P. M. (INDEPENDENT OF NO. OF CONDS.)	MULTIPLE CIRCUIT	Single	.1333	.200	.267	.333	.400	.467	.533
			Double ⊗	.0668	.100	.1333	.1667	.200	.233	.267
			Triple ⊗	.0445	.0667	.0888	.1111	.1333	.1555	.1778
		TWO CIRCUIT	Single	.267	.600	1.068	1.668	2.40	3.27	4.27
			Double ⊗	.1333	.300	.534	.834	1.200	1.635	2.14
			Triple ⊗	.0888	.200	.356	.556	.800	1.09	1.42

⊗ With Multiple Windings, the maximum Volts per bar is much more greatly in excess of the average Volts per bar than in Single Windings. This may be seen by a careful analysis of such Windings; which also shows that this may be more or less overcome by careful mutual adjustment of the position of the Brushes. This would not, however, be practicable with present methods.

NUMBER OF POLES	CONDUCTORS PER SLOT									VOLTS PER 100 CONDRS. PER 100 R.P.M. WITH FLUX = 1 MEGALINE	AVERAGE VOLTS BETWEEN COMR. SEGTS PER MEGALINE & PER 100 R.P.M. ⓟ
4	1	2		6		10		14		3.33	.267
6	1	2	4		8	10		14	16	5.00	.600
8	1	2		6		10		14		6.67	1.068
10	1	2	4	6	8		12	14	16	8.33	1.668
12	1	2				10		14		10.00	2.40
14	1	2	4	6	8	10	12		16	11.67	3.27
16	1	2		6		10		14		13.33	4.27

ⓟ Independent of number of Conductors

DATA FOR APPLYING TWO-CIRCUIT, SINGLE WINDINGS, FOR DRUM ARMATURES.

From the above Table the following Rule may be deduced:

In the ordinary two-circuit single winding, "C" is always such a number that the number of conductors per slot, and "n" the number of poles, cannot have a common factor greater than 2.

DATA FOR APPLYING TWO-CIRCUIT, DOUBLE WINDINGS, FOR DRUM ARMATURES.									VOLTS PER 100 CON DRS. PER 100 R.P.M. WITH FLUX = 1 MEGALINE	AVERAGE VOLTS BETWEEN COMR. SEGTS. PER MEGALINE & PER 100 R. P. M. ③
NUMBER OF POLES	CONDUCTORS PER SLOT									
	1	2	4	6	8	10	12	14	16		
4										1.667	.1333
6										2.50	300
8										3.33	.534
10										4.17	.834
12										5.00	1.200
14										5.83	1.635
16										6.67	2.14

② Independent of number of Conductors

③ Moreover, in multiple Windings this value is merely nominal, as a careful analysis of Multiple Windings shows that if this value can be approached at all, it is only by means of more careful mutual adjustment of the Brushes than is practicable with present methods.

NUMBER OF POLES	DATA FOR APPLYING TWO-CIRCUIT, TRIPLE WINDINGS, FOR DRUM ARMATURES. CONDUCTORS PER SLOT									VOLTS PER 100 CONDRS. PER 100 R.P.M. WITH FLUX = 1 MEGALINE	AVERAGE VOLTS BETWEEN COMMR. SEGTS. PER MEGALINE & PER 100 R. P. M. ⊗
	1	2	4	6	8	10	12	14	16		
4	⊗	⊗	OOO		⊗			⊗		1.111	.0888
6	⊗	⊗	⊗	⊗	⊗	⊗	⊗	⊗	⊗	1.667	.200
8	⊗	⊗	OOO		⊗			⊗		2.22	.356
10	⊗	⊗	⊗	OOO	⊗		OOO	⊗	⊗	2.78	.556
12	⊗	⊗		⊗		⊗		⊗		3.33	.800
14	⊗	⊗	⊗	OOO	⊗	⊗	OOO		⊗	3.89	1.09
16	⊗	⊗		OOO	⊗		⊗			4.44	1.42

⊗ Independent of number of Conductors

⊕ Moreover, in Multiple Windings this value is merely nominal, as a careful analysis of Multiple Windings shows that if this value can be approached at all, it is only by means of more careful mutual adjustment of the Brushes than is practicable with present methods.

WINDING TABLES FOR TWO-CIRCUIT, SINGLE WINDINGS
FOR DRUM ARMATURES.

TABLE OF TWO-CIRCUIT, SINGLE WINDINGS, FOR DRUM ARMATURES.

Nº OF CONDUCTORS	4 POLES		6 POLES		8 POLES		10 POLES		12 POLES		14 POLES		16 POLES		Nº OF CONDUCTORS
	F	B	F	B	F	B	F	B	F	B	F	B	F	B	
102					13	13	9	11							102
104			17	17											104
106			17	19	13	13			9	9					106
108							11	11							108
110			17	19	13	15			9	9	7	9	7	7	110
112			19	19			11	11							112
114					13	15					7	9	7	7	114
116			19	19											116
118			19	21	15	15	11	13	9	11					118
120															120
122			19	21	15	15	11	13	9	11					122
124			21	21							9	9			124
126					15	17							7	9	126
128			21	21			13	13			9	9			128
130			21	23	15	17			11	11			7	9	130
132							13	13							132
134			21	23	17	17			11	11					134
136			23	23											136
138					17	17	13	15			9	11			138
140			23	23											140
142			23	25	17	19	13	15	11	13	9	11	9	9	142
144															144
146			23	25	17	19			11	13			9	9	146
148			25	25			15	15							148
150					19	19									150
152			25	25			15	15			11	11			152
154			25	27	19	19			13	13					154
156											11	11			156
158			25	27	19	21	15	17	13	13			9	11	158
160			27	27											160
162					19	21	15	17					9	11	162
164			27	27											164
166			27	29	21	21			13	15	11	13			166
168							17	17							168
170			27	29	21	21			13	15	11	13			170
172			29	29			17	17							172
174					21	23							11	11	174
176			29	29											176
178			29	31	21	23	17	19	15	15			11	11	178
180											13	13			180
182			29	31	23	23	17	19	15	15					182
184			31	31							13	13			184
186					23	23									186
188			31	31			19	19							188
190			31	33	23	25			15	17			11	13	190
192							19	19							192
194			31	33	23	25			15	17	13	15	11	13	194
196			33	33											196
198					25	25	19	21			13	15			198
200			33	33											200

| 4 | 6 | 8 | 10 | 12 | 14 | 16 |

TABLE OF TWO-CIRCUIT, SINGLE WINDINGS, FOR DRUM ARMATURES.

FRONT AND BACK PITCHES.

No. OF CONDUCTORS	4 POLES F	4 POLES B	6 POLES F	6 POLES B	8 POLES F	8 POLES B	10 POLES F	10 POLES B	12 POLES F	12 POLES B	14 POLES F	14 POLES B	16 POLES F	16 POLES B	No. OF CONDUCTORS
202	49	51	33	35	25	25	19	21	17	17					202
204															204
206	51	51	33	35	25	27			17	17			13	13	206
208			35	35			21	21			15	15			208
210	51	51			25	27							13	13	210
212			35	35			21	21			15	15			212
214	53	53	35	37	27	27			17	19					214
216															216
218	53	53	35	37	27	27	21	23	17	19					218
220			37	37											220
222	53	53			27	29	21	23			15	17	13	15	222
224			37	37											224
226	55	55	37	39	27	29			19	19	15	17	13	15	226
228							23	23							228
230	55	55	37	39	29	29			19	19					230
232			39	39			23	23							232
234	55	55			29	29									234
236			39	39							17	17			236
238	57	57	39	41	29	31	23	25	19	21			15	15	238
240											17	17			240
242	57	57	39	41	29	31	23	25	19	21			15	15	242
244			41	41											244
246	57	57			31	31									246
248			41	41			25	25							248
250	59	59	41	43	31	31			21	21	17	19			250
252							25	25							252
254	59	59	41	43	31	33			21	21	17	19	15	17	254
256			43	43											256
258	59	59			31	33	25	27					15	17	258
260			43	43											260
262	61	61	43	45	33	33	25	27	21	23					262
264											19	19			264
266	61	61	43	45	33	33			21	23	19	19			266
268			45	45			27	27							268
270	61	61			33	35							17	17	270
272			45	45			27	27							272
274	63	63	45	47	33	35			23	23			17	17	274
276															276
278	63	63	45	47	35	35	27	29	23	23	19	21			278
280			47	47											280
282	63	63			35	35	27	29			19	21			282
284			47	47											284
286	65	65	47	49	35	37			23	25			17	19	286
288							29	29							288
290	65	65	47	49	35	37			23	25			17	19	290
292			49	49			29	29			21	21			292
294	65	65			37	37					21	21			294
296			49	49											296
298	67	67	49	51	37	37	29	31	25	25					298
300															300

| 4 | 6 | 8 | 10 | 12 | 14 | 16 |

TABLE OF TWO-CIRCUIT, SINGLE WINDINGS, FOR DRUM ARMATURES.

FRONT AND BACK PITCHES

No. of Conductors	4 POLES F	4 POLES B	6 POLES F	6 POLES B	8 POLES F	8 POLES B	10 POLES F	10 POLES B	12 POLES F	12 POLES B	14 POLES F	14 POLES B	16 POLES F	16 POLES B	No. of Conductors
302			49	51	37	39	29	31	25	25			19	19	302
304			51	51											304
306					37	39					21	23	19	19	306
308			51	51			31	31							308
310			51	53	39	39			25	27	21	23			310
312							31	31							312
314			51	53	39	39			25	27					314
316			53	53											316
318					39	41	31	33					19	21	318
320			53	53							23	23			320
322			53	55	39	41	31	33	27	27			19	21	322
324											23	23			324
326			53	55	41	41			27	27					326
328			55	55			33	33							328
330					41	41									330
332			55	55			33	33							332
334			55	57	41	43			27	29	23	25	21	21	334
336															336
338			55	57	41	43	33	35	27	29	23	25	21	21	338
340			57	57											340
342					43	43	33	35			—	—			342
344			57	57									—	—	344
346			57	59	43	43			29	29					346
348							35	35			25	25			348
350			57	59	43	45			29	29	25	25	21	23	350
352			59	59			35	35					21	23	352
354					43	45							21	23	354
356			59	59											356
358			59	61	45	45	35	37	29	31					358
360															360
362			59	61	45	45	35	37	29	31	25	27			362
364			61	61											364
366					45	47					25	27	23	23	366
368			61	61			37	37							368
370			61	63	45	47			31	31			23	23	370
372							37	37							372
374			61	63	47	47			31	31					374
376			63	63							27	27			376
378					47	47	37	39			27	27			378
380															380
382			63	65	47	49	37	39	31	33			23	25	382
384															384
386			63	65	47	49			31	33			23	25	386
388			65	65			39	39							388
390					49	49					27	29			390
392			65	65			39	39							392
394			65	67	49	49			33	33	27	29			394
396															396
398			66	67	49	51	39	41	33	33			25	25	398
400			67	67											400

| 4 | 6 | 8 | 10 | 12 | 14 | 16 |

TABLE OF TWO-CIRCUIT, SINGLE WINDINGS, FOR DRUM ARMATURES.

No. of Conductors	4 POLES		6 POLES		8 POLES		10 POLES		12 POLES		14 POLES		16 POLES		No. of Conductors
	F	B	F	B	F	B	F	B	F	B	F	B	F	B	
402					49	51	39	41					25	25	402
404			67	67							29	29			404
406			67	69	51	51			33	35					406
408							41	41			29	29			408
410			67	69	51	51			33	35					410
412			69	69			41	41							412
414					51	53							25	27	414
416			69	69											416
418			69	71	51	53	41	43	35	35	29	31	25	27	418
420															420
422			69	71	53	53	41	43	35	35	29	31			422
424			71	71											424
426					53	53									426
428			71	71			43	43							428
430			71	73	53	55			35	37			27	27	430
432							43	43			31	31			432
434			71	73	53	55			35	37			27	27	434
436			73	73							31	31			436
438					55	55	43	45							438
440			73	78											440
442			73	75	55	55	43	45	37	37					442
444															444
446			73	75	55	57			37	37	31	33	27	29	446
448			75	75			45	45							448
450					55	57					31	33	27	29	450
452			75	75			45	45							452
454			75	77	57	57			37	39					454
456															456
458			75	77	57	57	45	47	37	39					458
460			77	77							33	33			460
462					57	59	45	47					29	29	462
464			77	77							33	33			464
466			77	79	57	59			39	39			29	29	466
468							47	47							468
470			77	79	59	59			39	39					470
472			79	79			47	47							472
474					59	59					33	35			474
476			79	79											476
478			79	81	59	61	47	49	39	41	33	35	29	31	478
480															480
482			79	81	59	61	47	49	39	41			29	31	482
484			81	81											484
486					61	61									486
488			81	81			49	49			35	35			488
490			81	83	61	61			41	41					490
492							49	49			35	35			492
494			81	83	61	63			41	41			31	31	494
496			83	83											496
498					61	63	49	51					31	31	498
500			83	83											500

TABLE OF TWO CIRCUIT, SINGLE WINDINGS, FOR DRUM ARMATURES.

No. OF CONDUCTORS	FRONT AND BACK PITCHES														No. OF CONDUCTORS
	4 POLES		6 POLES		8 POLES		10 POLES		12 POLES		14 POLES		16 POLES		
	F	B	F	B	F	B	F	B	F	B	F	B	F	B	
502			83	85	63	63	49	51	41	43	35	37			502
504															504
506			83	85	63	63			41	43	35	37			506
508			85	85			51	51							508
510					63	65							31	33	510
512			85	85			51	51							512
514			85	87	63	65			43	43			31	33	514
516											37	37			516
518			85	87	65	65	51	53	43	43					518
520			87	87							37	37			520
522					65	65	51	53							522
524			87	87											524
526			87	89	65	67			43	45			33	33	526
528							53	53							528
530			87	89	65	67			43	45	37	39	33	33	530
532			89	89			53	53							532
534					67	67					37	39			534
536			89	89											536
538			89	91	67	67	53	55	45	45					538
540															540
542			89	91	67	69	53	55	45	45			33	35	542
544			91	91							39	39			544
546					67	69							33	35	546
548			91	91			55	55			39	39			548
550			91	93	69	69			45	47					550
552							55	55							552
554			91	93	69	69			45	47					554
556			93	93											556
558					69	71	55	57			39	41	35	35	558
560			93	93											560
562			93	95	69	71	55	57	47	47	39	41	35	35	562
564															564
566			93	95	71	71			47	47					566
568			95	95			57	57							568
570					71	71									570
572			95	95			57	57			41	41			572
574			95	97	71	73			47	49			35	37	574
576											41	41			576
578			95	97	71	73	57	59	47	49			35	37	578
580			97	97											580
582					73	73	57	59							582
584			97	97											584
586			97	99	73	73			49	49	41	43			586
588							59	59							588
590			97	99	73	75			49	49	41	43	37	37	590
592			99	99			59	59							592
594					73	75							37	37	594
596			99	99											596
598			99	101	75	75	59	61	49	51					598
600											43	43			600

| 4 | 6 | 8 | 10 | 12 | 14 | 16 |

TABLE OF TWO CIRCUIT SINGLE WINDINGS FOR DRUM ARMATURES.

No. of Conductors	4 POLES		6 POLES		8 POLES		10 POLES		12 POLES		14 POLES		16 POLES		No. of Conductors
	F	B	F	B	F	B	F	B	F	B	F	B	F	B	
602	…	…	99	101	75	75	59	61	49	51					602
604			101	101							43	43			604
606	…	…			75	77							37	39	606
608			101	101			61	61							608
610	…	…	101	103	75	77			51	51			37	39	610
612							61	61							612
614	…	…	101	103	77	77			51	51	43	45			614
616			103	103											616
618	…	…			77	77	61	63			43	45			618
620			103	103											620
622	…	…	103	105	77	79	61	63	51	53			39	39	622
624															624
626	…	…	103	105	77	79			51	53			39	39	626
628			105	105			63	63			45	45			628
630	…	…			79	79									630
632			105	105			63	63			45	45			632
634	…	…	105	107	79	79			53	53					634
636															636
638	…	…	105	107	79	81	63	65	53	53			39	41	638
640			107	107											640
642	…	…			79	81	63	65			45	47	39	41	642
644			107	107											644
646	…	…	107	109	81	81			53	55	45	47			646
648							65	65							648
650	…	…	107	109	81	81			53	55					650
652			109	109			65	65							652
654	…	…			81	83							41	41	654
656			109	109											656
658	…	…	109	111	81	83	65	67	55	55	47	47	41	41	658
660															660
662	…	…	109	111	83	83	65	67	55	55	47	47			662
664			111	111											664
666	…	…			83	83									666
668			111	111			67	67							668
670	…	…	111	113	83	85			55	57	47	49	41	43	670
672							67	67							672
674	…	…	111	113	83	85			55	57	47	49	41	43	674
676			113	113											676
678	…	…			85	85	67	69							678
680			113	113											680
682	…	…	113	115	85	85	67	69	57	57	49	49			682
684															684
686	…	…	113	115	85	87			57	57	49	49	43	43	686
688			115	115			69	69							688
690	…	…			85	87							43	43	690
692			115	115			69	69							692
694	…	…	115	117	87	87			57	59					694
696															696
698	…	…	115	117	87	87	69	71	57	59	49	51			698
700			117	117											700

4   6   8   10   12   14   16

TABLE OF TWO CIRCUIT SINGLE WINDINGS FOR DRUM ARMATURES.

No. of Conductors	4 POLES		6 POLES		8 POLES		10 POLES		12 POLES		14 POLES		16 POLES		No. of Conductors
	F	B	F	B	F	B	F	B	F	B	F	B	F	B	
702	*(illeg.)*	*(illeg.)*			87	89	69	71			49	51	43	45	702
704			117	117											704
706	*(illeg.)*	*(illeg.)*	117	119	87	89			59	59			43	45	706
708							71	71							708
710	*(illeg.)*	*(illeg.)*	117	119	89	89			59	59					710
712			119	119			71	71			51	51			712
714	*(illeg.)*	*(illeg.)*			89	89									714
716			119	119							51	51			716
718	*(illeg.)*	*(illeg.)*	119	121	89	91	71	73	59	61			45	45	718
720															720
722	*(illeg.)*	*(illeg.)*	119	121	89	91	71	73	59	61			45	45	722
724			121	121											724
726	*(illeg.)*	*(illeg.)*			91	91					51	53			726
728			121	121			73	73							728
730	*(illeg.)*	*(illeg.)*	121	123	91	91			61	61	51	53			730
732							73	73							732
734	*(illeg.)*	*(illeg.)*	121	123	91	93			61	61			45	47	734
736			123	123											736
738	*(illeg.)*	*(illeg.)*			91	93	73	75					45	47	738
740			123	123							53	53			740
742	*(illeg.)*	*(illeg.)*	123	125	93	93	73	75	61	63					742
744											53	53			744
746	*(illeg.)*	*(illeg.)*	123	125	93	93			61	63					746
748			125	125			75	75							748
750	*(illeg.)*	*(illeg.)*			93	95							47	47	750
752			125	125			75	75							752
754	*(illeg.)*	*(illeg.)*	125	127	93	95			63	63	53	55	47	47	754
756															756
758	*(illeg.)*	*(illeg.)*	125	127	95	95	75	77	63	63	53	55			758
760			127	127											760
762	*(illeg.)*	*(illeg.)*			95	95	75	77							762
764			127	127											764
766	*(illeg.)*	*(illeg.)*	127	129	95	97			63	65			47	49	766
768							77	77			55	55			768
770	*(illeg.)*	*(illeg.)*	127	129	95	97			63	65			47	49	770
772			129	129			77	77			55	55			772
774	*(illeg.)*	*(illeg.)*			97	97									774
776			129	129											776
778	*(illeg.)*	*(illeg.)*	129	131	97	97	77	79	65	65					778
780															780
782	*(illeg.)*	*(illeg.)*	129	131	97	99	77	79	65	65	55	57	49	49	782
784			131	131											784
786	*(illeg.)*	*(illeg.)*			97	99					55	57	49	49	786
788			131	131			79	79							788
790	*(illeg.)*	*(illeg.)*	131	133	99	99			65	67					790
792							79	79							792
794	*(illeg.)*	*(illeg.)*	131	133	99	99			65	67					794
796			133	133							57	57			796
798	*(illeg.)*	*(illeg.)*			99	101	79	81					49	51	798
800			133	133							57	57			800

4	6	8	10	12	14	16

WINDING TABLES FOR TWO-CIRCUIT, DOUBLE WINDINGS FOR DRUM ARMATURES.

TABLE OF TWO-CIRCUIT, DOUBLE WINDINGS, FOR DRUM ARMATURES.

N. OF CONDUCTORS	4 POLES			6 POLES			8 POLES			10 POLES			12 POLES			14 POLES			16 POLES			N. OF CONDUCTORS
	F	RE-ENTRANCY	B	F	RE-ENTRANCY	B	F	RE-ENTRANCY	B	F	RE-ENTRANCY	B	F	RE-ENTRANCY	B	F	RE-ENTRANCY	B	F	RE-ENTRANCY	B	
102																7	⊙	7				102
104		⊙		17	oo	19				9	oo	11	9	⊙	9							104
106				17	⊙	17				11	⊙	11										106
108		oo					1?	⊛	1?							7	oo	9	7	⊙	7	108
110				19	⊙	19																110
112		⊙		17	oo	19							9	⊙	9							112
114										11	⊙	11										114
116		oo		19	oo	21	1?	⊛	1?	11	oo	13	9	oo	11	7	oo	9	7	⊙	7	116
118				19	⊙	19																118
120		⊙																				120
122				21	⊙	21										9	⊙	9				122
124		oo		19	oo	21	1?	⊛	1?	11	oo	13	9	oo	11				7	oo	9	124
126										13	⊙	13										126
128		⊙		21	oo	23							11	⊙	11							128
130				21	⊙	21										9	⊙	9				130
132		oo					1?	⊛	1?							7	oo	9				132
134				23	⊙	23				13	⊙	13										134
136		⊙		21	oo	23				13	oo	15	11	⊙	11	9	oo	11				136
138																						138
140		oo		23	oo	25	?	⊛	1?				11	oo	13				9	⊙	9	140
142				23	⊙	23																142
144		⊙								13	oo	15				9	oo	11				144
146				25	⊙	25				15	⊙	15										146
148		oo		23	oo	25	1?	⊛	1?				11	oo	13				9	⊙	9	148
150																11	⊙	11				150
152		⊙		25	oo	27							13	⊙	13							152
154				25	⊙	25				15	⊙	15										154
156		oo					1?	⊛	1?	15	oo	17							9	oo	11	156
158				27	⊙	27										11	⊙	11				158
160		⊙		25	oo	27							13	⊙	13							160
162																						162
164		oo		27	oo	29	1?	⊛	1?	15	oo	17	13	oo	15	11	oo	13	9	oo	11	164
166				27	⊙	27				17	⊙	17										166
168		⊙																				168
170				29	⊙	29																170
172		oo		27	oo	29	?	⊛	?				13	oo	15	11	oo	13	11	⊙	11	172
174										17	⊙	17										174
176		⊙		29	oo	31				17	oo	19	15	⊙	15							176
178				29	⊙	29										13	⊙	13				178
180		oo					?	⊛	?										11	⊙	11	180
182				31	⊙	31																182
184		⊙		29	oo	31				17	oo	19	15	⊙	15							184
186										19	⊙	19				13	⊙	13				186
188		oo		31	oo	33	?	⊛	?				15	oo	17				11	oo	13	188
190				31	⊙	31																190
192		⊙														13	oo	15				192
194				33	⊙	33				19	⊙	19										194
196		oo		31	oo	33	?	⊛	?	19	oo	21	15	oo	17				11	oo	13	196
198																						198
200		⊙		33	oo	35							17	⊙	17	13	oo	15				200

4	6	8	10	12	14	16

TWO-CIRCUIT, DOUBLE WINDINGS, FOR DRUM ARMATURES.

FRONT AND BACK PITCHES

No. OF CONDUCTORS	4 POLES			6 POLES			8 POLES			10 POLES			12 POLES			14 POLES			16 POLES			No. OF CONDUCTORS
	F	••	B	F	••	B	F	••	B	F	••	B	F	••	B	F	••	B	F	••	B	
202				33	⊙	33																202
204		oo								19	oo	21		·					13	⊙	13	204
206				35	⊙	35				21	⊙	21				15	⊙	15				206
208		⊙		33	oo	35							17	⊙	17							208
210																						210
212		oo		35	oo	37							17	oo	19				13	⊙	13	212
214				35	⊙	35				21	⊙	21				15	⊙	15				214
216		⊙								21	oo	23										216
218				37	⊙	37																218
220		oo		35	oo	37							17	oo	19	15	oo	17	13	oo	15	220
222																						222
224		⊙		37	oo	39				21	oo	23	19	⊙	19							224
226				37	⊙	37				23	⊙	23										226
228		oo														15	oo	17	13	oo	15	228
230				39	⊙	39																230
232		⊙		37	oo	39							19	⊙	19							232
234										23	⊙	23				17	⊙	17				234
236		oo		39	oo	41				23	oo	25	19	oo	21				15	⊙	15	236
238				39	⊙	39																238
240		⊙																				240
242				41	⊙	41							17	⊙	17							242
244		oo		39	oo	41				23	oo	25							15	⊙	15	244
246										25	⊙	25										246
248		⊙		41	⊙	43							21	⊙	21	17	oo	19				248
250				41	⊙	41																250
252		oo																	15	oo	17	252
254				43	⊙	43				25	⊙	25										254
256		⊙		41	oo	43				25	oo	27	21	⊙	21	17	oo	19				256
258																						258
260		oo		43	oo	45							21	oo	23				15	oo	17	260
262				43	⊙	43										19	⊙	19				262
264		⊙								25	oo	27										264
266				45	⊙	45				27	⊙	27										266
268		oo		43	oo	45							21	oo	23	17	⊙	17				268
270				45	oo	47										19	⊙	19				270
272		⊙											23	⊙	23							272
274				45	⊙	45				27	⊙	27										274
276		oo								27	oo	29				19	oo	21	17	⊙	17	276
278				47	⊙	47																278
280		⊙		45	oo	47							23	⊙	23							280
282																						282
284		oo		47	oo	49				27	oo	29	23	oo	25	19	oo	21	17	oo	19	284
286				47	⊙	47				29	⊙	29										286
288		⊙																				288
290				49	⊙	49										21	⊙	21				290
292		oo		47	oo	49							23	oo	25				17	oo	19	292
294										29	⊙	29										294
296		⊙		49	oo	51				29	oo	31	25	⊙	25							296
298				49	⊙	49										21	⊙	21				298
300		oo		49	⊙	49													19	⊙	19	300
	4			6			8			10			12			14			16			

TWO-CIRCUIT, DOUBLE WINDINGS, FOR DRUM ARMATURES.

FRONT AND BACK PITCHES

No. of Conductors	4 POLES			6 POLES			8 POLES			10 POLES			12 POLES			14 POLES			16 POLES			No. of Conductors
	F	ent	B	F	ent	B	F	ent	B	F	ent	B	F	ent	B	F	ent	B	F	ent	B	
302				51	⊕	51																302
304	⅛	⊕	⅛	49	oo	51				29	oo	31	25	⊕	25	21	oo	23				304
306										31	⊕	31										306
308	⅛	oo	⅛	51	oo	53	⅜	⅜⅜	⅜				25	oo	27				19	⊕	19	308
310				51	⊕	51																310
312	⅛	⊕	⅛													21	oo	23				312
314				53	⊕	53				31	⊕	31										314
316	⅛	oo	⅛	51	oo	53	⅜	⅜⅜	⅜	31	oo	33	25	oo	27				19	oo	21	316
318																23	⊕	23				318
320	⅛	⊕	⅛	53	oo	55							27	⊕	27							320
322				53	⊕	53																322
324	⅛	oo	⅛				⅜	⅜⅜	⅜	31	oo	33							19	oo	21	324
326				55	⊕	55				33	⊕	33				23	⊕	23				326
328	⅛	⊕	⅛	53	oo	55							27	⊕	27							328
330																						330
332	⅛	oo	⅛	55	oo	57	⅜	⅜⅜	⅜				27	oo	29	23	oo	25	21	⊕	21	332
334				55	⊕	55				33	⊕	33										334
336	⅛	⊕	⅛							33	oo	35										336
338				57	⊕	57																338
340	⅛	oo	⅛	55	oo	57	⅜	⅜⅜	⅜				27	oo	29	23	oo	25	21	⊕	21	340
342																						342
344	⅛	⊕	⅛	57	oo	59				33	oo	35	29	⊕	29							344
346				57	⊕	57				35	⊕	35				25	⊕	25				346
348	⅛	oo	⅛				⅜	⅜⅜	⅜										21	oo	23	348
350				59	⊕	59																350
352	⅛	⊕	⅛	57	oo	59							29	⊕	29							352
354										35	⊕	35				25	⊕	25				354
356	⅛	oo	⅛	59	oo	61	⅜	⅜⅜	⅜	35	oo	37	29	oo	31				21	oo	23	356
358				59	⊕	59																358
360	⅛	⊕	⅛													25	oo	27				360
362				61	⊕	61																362
364	⅛	oo	⅛	59	oo	61	⅜	⅜⅜	⅜	35	oo	37	29	oo	31				23	⊕	23	364
366										37	⊕	37										366
368	⅛	⊕	⅛	61	oo	63							31	⊕	31	25	oo	27				368
370				61	⊕	61																370
372	⅛	oo	⅛				⅜	⅜⅜	⅜										23	⊕	23	372
374				63	⊕	63				37	⊕	37				27	⊕	27				374
376	⅛	⊕	⅛	61	oo	63				37	oo	39	31	⊕	31							376
378																						378
380	⅛	oo	⅛	63	oo	65	⅜	⅜⅜	⅜				31	oo	33				23	oo	25	380
382				63	⊕	63										27	⊕	27				382
384	⅛	⊕	⅛							37	oo	39										384
386				65	⊕	65				39	⊕	39										386
388	⅛	oo	⅛	63	oo	65	⅜	⅜⅜	⅜				31	oo	33	27	oo	29	23	oo	25	388
390																						390
392	⅛	⊕	⅛	65	oo	67							33	⊕	33							392
394				65	⊕	65				39	⊕	39										394
396	⅛	oo	⅛				⅜	⅜⅜	⅜	39	oo	41				27	oo	29	25	⊕	25	396
398				67	⊕	67																398
400	⅛	⊕	⅛	65	oo	67							33	⊕	33							400

4	6	8	10	12	14	16

TWO-CIRCUIT, DOUBLE WINDINGS, FOR DRUM ARMATURES.

FRONT AND BACK PITCHES

No. of Conductors	4 POLES			6 POLES			8 POLES			10 POLES			12 POLES			14 POLES			16 POLES			No. of Conductors
	F	ent.	B	F	ent.	B	F	ent.	B	F	ent.	B	F	ent.	B	F	ent.	B	F	ent.	B	
402													29	⊕	29							402
404	…	o o	…	67	o o	69	…	o o	…	39	o o	41	33	o o	35				25	⊕	25	404
406				67	⊕	67				41	⊕	41										406
408	…	⊕	…																			408
410				69	⊕	69							29	⊕	29							410
412	…	o o	…	67	o o	69	…	o o	…				33	o o	35				25	o o	27	412
414										41	⊕	41										414
416	…	⊕	…	69	o o	71				41	o o	43	35	⊕	35	29	o o	31				416
418				69	⊕	69																418
420	…	o o	…																25	o o	27	420
422				71	⊕	71																422
424	…	⊕	…	69	o o	71				41	o o	43	35	⊕	35	29	o o	31				424
426										43	⊕	43										426
428	…	o o	…	71	o o	73	…	o o	…				35	o o	37				27	⊕	27	428
430				71	⊕	71										31	⊕	31				430
432	…	⊕	…																			432
434				73	⊕	73				43	⊕	43										434
436	…	o o	…	71	o o	73	…	o o	…	43	o o	45	35	o o	37				27	⊕	27	436
438																31	⊕	31				438
440	…	⊕	…	73	o o	75							37	⊕	37							440
442				73	⊕	73																442
444	…	o o	…				…	o o	…	43	o o	45				31	o o	33	27	o o	29	444
446				75	⊕	75				45	⊕	45										446
448	…	⊕	…	73	o o	75							37	⊕	37							448
450																						450
452	…	o o	…	75	o o	77	…	o o	…				37	o o	39	31	o o	33	27	o o	29	452
454				75	⊕	75				45	⊕	45										454
456	…	⊕	…							45	o o	47										456
458				77	⊕	77										33	⊕	33				458
460	…	o o	…	75	o o	77	…	o o	…				37	o o	39				29	⊕	29	460
462																						462
464	…	⊕	…	77	o o	79				45	o o	47	39	⊕	39							464
466				77	⊕	77				47	⊕	47				33	⊕	33				466
468	…	o o	…				…	o o	…										29	⊕	29	468
470				79	⊕	79																470
472	…	⊕	…	77	o o	79							39	⊕	39	33	o o	35				472
474										47	⊕	47										474
476	…	o o	…	79	o o	81	…	o o	…	47	o o	49	39	o o	41				29	o o	31	476
478				79	⊕	79																478
480	…	⊕	…													33	o o	35				480
482				81	⊕	81																482
484	…	o o	…	79	o o	81	…	o o	…	47	o o	49	39	o o	41				29	o o	31	484
486										49	⊕	49				35	⊕	35				486
488	…	⊕	…	81	o o	83							41	⊕	41							488
490				81	⊕	81																490
492	…	o o	…				…	o o	…										31	⊕	31	492
494				83	⊕	83				49	⊕	49				35	⊕	35				494
496	…	⊕	…	81	o o	83				49	o o	51	41	⊕	41							496
498																						498
500	…	o o	…	83	o o	85	…	o o	…				41	o o	43	35	o o	37	31	⊕	31	500

No. of Conductors	4 POLES			6 POLES			8 POLES			10 POLES			12 POLES			14 POLES			16 POLES			No. of Conductors
	F	RE-ENTRANCY	B	F	RE-ENTRANCY	B	F	RE-ENTRANCY	B	F	RE-ENTRANCY	B	F	RE-ENTRANCY	B	F	RE-ENTRANCY	B	F	RE-ENTRANCY	B	
502				83	⊕	83																502
504		⊕								49	o o	51										504
506				85	⊕	85				51	⊕	51										506
508		o o		83	o o	85		⊕					41	o o	43	35	o o	37	31	o o	33	508
510																						510
512		⊕		95	o o	87							43	⊕	43							512
514				85	⊕	85				51	⊕	51				37	⊕	37				514
516		o o						⊕		51	o o	53							31	o o	33	516
518				87	⊕	87																518
520		⊕		85	o o	87							43	⊕	43							520
522																37	⊕	37				522
524		o o		87	o o	89		⊕		51	o o	53	43	o o	45				33	⊕	33	524
526				87	⊕	87				53	⊕	53										526
528		⊕														37	o o	39				528
530				89	⊕	89																530
532		o o		87	o o	89		⊕					43	o o	45				33	⊕	33	532
534										53	⊕	53										534
536		⊕		89	o o	91				53	o o	55	45	⊕	45	37	o o	39				536
538				89	⊕	89																538
540		o o						⊕											33	o o	35	540
542				91	⊕	91										39	⊕	39				542
544		⊕		89	o o	91				53	o o	55	45	⊕	45							544
546										55	⊕	55										546
548		o o		91	o o	93		⊕					45	o o	47				33	o o	35	548
550				91	⊕	91										39	⊕	39				550
552		⊕																				552
554				93	⊕	93				55	⊕	55										554
556		o o		91	o o	93		⊕		55	o o	57	45	o o	47	39	o o	41	35	⊕	35	556
558																						558
560		⊕		93	o o	95							47	⊕	47							560
562				93	⊕	93																562
564		o o						⊕		55	o o	57				39	o o	41	35	⊕	35	564
566				95	⊕	95				57	⊕	57										566
568		⊕		93	o o	95							47	⊕	47							568
570																41	⊕	41				570
572		o o		95	o o	97		⊕					47	o o	49				35	o o	37	572
574				95	⊕	95				57	⊕	57										574
576		⊕								57	o o	59										576
578				97	⊕	97										41	⊕	41				578
580		o o		95	o o	97		⊕					47	o o	49				35	o o	37	580
582																						582
584		⊕		97	o o	99				57	o o	59	49	⊕	49	41	o o	43				584
586				97	⊕	97				59	⊕	59										586
588		o o						⊕											37	⊕	37	588
590				99	⊕	99																590
592		⊕		97	o o	99							49	⊕	49	41	o o	43				592
594										59	⊕	59										594
596		o o		99	o o	101		⊕		59	o o	61	49	o o	51				37	⊕	37	596
598				99	⊕	99										43	⊕	43				598
600		⊕																				600
	4			6			8			10			12			14			16			

TWO-CIRCUIT, DOUBLE WINDINGS, FOR DRUM ARMATURES.

FRONT AND BACK PITCHES

No. of Cond.	4 POLES F	Re-entrancy	B	6 POLES F	Re-entrancy	B	8 POLES F	Re-entrancy	B	10 POLES F	Re-entrancy	B	12 POLES F	Re-entrancy	B	14 POLES F	Re-entrancy	B	16 POLES F	Re-entrancy	B	No. of Cond.
602				101	Ⓓ	101																602
604	1##	o o	1##	99	o o	101	½	##	##	59	o o	61	49	o o	51				37	o o	39	604
606										61	Ⓓ	61				43	Ⓓ	43				606
608	1##	Ⓓ	1##	101	o o	103							51	Ⓓ	51							608
610				101	Ⓓ	101																610
612	1##	o o	1##				##	##	##							43	o o	45	37	o o	39	612
614				103	Ⓓ	103																614
616	1##	Ⓓ	1##	101	o o	103				61	o o	63	51	Ⓓ	51							616
618																						618
620	1##	o o	1##	103	o o	105	##	##	##				51	o o	53	43	o o	45	39	Ⓓ	39	620
622				103	Ⓓ	103																622
624	1##	Ⓓ	1##							61	o o	63										624
626				105	Ⓓ	105				63	Ⓓ	63				45	Ⓓ	45				626
628	1##	o o	1##	103	o o	105	##	##	##				51	o o	53				39	Ⓓ	39	628
630																						630
632	1##	Ⓓ	1##	105	o o	107				63	Ⓓ	63	53	Ⓓ	53							632
634				105	Ⓓ	105										45	Ⓓ	45				634
636	1##	o o	1##				##	##	##	63	o o	65							39	o o	41	636
638				107	Ⓓ	107																638
640	1##	Ⓓ	1##	105	o o	107							53	Ⓓ	53	45	o o	47				640
642																						642
644	1##	o o	1##	107	o o	109	##	##	##	63	o o	65	53	o o	55				39	o o	41	644
646				107	Ⓓ	107				65	Ⓓ	65										646
648	1##	Ⓓ	1##													45	o o	47				648
650				109	Ⓓ	109																650
652	1##	o o	1##	107	o o	109	##	##	##				53	o o	55				41	Ⓓ	41	652
654										65	Ⓓ	65				47	Ⓓ	47				654
656	1##	Ⓓ	1##	109	o o	111				65	o o	67	55	Ⓓ	55							656
658				109	Ⓓ	109																658
660	1##	o o	1##				##	##	##										41	Ⓓ	41	660
662				111	Ⓓ	111										47	Ⓓ	47				662
664	1##	Ⓓ	1##	109	o o	111				65	o o	67	55	Ⓓ	55							664
666										67	Ⓓ	67										666
668	1##	o o	1##	111	o o	113	##	##	##				55	o o	57	47	o o	49	41	o o	43	668
670				111	Ⓓ	111																670
672	1##	Ⓓ	1##																			672
674				113	Ⓓ	113				67	Ⓓ	67										674
676	1##	o o	1##	111	o o	113	##	##	##	67	o o	69	55	o o	57	47	o o	49	41	o o	43	676
678																						678
680	1##	Ⓓ	1##	113	o o	115							57	Ⓓ	57	49	Ⓓ	49				680
682				113	Ⓓ	113																682
684	1##	o o	1##				##	##	##	67	o o	69							43	Ⓓ	43	684
686				115	Ⓓ	115				69	Ⓓ	69										686
688	1##	Ⓓ	1##	113	o o	115							57	Ⓓ	57	49	Ⓓ	49				688
690																						690
692	1##	o o	1##	115	o o	117	##	##	##				57	o o	59				43	Ⓓ	43	692
694				115	Ⓓ	115				69	Ⓓ	69										694
696	1##	Ⓓ	1##							69	o o	71				49	o o	51				696
698				117	Ⓓ	117																698
700	1##	o o	1##	115	o o	117	##	##	##				57	o o	59				43	o o	45	700

| 4 | 6 | 8 | 10 | 12 | 14 | 16 |

TWO-CIRCUIT, DOUBLE WINDINGS, FOR DRUM ARMATURES.

FRONT AND BACK PITCHES

No. OF CONDUCTORS	4 POLES			6 POLES			8 POLES			10 POLES			12 POLES			14 POLES			16 POLES			No. OF CONDUCTORS
	F	ENTRANCE	B	F	ENTRANCE	B	F	ENTRANCE	B	F	ENTRANCE	B	F	ENTRANCE	B	F	ENTRANCE	B	F	ENTRANCE	B	
702																						702
704		⊙		117	oo	119				69	oo	71	59	⊙	59	49	oo	51				704
706				117	⊙	117				71	⊙	71										706
708		oo																	43	oo	45	708
710				119	⊙	119										51	⊙	51				710
712		⊙		117	oo	119							59	⊙	59							712
714										71	⊙	71										714
716		oo		119	oo	121				71	oo	73	59	oo	61				45	⊙	45	716
718				119	⊙	119										51	⊙	51				718
720		⊙																				720
722				121	⊙	121																722
724		oo		119	oo	121				71	oo	73	59	oo	61	51	oo	53	45	⊙	45	724
726										73	⊙	73										726
728		⊙		121	oo	123							61	⊙	61							728
730				121	⊙	121																730
732		oo														51	oo	53	45	oo	47	732
734				123	⊙	123				73	⊙	73										734
736		⊙		121	oo	123				73	oo	75	61	⊙	61							736
738																53	⊙	53				738
740		oo		123	oo	125							61	oo	63				45	oo	47	740
742				123	⊙	123																742
744		⊙								73	oo	75										744
746				125	⊙	125				75	⊙	75				53	⊙	53				746
748		oo		123	oo	125							61	oo	63				47	⊙	47	748
750																						750
752		⊙		125	oo	127							63	⊙	63	53	oo	55				752
754				125	⊙	125				75	⊙	75										754
756		oo								75	oo	77							47	⊙	47	756
758				127	⊙	127																758
760		⊙		125	oo	127							63	⊙	63	53	oo	55				760
762																						762
764		oo		127	oo	129				75	oo	77	63	oo	65				47	oo	49	764
766				127	⊙	127				77	⊙	77				55	⊙	55				766
768		⊙																				768
770				129	⊙	129																770
772		oo		127	oo	129							63	oo	65	55	⊙	55	47	oo	49	772
774										77	⊙	77										774
776		⊙		129	oo	131				77	oo	79	65	⊙	65							776
778				129	⊙	129																778
780		oo														55	oo	57	49	⊙	49	780
782				131	⊙	131																782
784		⊙		129	oo	131				77	oo	79	65	⊙	65							784
786										79	⊙	79										786
788		oo		131	oo	133							65	oo	67	55	oo	57	49	⊙	49	788
790				131	⊙	131																790
792		⊙																				792
794				133	⊙	133				79	⊙	79				57	⊙	57				794
796		oo		131	oo	133				79	oo	81	65	oo	67				49	oo	51	796
798																						798
800		⊙		133	oo	135																800

| 4 | 6 | 8 | 10 | 12 | 14 | 16 |

WINDING TABLES FOR TWO-CIRCUIT, TRIPLE WINDINGS FOR DRUM ARMATURES.

TABLE OF TWO-CIRCUIT, TRIPLE WINDINGS, FOR DRUM ARMATURES.

FRONT AND BACK PITCHES

No. of Conductors	4 POLES			6 POLES			8 POLES			10 POLES			12 POLES			14 POLES			16 POLES			No. of Conductors
	F		B	F		B	F		B	F		B	F		B	F		B	F		B	
102		ooo			₽₽₽		11	ooo	13					₽₽₽						5	ooo	7	102
104		⊚								11	⊚	11				7	⊚	7				104
106		⊚			⊚		13	⊚	15	9	⊚	11				7	⊚	9	7	⊚	7	106
108					⊚																	108
110		⊚					13	⊚	13													110
112																						112
114		ooo			₽₽₽		15	ooo	15	11	ooo	13		₽₽₽								114
116										11	⊚	11										116
118		⊚					13	⊚	15							7	⊚	9	7	⊚	7	118
120					₽₽₽											9	ooo	9				120
122		⊚					15	⊚	17										7	⊚	9	122
124										13	⊚	13										124
126		ooo			⊚		15	ooo	15	11	ooo	13		₽₽₽								126
128																						128
130		⊚					17	⊚	17													130
132					₽₽₽											9	ooo	9				132
134		⊚					15	⊚	17	13	⊚	15				9	⊚	11	7	⊚	9	134
136										13	⊚	13										136
138		ooo			₽₽₽		17	ooo	19					₽₽₽					9	ooo	9	138
140																						140
142		⊚					17	⊚	17													142
144					⊚					15	ooo	15										144
146		⊚					19	⊚	19	13	⊚	15				9	⊚	11				146
148																11	⊚	11				148
150		ooo			₽₽₽		17	ooo	19					₽₽₽					9	ooo	9	150
152																						152
154		⊚					19	⊚	21	15	⊚	17							9	⊚	11	154
156					₽₽₽					15	ooo	15										156
158		⊚					19	⊚	19													158
160																11	⊚	11				160
162		ooo			⊚		21	ooo	21					₽₽₽		11	ooo	13				162
164										17	⊚	17										164
166		⊚					19	⊚	21	15	⊚	17							9	⊚	11	166
168					₽₽₽																	168
170		⊚					21	⊚	23										11	⊚	11	170
172																						172
174		ooo			₽₽₽		21	ooo	21	17	ooo	19		₽₽₽		11	ooo	13				174
176		⊚								17	⊚	17				13	⊚	13				176
178					⊚		23	⊚	23													178
180																						180
182		⊚					21	⊚	23										11	⊚	11	182
184										19	⊚	19										184
186		ooo			₽₽₽		23	ooo	25	17	ooo	19		₽₽₽		11	ooo	13				186
188																13	⊚	13				188
190		⊚			₽₽₽		23	⊚	23							13	⊚	15				190
192																						192
194		⊚					25	⊚	25	19	⊚	21										194
196										19	⊚	19										196
198		ooo			⊚		23	ooo	25					₽₽₽					11	ooo	13	198
200																						200

TABLE OF TWO-CIRCUIT, TRIPLE WINDINGS, FOR DRUM ARMATURES.

No. OF CONDUCTORS	4 POLES		6 POLES		8 POLES		10 POLES		12 POLES		14 POLES		16 POLES		No. OF CONDUCTORS
	F	B	F	B	F	B	F	B	F	B	F	B	F	B	
202					25	27					13	15	13	13	202
204							21	21			15	15			204
206					25	25	19	21							206
208															208
210					27	27									210
212															212
214					25	27	21	23			13	13			214
216							21	21					13	13	216
218					27	29					15	15			218
220											15	17	13	15	220
222					27	27									222
224							23	23							224
226					29	29	21	23							226
228															228
230					27	29					15	17	13	15	230
232											17	17			232
234					29	31	23	25					15	15	234
236							23	23							236
238					29	29									238
240															240
242					31	31					17	17			242
244							25	25							244
246					29	31	23	25			17	19	15	15	246
248															248
250					31	33							15	17	250
252															252
254					31	31	25	27							254
256							25	25							256
258					33	33					17	19			258
260											19	19			260
262					31	33							15	17	262
264							27	27							264
266					33	35	25	27					17	17	266
268															268
270					33	33					19	19			270
272											19	21			272
274					35	35	27	29							274
276							27	27							276
278					33	35							17	17	278
280															280
282					35	37							17	19	282
284							29	29							284
286					35	35	27	29			19	21			286
288											21	21			288
290					37	37									290
292															292
294					35	37	29	31					17	19	294
296							29	29							296
298					37	39							19	19	298
300											21	21			300

TABLE OF TWO-CIRCUIT, TRIPLE WINDINGS, FOR DRUM ARMATURES.

FRONT AND BACK PITCHES

No. of Conductors	4 POLES F	RE-ENTRANCY	B	6 POLES F	RE-ENTRANCE	B	8 POLES F	RE-ENTRANCY	B	10 POLES F	RE-ENTRANCY	B	12 POLES F	RE-ENTRANCY	B	14 POLES F	RE-ENTRANCY	B	16 POLES F	RE-ENTRANCY	B	No. of Conductors
302	¾	(QQ)	¾				37	(QQ)	37							21	(QQ)	23				302
304										31	(QQ)	31										304
306	¾	ooo	¾	¾	(QQ)	¾	39	ooo	39	29	ooo	31	¾	88	¾							306
308																						308
310	¾	(QQ)	¾				37	(QQ)	39										19	(QQ)	19	310
312				¾	888	¾																312
314	¾	(QQ)	¾				39	(QQ)	41	31	(QQ)	33				21	(QQ)	23	19	(QQ)	21	314
316										31	(QQ)	31				23	(QQ)	23				316
318	¾	ooo	¾	¾	888	¾	39	ooo	39				¾	888	¾							318
320																						320
322	¾	(QQ)	¾				41	(QQ)	41													322
324				¾	(QQ)	¾				33	ooo	33										324
326	¾	(QQ)	¾				39	(QQ)	41	31	(QQ)	33							19	(QQ)	21	326
328																23	(QQ)	23				328
330	¾	ooo	¾	¾	888	¾	41	ooo	43				¾	888	¾	23	ooo	25	21	ooo	21	330
332																						332
334	¾	(QQ)	¾				41	(QQ)	41	33	(QQ)	35										334
336				¾	888	¾				33	ooo	33										336
338	¾	(QQ)	¾				43	(QQ)	43													338
340																						340
342	¾	ooo	¾	¾	(QQ)	¾	41	ooo	43				¾	88	¾	23	ooo	25	21	ooo	21	342
344										35	(QQ)	35				25	(QQ)	25				344
346	¾	(QQ)	¾				43	(QQ)	45	33	(QQ)	35							21	(QQ)	23	346
348				¾	888	¾																348
350	¾	(QQ)	¾				43	(QQ)	43													350
352																						352
354	¾	ooo	¾	¾	888	¾	45	ooo	45	35	ooo	37	¾	888	¾							354
356										35	(QQ)	35				25	(QQ)	25				356
358	¾	(QQ)	¾				43	(QQ)	45							25	(QQ)	27	21	(QQ)	23	358
360				¾	(QQ)	¾																360
362	¾	(QQ)	¾				45	(QQ)	47										23	(QQ)	23	362
364										37	(QQ)	37										364
366	¾	ooo	¾	¾	888	¾	45	ooo	45	35	ooo	37	¾	888	¾							366
368																						368
370	¾	(QQ)	¾				47	(QQ)	47							25	(QQ)	27				370
372				¾	888	¾										27	ooo	27				372
374	¾	(QQ)	¾				45	(QQ)	47	37	(QQ)	39							23	(QQ)	23	374
376										37	(QQ)	37										376
378	¾	ooo	¾	¾	(QQ)	¾	47	ooo	49				¾	88	¾				23	ooo	25	378
380																						380
382	¾	(QQ)	¾				47	(QQ)	47													382
384				¾	888	¾				39	ooo	39				27	ooo	27				384
386	¾	(QQ)	¾				49	(QQ)	49	37	(QQ)	39				27	(QQ)	29				386
388																						388
390	¾	ooo	¾	¾	888	¾	47	ooo	49				¾	888	¾				23	ooo	25	390
392																						392
394	¾	(QQ)	¾				49	(QQ)	51	39	(QQ)	41							25	(QQ)	25	394
396				¾	(QQ)	¾				39	ooo	39										396
398	¾	(QQ)	¾				49	(QQ)	49							27	(QQ)	29				398
400																29	(QQ)	29				400

4	6	8	10	12	14	16

TABLE OF TWO-CIRCUIT, TRIPLE WINDINGS, FOR DRUM ARMATURES.

No. OF CONDUCTORS	4 POLES		6 POLES		8 POLES		10 POLES		12 POLES		14 POLES		16 POLES		No. OF CONDUCTORS
	F	B	F	B	F	B	F	B	F	B	F	B	F	B	
402					51	51									402
404							41	41							404
406					49	51	39	41					25	25	406
408															408
410					51	53							25	27	410
412											29	29			412
414					51	51	41	43			29	31			414
416							41	41							416
418					53	53									418
420															420
422					51	53							25	27	422
424							43	43							424
426					53	55	41	43			29	31	27	27	426
428											31	31			428
430					53	53									430
432															432
434					55	55	43	45							434
436							43	43							436
438					53	55							27	27	438
440											31	31			440
442					55	57					31	33	27	29	442
444							45	45							444
446					55	55	43	45							446
448															448
450					57	57									450
452															452
454					55	57	45	47			31	33	27	29	454
456							45	45			33	33			456
458					57	59							29	29	458
460															460
462					57	57									462
464							47	47							464
466					59	59	45	47							466
468											33	33			468
470					57	59					33	35	29	29	470
472															472
474					59	61	47	49					29	31	474
476							47	47							476
478					59	59									478
480															480
482					61	61					33	35			482
484							49	49			35	35			484
486					59	61	47	49					29	31	486
488															488
490					61	63							31	31	490
492															492
494					61	61	49	51					31	31	494
496							49	49			35	35			496
498					63	63					35	37			498
500															500

| 4 | 6 | 8 | 10 | 12 | 14 | 16 |

TABLE OF TWO-CIRCUIT, TRIPLE WINDINGS, FOR DRUM ARMATURES.

No. of Conductors	4 POLES			6 POLES			8 POLES			10 POLES			12 POLES			14 POLES			16 POLES			No. of Conductors
	F	RE-ENTRANCY	B	F	RE-ENTRANCY	B	F	RE-ENTRANCY	B	F	RE-ENTRANCY	B	F	RE-ENTRANCY	B	F	RE-ENTRANCY	B	F	RE-ENTRANCY	B	
502		⊛					61	⊛	63										31	⊛	31	502
504					⊛					51	ooo	51										504
506		⊛			⊛		63	⊛	65	49	⊛	51							31	⊛	33	506
508																						508
510		ooo			⊛		63	ooo	63					⊛		35	ooo	37				510
512																37	⊛	37				512
514		⊛					65	⊛	65	51	⊛	53										514
516					⊛					51	ooo	51										516
518		⊛					63	⊛	65										31	⊛	33	518
520																						520
522		ooo			⊛		65	ooo	67					⊛					33	ooo	33	522
524										53	⊛	53				37	⊛	37				524
526		⊛					65	⊛	65	51	⊛	53				37	⊛	39				526
528					⊛																	528
530		⊛					67	⊛	67													530
532																						532
534		ooo			⊛		65	ooo	67	53	ooo	55		⊛					33	ooo	33	534
536										53	⊛	53										536
538		⊛					67	⊛	69							37	⊛	39	33	⊛	35	538
540					⊛											39	ooo	39				540
542		⊛					67	⊛	67													542
544																						544
546		ooo			⊛		69	ooo	69	53	ooo	55		⊛								546
548																						548
550		⊛					67	⊛	69										33	⊛	35	550
552					⊛											39	ooo	39				552
554		⊛					69	⊛	71	55	⊛	57				39	⊛	41	35	⊛	35	554
556										55	⊛	55										556
558		ooo			⊛		69	ooo	69					⊛								558
560																						560
562		⊛					71	⊛	71													562
564					⊛					57	ooo	57										564
566		⊛					69	⊛	71	55	⊛	57				39	⊛	41	35	⊛	35	566
568																41	⊛	41				568
570		ooo			⊛		71	ooo	73					⊛					35	ooo	37	570
572																						572
574		⊛					71	⊛	71	57	⊛	59										574
576					⊛					57	ooo	57										576
578		⊛					73	⊛	73													578
580																41	⊛	41				580
582		ooo			⊛		71	ooo	73					⊛		41	ooo	43	35	ooo	37	582
584										59	⊛	59										584
586		⊛			⊛		73	⊛	75	57	⊛	59							37	⊛	37	586
588					⊛																	588
590		⊛					73	⊛	73													590
592																						592
594		ooo			⊛		75	ooo	75	59	ooo	61		⊛		41	ooo	43				594
596										59	⊛	59				43	⊛	43				596
598		⊛					73	⊛	75										37	⊛	37	598
600					⊛																	600

| 4 | 6 | 8 | 10 | 12 | 14 | 16 |

TABLE OF TWO-CIRCUIT, TRIPLE WINDINGS, FOR DRUM ARMATURES.

FRONT AND BACK PITCHES

No. OF CONDUCTORS	4 POLES F	B	6 POLES F	B	8 POLES F	B	10 POLES F	B	12 POLES F	B	14 POLES F	B	16 POLES F	B	No. OF CONDUCTORS
602					75	77							37	39	602
604							61	61							604
606					75	75	59	61							606
608											43	43			608
610					77	77					43	45			610
612															612
614					75	77	61	63					37	39	614
616							61	61							616
618					77	79							39	39	618
620															620
622					77	77					43	45			622
624							63	63			45	45			624
626					79	79	61	63							626
628															628
630					77	79							39	?0	630
632															632
634					79	81	63	65					39	41	634
636							63	63							636
638					79	79					45	47			638
640															640
642					81	81									642
644							65	65							644
646					79	81	63	65					39	41	646
648															648
650					81	83					45	47	41	41	650
652											47	47			652
654					81	81	65	67							654
656							65	65							656
658					83	83									658
660															660
662					81	83							41	41	662
664							67	67			47	47			664
666					83	85	65	67			47	49	41	43	666
668															668
670					83	83									670
672															672
674					85	85	67	69							674
676							67	67							676
678					83	85					47	49	41	43	678
680											49	49			680
682					85	87							43	43	682
684							69	69							684
686					85	85	67	69							686
688															688
690					87	87									690
692											49	49			692
694					85	87	69	71			49	51	43	43	694
696							69	69							696
698					87	89							43	45	698
700															700

4   6   8   10   12   14   16

TABLE OF TWO-CIRCUIT, TRIPLE WINDINGS, FOR DRUM ARMATURES.

FRONT AND BACK PITCHES

No. OF CONDUCTORS	4 POLES F	Re-entrancy	B	6 POLES F	Re-entrancy	B	8 POLES F	Re-entrancy	B	10 POLES F	Re-entrancy	B	12 POLES F	Re-entrancy	B	14 POLES F	Re-entrancy	B	16 POLES F	Re-entrancy	B	No. OF CONDUCTORS
702		ooo			⊕		87	ooo	87					⊞								702
704										71	⊕	71										704
706		⊕					80	⊕	89	69	⊕	71				49	⊕	51				706
708					⊕											51	ooo	51				708
710		⊕			⊕		87	⊕	89										43	⊕	45	710
712																						712
714		ooo			⊕		89	ooo	91	71	ooo	73		⊕					45	ooo	45	714
716										71	⊕	71										716
718		⊕					80	⊕	89													718
720					⊕											51	ooo	51				720
722		⊕			⊕		91	⊕	91							51	⊕	53				722
724										73	⊕	73										724
726		ooo			⊕		89	ooo	91	71	ooo	73		⊕					45	ooo	45	726
728																						728
730		⊕			⊕		91	⊕	93										45	⊕	47	730
732					⊕																	732
734		⊕			⊕		91	⊕	91	73	⊕	75				51	⊕	53				734
736										73	⊕	73				53	⊕	53				736
738		ooo			⊕		93	ooo	93					⊕								738
740																						740
742		⊕			⊕		91	⊕	93										45	⊕	47	742
744					⊕					75	ooo	75										744
746		⊕			⊕		93	⊕	95	73	⊕	75							47	⊕	47	746
748																53	⊕	53				748
750		ooo			⊕		93	ooo	93					⊕		53	ooo	55				750
752																						752
754		⊕			⊕		95	⊕	95	75	⊕	77										754
756					⊕					75	ooo	75										756
758		⊕					93	⊕	95										47	⊕	47	758
760																						760
762		ooo			⊕		95	ooo	97					⊕		53	ooo	55	47	ooo	49	762
764										77	⊕	77				55	⊕	55				764
766		⊕			⊕		95	⊕	95	75	⊕	77										766
768					⊕																	768
770		⊕					97	⊕	97													770
772																						772
774		ooo			⊕		95	ooo	97	77	ooo	79		⊕					47	ooo	49	774
776										77	⊕	77				55	⊕	55				776
778		⊕			⊕		97	⊕	99							55	⊕	57	49	⊕	49	778
780					⊕																	780
782		⊕			⊕		97	⊕	97													782
784										79	⊕	79										784
786		ooo			⊕		99	ooo	99	77	ooo	79		⊕								786
788																						788
790		⊕			⊕		97	⊕	99							55	⊕	57	49	⊕	49	790
792					⊕											57	ooo	57				792
794		⊕			⊕		99	⊕	101	79	⊕	81							49	⊕	51	794
796										79	⊕	79										796
798		ooo			⊕		99	ooo	99					⊞								798
800																						800

4 6 8 10 12 14 16

WINDING TABLES FOR MULTIPLE–CIRCUIT, SINGLE WINDINGS FOR DRUM ARMATURES.

MULTIPLE-CIRCUIT, SINGLE WINDINGS, FOR DRUM ARMATURES.

No. OF CONDUCTORS	FRONT AND BACK PITCHES													No. OF CONDUCTORS
	4 POLES		6 POLES		8 POLES		10 POLES		12 POLES		14 POLES		16 POLES		
	F	B	F	B	F	B	F	B	F	B	F	B	F	B	
202	49	51	33	35	25	27	19	21	15	17	13	15	11	13	202
204	49	51	33	35	25	27	19	21	15	17	13	15	11	13	204
206	51	53	33	35	25	27	19	21	17	19	13	15	11	13	206
208	51	53	33	35	25	27	19	21	17	19	13	16	11	13	208
210	51	53	33	35	25	27	19	21	17	19	13	15	13	15	210
212	51	53	35	37	25	27	21	23	17	19	15	17	13	15	212
214	53	55	35	37	25	27	21	23	17	19	15	17	13	15	214
216	53	55	35	37	25	27	21	23	17	19	15	17	13	15	216
218	53	55	35	37	27	29	21	23	17	19	15	17	13	15	218
220	53	55	35	37	27	29	21	23	17	19	15	17	13	15	220
222	55	57	35	37	27	29	21	23	17	19	15	17	13	15	222
224	55	57	37	39	27	29	21	23	17	19	15	17	13	15	224
226	55	57	37	39	27	29	21	23	17	19	15	17	13	15	226
228	55	57	37	39	27	29	21	23	17	19	15	17	13	15	228
230	57	59	37	39	27	29	21	23	19	21	15	17	13	15	230
232	57	59	37	39	27	29	23	25	19	21	15	17	13	15	232
234	57	59	37	39	29	31	23	25	19	21	15	17	13	15	234
236	57	59	39	41	29	31	23	25	19	21	15	17	13	15	236
238	59	61	39	41	29	31	23	25	19	21	15	17	13	15	238
240	59	61	39	41	29	31	23	25	19	21	17	19	13	15	240
242	59	61	39	41	29	31	23	25	19	21	17	19	15	17	242
244	59	61	39	41	29	31	23	25	19	21	17	19	15	17	244
246	61	63	39	41	29	31	23	25	19	21	17	19	15	17	246
248	61	63	41	43	29	31	23	25	19	21	17	19	15	17	248
250	61	63	41	43	31	33	23	25	19	21	17	19	15	17	250
252	61	63	41	43	31	33	25	27	19	21	17	19	15	17	252
254	63	65	41	43	31	33	25	27	21	23	17	19	15	17	254
256	63	65	41	43	31	33	25	27	21	23	17	19	15	17	256
258	63	65	41	43	31	33	25	27	21	23	17	19	15	17	258
260	63	65	43	45	31	33	25	27	21	23	17	19	15	17	260
262	65	67	43	45	31	33	25	27	21	23	17	19	15	17	262
264	65	67	43	45	31	33	25	27	21	23	17	19	15	17	264
266	65	67	43	45	33	35	25	27	21	23	17	19	15	17	266
268	65	67	43	45	33	35	25	27	21	23	19	21	15	17	268
270	67	69	43	45	33	35	25	27	21	23	19	21	15	17	270
272	67	69	45	47	33	35	27	29	21	23	19	21	15	17	272
274	67	69	45	47	33	35	27	29	21	23	19	21	17	19	274
276	67	69	45	47	33	35	27	29	21	23	19	21	17	19	276
278	69	71	45	47	33	35	27	29	23	25	19	21	17	19	278
280	69	71	45	47	33	35	27	29	23	25	19	21	17	19	280
282	69	71	45	47	35	37	27	29	23	25	19	21	17	19	282
284	69	71	47	49	35	37	27	29	23	25	19	21	17	19	284
286	71	73	47	49	35	37	27	29	23	25	19	21	17	19	286
288	71	73	47	49	35	37	27	29	23	25	19	21	17	19	288
290	71	73	47	49	35	37	27	29	23	25	19	21	17	19	290
292	71	73	47	49	35	37	29	31	23	25	19	21	17	19	292
294	73	75	47	49	35	37	29	31	23	25	19	21	17	19	294
296	73	75	49	51	35	37	29	31	23	25	21	23	17	19	296
298	73	75	49	51	37	39	29	31	23	25	21	23	17	19	298
300	73	75	49	51	37	39	29	31	23	25	21	23	17	19	300

Above choice of Pitches will prove most satisfactory, although, as stated in text, the absolute magnitude of average pitch may be varied within reasonable limits.

MULTIPLE-CIRCUIT, SINGLE WINDINGS, FOR DRUM ARMATURES.

FRONT AND BACK PITCHES.

No. OF CONDUCTORS	4 POLES		6 POLES		8 POLES		10 POLES		12 POLES		14 POLES		16 POLES		No. OF CONDUCTORS
	F	B	F	B	F	B	F	B	F	B	F	B	F	B	
302	75	77	49	51	37	39	29	31	25	27	21	23	17	19	302
304	75	77	49	51	37	39	29	31	25	27	21	23	17	19	304
306	75	77	49	51	37	39	29	31	25	27	21	23	19	21	306
308	75	77	51	53	37	39	29	31	25	27	21	23	19	21	308
310	77	79	51	53	37	39	29	31	25	27	21	23	19	21	310
312	77	79	51	53	37	39	31	33	25	27	21	23	19	21	312
314	77	79	51	53	39	41	31	33	25	27	21	23	19	21	314
316	77	79	51	53	39	41	31	33	25	27	21	23	19	21	316
318	79	81	51	53	39	41	31	33	25	27	21	23	19	21	318
320	79	81	53	55	39	41	21	33	25	27	21	23	19	21	320
322	79	81	53	55	39	41	31	33	25	27	21	23	19	21	322
324	79	81	53	55	39	41	31	33	25	27	23	25	19	21	324
326	81	83	53	55	39	41	31	33	27	29	23	25	19	21	326
328	81	83	53	55	39	41	31	33	27	29	23	25	19	21	328
330	81	83	53	55	41	43	31	33	27	29	23	25	19	21	330
332	81	83	55	57	41	43	33	35	27	29	23	25	19	21	332
334	83	85	55	57	41	43	33	35	27	29	23	25	19	21	334
336	83	85	55	57	41	43	33	35	27	29	23	25	19	21	336
338	83	85	55	57	41	43	33	35	27	29	23	25	21	23	338
340	83	85	55	57	41	43	33	35	27	29	23	25	21	23	340
342	85	87	55	57	41	43	33	35	27	29	23	25	21	23	342
344	85	87	57	59	41	43	33	35	27	29	23	25	21	23	344
346	85	87	57	59	43	45	33	35	27	29	23	25	21	23	346
348	85	87	57	59	43	45	33	35	27	29	23	25	21	23	348
350	87	89	57	59	43	45	33	35	29	31	23	25	21	23	350
352	87	89	57	59	43	45	35	37	29	31	25	27	21	23	352
354	87	89	57	59	43	45	35	87	29	31	25	27	21	23	354
356	87	89	59	61	43	45	35	37	29	31	25	27	21	23	356
358	89	91	59	61	43	45	35	37	29	31	25	27	21	23	358
360	89	91	59	61	43	45	35	37	29	31	25	27	21	23	360
362	89	91	59	61	45	47	35	37	29	31	25	27	21	23	362
364	89	91	59	61	45	47	35	37	29	31	25	27	21	23	364
366	91	93	59	61	45	47	35	37	29	31	25	27	21	23	366
368	91	93	61	63	45	47	35	37	29	31	25	27	21	23	368
370	91	93	61	63	45	47	35	37	29	31	25	27	23	25	370
372	91	93	61	63	45	47	37	39	29	31	25	27	23	25	372
374	93	95	61	63	45	47	37	39	31	33	25	27	23	25	374
376	93	95	61	63	47	49	37	39	31	33	25	27	23	25	376
378	93	95	61	63	47	49	37	39	31	33	25	27	23	25	378
380	93	95	63	65	47	49	37	39	31	33	27	29	23	25	380
382	95	97	63	65	47	49	37	39	31	33	27	29	23	25	382
384	95	97	63	65	47	49	37	39	31	33	27	29	23	25	384
386	95	97	63	65	47	49	37	39	31	33	27	29	23	25	386
388	95	97	63	65	47	49	37	39	31	33	27	29	23	25	388
390	97	99	63	65	47	49	37	39	31	33	27	29	23	25	390
392	97	99	65	67	47	49	39	41	31	33	27	29	23	25	392
394	97	99	65	67	49	51	39	41	31	33	27	29	23	25	394
396	97	99	65	67	49	51	39	41	31	33	27	29	23	25	396
398	99	101	65	67	49	51	39	41	33	35	27	29	23	25	398
400	99	101	65	67	49	51	39	41	33	35	27	29	23	25	400

Above choice of Pitches will prove most satisfactory, although, as stated in text, the absolute magnitude of average pitch may be varied within reasonable limits.

MULTIPLE-CIRCUIT, SINGLE WINDINGS, FOR DRUM ARMATURES.

No. OF CONDUCTORS	FRONT AND BACK PITCHES														No. OF CONDUCTORS
	4 POLES		6 POLES		8 POLES		10 POLES		12 POLES		14 POLES		16 POLES		
	F	B	F	B	F	B	F	B	F	B	F	B	F	B	
402	99	101	65	67	49	51	39	41	33	35	27	29	25	27	402
404	99	101	67	69	49	51	39	41	33	35	27	29	25	27	404
406	101	103	67	69	49	51	39	41	33	35	27	29	25	27	406
408	101	103	67	69	49	51	39	41	33	35	29	31	25	27	408
410	101	103	67	69	51	53	39	41	33	35	29	31	25	27	410
412	101	103	67	69	51	53	41	43	33	35	29	31	25	27	412
414	103	105	67	69	51	53	41	43	33	35	29	31	25	27	414
416	103	105	69	71	51	53	41	43	33	35	29	31	25	27	416
418	103	105	69	71	51	53	41	43	33	35	29	31	25	27	418
420	103	105	69	71	51	53	41	43	33	35	29	31	25	27	420
422	105	107	69	71	51	53	41	43	33	35	29	31	25	27	422
424	105	107	69	71	51	53	41	43	35	37	29	31	25	27	424
426	105	107	69	71	53	55	41	43	35	37	29	31	25	27	426
428	105	107	71	73	53	55	41	43	35	37	29	31	25	27	428
430	107	109	71	73	53	55	41	43	35	37	29	31	25	27	430
432	107	109	71	73	53	55	43	45	35	37	29	31	25	27	432
434	107	109	71	73	53	55	43	45	35	37	29	31	27	29	434
436	107	109	71	73	53	55	43	45	35	37	31	33	27	29	436
438	109	111	71	73	53	55	43	45	35	37	31	33	27	29	438
440	109	111	73	75	53	55	43	45	35	37	31	33	27	29	440
442	109	111	73	75	55	57	43	45	35	37	31	33	27	29	442
444	109	111	73	75	55	57	43	45	35	37	31	33	27	29	444
446	111	113	73	75	55	57	43	45	37	39	31	33	27	29	446
448	111	113	73	75	55	57	43	45	37	39	31	33	27	29	448
450	111	113	73	75	55	57	43	45	37	39	31	33	27	29	450
452	111	113	75	77	55	57	45	47	37	39	31	33	27	29	452
454	113	115	75	77	55	57	45	47	37	39	31	33	27	29	454
456	113	115	75	77	55	57	45	47	37	39	31	33	27	29	456
458	113	115	75	77	57	59	45	47	37	39	31	33	27	29	458
460	113	115	75	77	57	59	45	47	37	39	31	33	27	29	460
462	115	117	75	77	57	59	45	47	37	39	31	33	27	29	462
464	115	117	77	79	57	59	45	47	37	39	33	35	27	29	464
466	115	117	77	79	57	59	45	47	37	39	33	35	29	31	466
468	115	117	77	79	57	59	45	47	37	39	33	35	29	31	468
470	117	119	77	79	57	59	45	47	39	41	33	35	29	31	470
472	117	119	77	79	57	59	47	49	39	41	33	35	29	31	472
474	117	119	77	79	59	61	47	49	39	41	33	35	29	31	474
476	117	119	79	81	59	61	47	49	39	41	33	35	29	31	476
478	119	121	79	81	59	61	47	49	39	41	33	35	29	31	478
480	119	121	79	81	59	61	47	49	39	41	33	35	29	31	480
482	119	121	79	81	59	61	47	49	39	41	33	35	29	31	482
484	119	121	79	81	59	61	47	49	39	41	33	35	29	31	484
486	121	123	79	81	59	61	47	49	39	41	33	35	29	31	486
488	121	123	81	83	59	61	47	49	39	41	33	35	29	31	488
490	121	123	81	83	61	63	47	49	39	41	33	35	29	31	490
492	121	123	81	83	61	63	49	51	39	41	35	37	29	31	492
494	123	125	81	83	61	63	49	51	41	43	35	37	29	31	494
496	123	125	81	83	61	63	49	51	41	43	35	37	29	31	496
498	123	125	81	83	61	63	49	51	41	43	35	37	31	33	498
500	123	125	83	85	61	63	49	51	41	43	35	37	31	33	500

Above choice of Pitches will prove most satisfactory, although, as stated in text, the absolute magnitude of average pitch may be varied within reasonable limits.

MULTIPLE-CIRCUIT, SINGLE WINDINGS, FOR DRUM ARMATURES.

No. OF CONDUCTORS	FRONT AND BACK PITCHES														No. OF CONDUCTORS
	4 POLES		6 POLES		8 POLES		10 POLES		12 POLES		14 POLES		16 POLES		
	F	B	F	B	F	B	F	B	F	B	F	B	F	B	
502	125	127	83	85	61	63	49	51	41	43	35	37	31	33	502
504	125	127	83	85	61	63	49	51	41	43	35	37	31	33	504
506	125	127	83	85	63	65	49	51	41	43	35	37	31	33	506
508	125	127	83	85	63	65	49	51	41	43	35	37	31	33	508
510	127	129	83	85	63	65	49	51	41	43	35	37	31	33	510
512	127	129	85	87	63	65	51	53	41	43	35	37	31	33	512
514	127	129	85	87	63	65	51	53	41	43	35	37	31	33	514
516	127	129	85	87	63	65	51	53	41	43	35	37	31	33	516
518	129	131	85	87	63	65	51	53	43	45	35	37	31	33	518
520	129	131	85	87	63	65	51	53	43	45	37	39	31	33	520
522	129	131	85	87	65	67	51	53	43	45	37	39	31	33	522
524	129	131	87	89	65	67	51	53	43	45	37	39	31	33	524
526	131	133	87	89	65	67	51	53	43	45	37	39	31	33	526
528	131	133	87	89	65	67	51	53	43	45	37	39	31	33	528
530	131	133	87	89	65	67	51	53	43	45	37	39	33	35	530
532	131	133	87	89	65	67	53	55	43	45	37	39	33	35	532
534	133	135	87	89	65	67	53	55	43	45	37	39	33	35	534
536	133	135	89	91	65	67	53	55	43	45	37	39	33	35	536
538	133	135	89	91	67	69	53	55	43	45	37	39	33	35	538
540	133	135	89	91	67	69	53	55	43	45	37	39	33	35	540
542	135	137	89	91	67	69	53	55	45	47	37	39	33	35	542
544	135	137	89	91	67	69	53	55	45	47	37	39	33	35	544
546	135	137	89	91	67	69	53	55	45	47	37	39	33	35	546
548	135	137	91	93	67	69	53	55	45	47	39	41	33	35	548
550	137	139	91	93	67	69	53	55	45	47	39	41	33	35	550
552	137	139	91	93	67	69	53	57	45	47	39	41	33	35	552
554	137	139	91	93	69	71	55	57	45	47	39	41	33	35	554
556	137	139	91	93	69	71	55	57	45	47	39	41	33	35	556
558	139	141	91	93	69	71	55	57	45	47	39	41	33	35	558
560	139	141	93	95	69	71	55	57	45	47	39	41	33	35	560
562	139	141	93	95	69	71	55	57	45	47	39	41	35	37	562
564	139	141	93	95	69	71	55	57	45	47	39	41	35	37	564
566	141	143	93	95	69	71	55	57	47	49	39	41	35	37	566
568	141	143	93	95	69	71	55	57	47	49	39	41	35	37	568
570	141	143	93	95	71	73	55	57	47	49	39	41	35	37	570
572	141	143	95	97	71	73	57	59	47	49	39	41	35	37	572
574	143	145	95	97	71	73	57	59	47	49	39	41	35	37	574
576	143	145	95	97	71	73	57	59	47	49	41	43	35	37	576
578	143	145	95	97	71	73	57	59	47	49	41	43	35	37	578
580	143	145	95	97	71	73	57	59	47	49	41	43	35	37	580
582	145	147	95	97	71	73	57	59	47	49	41	43	35	37	582
584	145	147	97	99	71	73	57	59	47	49	41	43	35	37	584
586	145	147	97	99	73	75	57	59	47	49	41	43	35	37	586
588	145	147	97	99	73	75	57	59	47	49	41	43	35	37	588
590	147	149	97	99	73	75	57	59	49	51	41	43	35	37	590
592	147	149	97	99	73	75	59	61	49	51	41	43	35	37	592
594	147	149	97	99	73	75	59	61	49	51	41	43	37	39	594
596	147	149	99	101	73	75	59	61	49	51	41	43	37	39	596
598	149	151	99	101	73	75	59	61	49	51	41	43	37	39	598
600	149	151	99	101	73	75	59	61	49	51	41	43	37	39	600

Above choice of Pitches will prove most satisfactory, although, as stated in text, the absolute magnitude of average pitch may be varied within reasonable limits.

MULTIPLE-CIRCUIT, SINGLE WINDINGS, FOR DRUM ARMATURES.

No. OF CONDUCTORS	FRONT AND BACK PITCHES													No. OF CONDUCTORS
	4 POLES		6 POLES		8 POLES		10 POLES		12 POLES		14 POLES		16 POLES	
	F	B	F	B	F	B	F	B	F	B	F	B	F	B	
602	149	151	99	101	75	77	59	61	49	51	41	43	37	39	602
604	149	151	99	101	75	77	59	61	49	51	43	45	37	39	604
606	151	153	99	101	75	77	59	61	49	51	43	45	37	39	606
608	151	153	101	103	75	77	59	61	49	51	43	45	37	39	608
610	151	153	101	103	75	77	59	61	49	51	43	45	37	39	610
612	151	153	101	103	75	77	61	63	49	51	43	45	37	39	612
614	153	155	101	103	75	77	61	63	51	53	43	45	37	39	614
616	153	155	101	103	75	77	61	63	51	53	43	45	37	39	616
618	153	155	101	103	77	79	61	63	51	53	43	45	37	39	618
620	153	155	103	105	77	79	61	63	51	53	43	45	37	39	620
622	155	157	103	105	77	79	61	63	51	53	43	45	37	39	622
624	155	157	103	105	77	79	61	63	51	53	43	45	37	39	624
626	155	157	103	105	77	79	61	63	51	53	43	45	39	41	626
628	155	157	103	105	77	79	61	63	51	53	43	45	39	41	628
630	157	159	103	105	77	79	61	63	51	53	43	45	39	41	630
632	157	159	105	107	77	79	63	65	51	53	45	47	39	41	632
634	157	159	105	107	79	81	63	65	51	53	45	47	39	41	634
636	157	159	105	107	79	81	63	65	51	53	45	47	39	41	636
638	159	161	105	107	79	81	63	65	53	55	45	47	39	41	638
640	159	161	105	107	79	81	63	65	53	55	45	47	39	41	640
642	159	161	105	107	79	81	63	65	53	55	45	47	39	41	642
644	159	161	107	109	79	81	63	65	53	55	45	47	39	41	644
646	161	163	107	109	79	81	63	65	53	55	45	47	39	41	646
648	161	163	107	109	81	83	63	65	53	55	45	47	39	41	648
650	161	163	107	109	81	83	63	65	53	55	45	47	39	41	650
652	161	163	107	109	81	83	65	67	53	55	45	47	39	41	652
654	163	165	107	109	81	83	65	67	53	55	45	47	39	41	654
656	163	165	109	111	81	83	65	67	53	55	45	47	39	41	656
658	163	165	109	111	81	83	65	67	53	55	45	47	41	43	658
660	163	165	109	111	81	83	65	67	53	55	47	49	41	43	660
662	165	167	109	111	81	83	65	67	55	57	47	49	41	43	662
664	165	167	109	111	81	83	65	67	55	57	47	49	41	43	664
666	165	167	109	111	83	85	65	67	55	57	47	49	41	43	666
668	165	167	111	113	83	85	65	67	55	57	47	49	41	43	668
670	167	169	111	113	83	85	65	67	55	57	47	49	41	43	670
672	167	169	111	113	83	85	67	69	55	57	47	49	41	43	672
674	167	169	111	113	83	85	67	69	55	57	47	49	41	43	674
676	167	169	111	113	83	85	67	69	55	57	47	49	41	43	676
678	169	171	111	113	83	85	67	69	55	57	47	49	41	43	678
680	169	171	113	115	83	85	67	69	55	57	47	49	41	43	680
682	169	171	113	115	85	87	67	69	55	57	47	49	41	43	682
684	169	171	113	115	85	87	67	69	55	57	47	49	41	43	684
686	171	173	113	116	85	87	67	69	57	59	47	49	41	43	686
688	171	173	113	115	85	87	67	69	57	59	49	51	41	43	688
690	171	173	113	115	85	87	67	69	57	59	49	51	43	45	690
692	171	173	115	117	85	87	69	71	57	59	49	51	43	45	692
694	173	175	115	117	85	87	69	71	57	59	49	51	43	45	694
696	173	175	115	117	85	87	69	71	57	59	49	51	43	45	696
698	173	175	115	117	87	89	69	71	57	59	49	51	43	45	698
700	173	175	115	117	87	89	69	71	57	59	49	51	43	45	700

Above choice of Pitches will prove most satisfactory, although, as stated in text, the absolute magnitude of average pitch may be varied within reasonable limits.

MULTIPLE-CIRCUIT, SINGLE WINDINGS, FOR DRUM ARMATURES.

No. of Conductors	FRONT AND BACK PITCHES														No. of Conductors
	4 POLES		6 POLES		8 POLES		10 POLES		12 POLES		14 POLES		16 POLES		
	F	B	F	B	F	B	F	B	F	B	F	B	F	B	
702	175	177	115	117	87	89	69	71	57	59	49	51	43	45	702
704	175	177	117	119	87	89	69	71	57	59	49	51	43	45	704
706	175	177	117	119	87	89	69	71	57	59	49	51	43	45	706
708	175	177	117	119	87	89	69	71	57	59	49	51	43	45	708
710	177	179	117	119	87	89	69	71	59	61	49	51	43	45	710
712	177	179	117	119	87	89	71	73	59	61	49	51	43	45	712
714	177	179	117	119	89	91	71	73	59	61	49	51	43	45	714
716	177	179	119	121	89	91	71	73	59	61	51	53	43	45	716
718	179	181	119	121	89	91	71	73	59	61	51	53	43	45	718
720	179	181	119	121	89	91	71	73	59	61	51	53	43	45	720
722	179	181	119	121	89	91	71	73	59	61	51	53	45	47	722
724	179	181	119	121	89	91	71	73	59	61	51	53	45	47	724
726	181	183	119	121	89	91	71	73	59	61	51	53	45	47	726
728	181	183	121	123	89	91	71	73	59	61	51	53	45	47	728
730	181	183	121	123	91	93	71	73	59	61	51	53	45	47	730
732	181	183	121	123	91	93	73	75	59	61	51	53	45	47	732
734	183	185	121	123	91	93	73	75	61	63	51	53	45	47	734
736	183	185	121	123	91	93	73	75	61	63	51	53	45	47	736
738	183	185	121	123	91	93	73	75	61	63	51	53	45	47	738
740	183	185	123	125	91	93	73	75	61	63	51	53	45	47	740
742	185	187	123	125	91	93	73	75	61	63	51	53	45	47	742
744	185	187	123	125	91	93	73	75	61	63	53	55	45	47	744
746	185	187	123	125	93	95	73	75	61	63	53	55	45	47	746
748	185	187	123	125	93	95	73	75	61	63	53	55	45	47	748
750	187	189	123	125	93	95	73	75	61	63	53	55	45	47	750
752	187	189	125	127	93	95	75	77	61	63	53	55	45	47	752
754	187	189	125	127	93	95	75	77	61	63	53	55	47	49	754
756	187	189	125	127	93	95	75	77	61	63	53	55	47	49	756
758	189	191	125	127	93	95	75	77	63	65	53	55	47	49	758
760	189	191	125	127	93	95	75	77	63	65	53	55	47	49	760
762	189	191	125	127	95	97	75	77	63	65	53	55	47	49	762
764	189	191	127	129	95	97	75	77	63	65	53	55	47	49	764
766	191	193	127	129	95	97	75	77	63	65	53	55	47	49	766
768	191	193	127	129	95	97	75	77	63	65	53	55	47	49	768
770	191	193	127	129	95	97	75	77	63	65	53	55	47	49	770
772	191	193	127	129	95	97	77	79	63	65	55	57	47	49	772
774	193	195	127	129	95	97	77	79	63	65	55	57	47	49	774
776	193	195	129	131	95	97	77	79	63	65	55	57	47	49	776
778	193	195	129	131	97	99	77	79	63	65	55	57	47	49	778
780	193	195	129	131	97	99	77	79	63	65	55	57	47	49	780
782	195	197	129	131	97	99	77	79	65	67	55	57	47	49	782
784	195	197	129	131	97	99	77	79	65	67	55	57	47	49	784
786	195	197	129	131	97	99	77	79	65	67	55	57	49	51	786
788	195	197	131	133	97	99	77	79	65	67	55	57	49	51	788
790	197	199	131	133	97	99	77	79	65	67	55	57	49	51	790
792	197	199	131	133	97	99	79	81	65	67	55	57	49	51	792
794	197	199	131	133	99	101	79	81	65	67	55	57	49	51	794
796	197	199	131	133	99	101	79	81	65	67	55	57	49	51	796
798	199	201	131	133	99	101	79	81	65	67	55	57	49	51	798
800	199	201	133	135	99	101	79	81	65	67	57	59	49	51	800

Above choice of Pitches will prove most satisfactory, although, as stated in text, the absolute magnitude of average pitch may be varied within reasonable limits.

MULTIPLE-CIRCUIT SINGLE WINDINGS, FOR DRUM ARMATURES.

FRONT AND BACK PITCHES

No. OF CONDUCTORS	4 POLES		6 POLES		8 POLES		10 POLES		12 POLES		14 POLES		16 POLES		No. OF CONDUCTORS
	F	B	F	B	F	B	F	B	F	B	F	B	F	B	
802	199	201	133	135	99	101	79	81	65	67	57	59	49	51	802
804	199	201	133	135	99	101	79	81	65	67	57	59	49	51	804
806	201	203	133	135	99	101	79	81	67	69	57	59	49	51	806
808	201	203	133	135	99	101	79	81	67	69	57	59	49	51	808
810	201	203	133	135	101	103	79	81	67	69	57	59	49	51	810
812	201	203	135	137	101	103	81	83	67	69	57	59	49	51	812
814	203	205	135	137	101	103	81	83	67	69	57	59	49	51	814
816	203	205	135	137	101	103	81	83	67	69	57	59	49	51	816
818	203	205	135	137	101	103	81	83	67	69	57	59	51	53	818
820	203	205	135	137	101	103	81	83	67	69	57	59	51	53	820
822	205	207	135	137	101	103	81	83	67	69	57	59	51	53	822
824	205	207	137	139	101	103	81	83	67	69	57	59	51	53	824
826	205	207	137	139	103	105	81	83	67	69	57	59	51	53	826
828	205	207	137	139	103	105	81	83	67	69	59	61	51	53	828
830	207	209	137	139	103	105	81	83	69	71	59	61	51	53	830
832	207	209	137	139	103	105	83	85	69	71	59	61	51	53	832
834	207	209	137	139	103	105	83	85	69	71	59	61	51	53	834
836	207	209	139	141	103	105	83	85	69	71	59	61	51	53	836
838	209	211	139	141	103	105	83	85	69	71	59	61	51	53	838
840	209	211	139	141	103	105	83	85	69	71	59	61	51	53	840
842	209	211	139	141	105	107	83	85	69	71	59	61	51	53	842
844	209	211	139	141	105	107	83	85	69	71	59	61	51	53	844
846	211	213	139	141	105	107	83	85	69	71	59	61	51	53	846
848	211	213	141	143	105	107	83	85	69	71	59	61	51	53	848
850	211	213	141	143	105	107	83	85	69	71	59	61	53	55	850
852	211	213	141	143	105	107	83	85	69	71	59	61	53	55	852
854	213	215	141	143	105	107	85	87	71	73	59	61	53	55	854
856	213	215	141	143	105	107	85	87	71	73	61	63	53	55	856
858	213	215	141	143	107	109	85	87	71	73	61	63	53	55	858
860	213	215	143	145	107	109	85	87	71	73	61	63	53	55	860
862	215	217	143	145	107	109	85	87	71	73	61	63	53	55	862
864	215	217	143	145	107	109	85	87	71	73	61	63	53	55	864
866	215	217	143	145	107	109	85	87	71	73	61	63	53	55	866
868	215	217	143	145	107	109	85	87	71	73	61	63	53	55	868
870	217	219	143	145	107	109	85	87	71	73	61	63	53	55	870
872	217	219	145	147	107	109	87	89	71	73	61	63	53	56	872
874	217	219	145	147	109	111	87	89	71	73	61	63	53	55	874
876	217	219	145	147	109	111	87	89	71	73	61	63	53	55	876
878	219	221	145	147	109	111	87	89	73	75	61	63	53	55	878
880	219	221	145	147	109	111	87	89	73	75	61	63	53	55	880
882	219	221	145	147	109	111	87	89	73	75	61	63	55	57	882
884	219	221	147	149	109	111	87	89	73	75	63	65	55	57	884
886	221	223	147	149	109	111	87	89	73	75	63	65	55	57	886
888	221	223	147	149	109	111	87	89	73	75	63	65	55	57	888
890	221	223	147	149	111	113	87	89	73	75	63	65	55	57	890
892	221	223	147	149	111	113	89	91	73	75	63	65	55	57	892
894	223	225	147	149	111	113	89	91	73	75	63	65	55	57	894
896	223	225	149	151	111	113	89	91	73	75	63	65	55	57	896
898	223	225	149	151	111	113	89	91	73	75	63	65	55	57	898
900	223	225	149	151	111	113	89	91	73	75	63	65	55	57	900

Above choice of Pitches will prove most satisfactory, although, as stated in text, the absolute magnitude of average pitch may be varied within reasonable limits.

MULTIPLE-CIRCUIT, SINGLE WINDINGS, FOR DRUM ARMATURES.

No. OF CONDUCTORS	4 POLES		6 POLES		8 POLES		10 POLES		12 POLES		14 POLES		16 POLES		No. OF CONDUCTORS
	F	B	F	B	F	B	F	B	F	B	F	B	F	B	
902	225	227	149	151	111	113	89	91	75	77	63	65	55	57	902
904	225	227	149	151	111	113	89	91	75	77	63	65	55	57	904
906	225	227	149	151	113	115	89	91	75	77	63	65	55	57	906
908	225	227	151	153	113	115	89	91	75	77	63	65	55	57	908
910	227	229	151	153	113	115	89	91	75	77	63	65	55	57	910
912	227	229	151	153	113	115	91	93	75	77	65	67	55	57	912
914	227	229	151	153	113	115	91	93	75	77	65	67	57	59	914
916	227	229	151	153	113	115	91	93	75	77	65	67	57	59	916
918	229	231	151	153	113	115	91	93	75	77	65	67	57	59	918
920	229	231	153	155	113	115	91	93	75	77	65	67	57	59	920
922	229	231	153	155	115	117	91	93	75	77	65	67	57	59	922
924	229	231	153	155	115	117	91	93	75	77	65	67	57	59	924
926	231	233	153	155	115	117	91	93	77	79	65	67	57	59	926
928	231	233	153	155	115	117	91	93	77	79	65	67	57	59	928
930	231	233	155	157	115	117	91	93	77	79	65	67	57	59	930
932	231	233	155	157	115	117	93	95	77	79	65	67	57	59	932
934	233	235	155	157	115	117	93	95	77	79	65	67	57	59	934
936	233	235	155	157	115	117	93	95	77	79	65	67	57	59	936
938	233	235	155	157	117	119	93	95	77	79	65	67	57	59	938
940	233	235	155	157	117	119	93	95	77	79	67	69	57	59	940
942	235	237	155	157	117	119	93	95	77	79	67	69	57	59	942
944	235	237	157	159	117	119	93	95	77	79	67	69	57	59	944
946	235	237	157	159	117	119	93	95	77	79	67	69	59	61	946
948	235	237	157	159	117	119	93	95	77	79	67	69	59	61	948
950	237	239	157	159	117	119	93	95	79	81	67	69	59	61	950
952	237	239	157	159	117	119	95	97	79	81	67	69	59	61	952
954	237	239	157	159	119	121	95	97	79	81	67	69	59	61	954
956	237	239	159	161	119	121	95	97	79	81	67	69	59	61	956
958	239	241	159	161	119	121	95	97	79	81	67	69	59	61	958
960	239	241	159	161	119	121	95	97	79	81	67	69	59	61	960
962	239	241	159	161	119	121	95	97	79	81	67	69	59	61	962
964	239	241	159	161	119	121	95	97	79	81	67	69	59	61	964
966	241	243	159	161	119	121	95	97	79	81	67	69	59	61	966
968	241	243	161	163	119	121	95	97	79	81	69	71	59	61	968
970	241	243	161	163	121	123	95	97	79	81	69	71	59	61	970
972	241	243	161	163	121	123	97	99	79	81	69	71	59	61	972
974	243	245	161	163	121	123	97	99	81	83	69	71	59	61	974
976	243	245	161	163	121	123	97	99	81	83	69	71	59	61	976
978	243	245	161	163	121	123	97	99	81	83	69	71	61	63	978
980	243	245	163	165	121	123	97	99	81	83	69	71	61	63	980
982	245	247	163	165	121	123	97	99	81	83	69	71	61	63	982
984	245	247	163	165	121	123	97	99	81	83	69	71	61	63	984
986	245	247	163	165	123	125	97	99	81	83	69	71	61	63	986
988	245	247	163	165	123	125	97	99	81	83	69	71	61	63	988
990	247	249	163	165	123	125	97	99	81	83	69	71	61	63	990
992	247	249	165	167	123	125	99	101	81	83	69	71	61	63	992
994	247	249	165	167	123	125	99	101	81	83	69	71	61	63	994
996	247	249	165	167	123	125	99	101	81	83	71	73	61	63	996
998	249	251	165	167	123	125	99	101	83	85	71	73	61	63	998
1000	249	251	165	167	123	125	99	101	83	85	71	73	61	63	1000

Above choice of Pitches will prove most satisfactory, although, as stated in text, the absolute magnitude of average pitch may be varied within reasonable limits.

MULTIPLE CIRCUIT, SINGLE WINDINGS, FOR DRUM ARMATURES.

FRONT AND BACK PITCHES

No. of Conductors	4 POLES		6 POLES		8 POLES		10 POLES		12 POLES		14 POLES		16 POLES		No. of Conductors
	F	B	F	B	F	B	F	B	F	B	F	B	F	B	
1002	249	251	165	167	125	127	99	101	83	85	71	73	61	63	1002
1004	249	251	167	169	125	127	99	101	83	85	71	73	61	63	1004
1006	251	253	167	169	125	127	99	101	83	85	71	73	61	63	1006
1008	251	253	167	169	125	127	99	101	83	85	71	73	61	63	1008
1010	251	253	167	169	125	127	99	101	83	85	71	73	63	65	1010
1012	251	253	167	169	125	127	101	103	83	85	71	73	63	65	1012
1014	253	255	167	169	125	127	101	103	83	85	71	73	63	65	1014
1016	253	255	169	171	125	127	101	103	83	85	71	73	63	65	1016
1018	253	255	169	171	127	129	101	103	83	85	71	73	63	65	1018
1020	253	255	169	171	127	129	101	103	83	85	71	73	63	65	1020
1022	255	257	169	171	127	129	101	103	85	87	71	73	63	65	1022
1024	255	257	169	171	127	129	101	103	85	87	73	75	63	65	1024
1026	255	257	169	171	127	129	101	103	85	87	73	75	63	65	1026
1028	255	257	171	173	127	129	101	103	85	87	73	75	63	65	1028
1030	257	259	171	173	127	129	101	103	85	87	73	75	63	65	1030
1032	257	259	171	173	127	129	103	105	85	87	73	75	63	65	1032
1034	257	259	171	173	129	131	103	105	85	87	73	75	63	65	1034
1036	257	259	171	173	129	131	103	105	85	87	73	75	63	65	1036
1038	259	261	171	173	129	131	103	105	85	87	73	75	63	65	1038
1040	259	261	173	175	129	131	103	105	85	87	73	75	63	65	1040
1042	259	261	173	175	129	131	103	105	85	87	73	75	65	67	1042
1044	259	261	173	175	129	131	103	105	85	87	73	75	65	67	1044
1046	261	263	173	175	129	131	103	105	87	89	73	75	65	67	1046
1048	261	263	173	175	129	131	103	105	87	89	73	75	65	67	1048
1050	261	263	173	175	131	133	103	105	87	89	73	75	65	67	1050
1052	261	263	175	177	131	133	103	105	87	89	75	77	65	67	1052
1054	263	265	175	177	131	133	105	107	87	89	75	77	65	67	1054
1056	263	265	175	177	131	133	105	107	87	89	75	77	65	67	1056
1058	263	265	175	177	131	133	105	107	87	89	75	77	65	67	1058
1060	263	265	175	177	131	133	105	107	87	89	75	77	65	67	1060
1062	265	267	175	177	131	133	105	107	87	89	75	77	65	67	1062
1064	265	267	177	179	131	133	105	107	87	89	75	77	65	67	1064
1066	265	267	177	179	133	135	105	107	87	89	75	77	65	67	1066
1068	265	267	177	179	133	135	105	107	87	89	75	77	65	67	1068
1070	267	269	177	179	133	135	105	107	89	91	75	77	65	67	1070
1072	267	269	177	179	133	135	107	109	89	91	75	77	65	67	1072
1074	267	269	177	179	133	135	107	109	89	91	75	77	67	69	1074
1076	267	269	179	181	133	135	107	109	89	91	75	77	67	69	1076
1078	269	271	179	181	133	135	107	109	89	91	75	77	67	69	1078
1080	269	271	179	181	133	135	107	109	89	91	77	79	67	69	1080
1082	269	271	179	181	135	137	107	109	89	91	77	79	67	69	1082
1084	269	271	179	181	135	137	107	109	89	91	77	79	67	69	1084
1086	271	273	179	181	135	137	107	109	89	91	77	79	67	69	1086
1088	271	273	181	183	135	137	107	109	89	91	77	79	67	69	1088
1090	271	273	181	183	135	137	107	109	89	91	77	79	67	69	1090
1092	271	273	181	183	135	137	109	111	89	91	77	79	67	69	1092
1094	273	275	181	183	135	137	109	111	91	93	77	79	67	69	1094
1096	273	275	181	183	135	137	109	111	91	93	77	79	67	69	1096
1098	273	275	181	183	137	139	109	111	91	93	77	79	67	69	1098
1100	273	275	183	185	137	139	109	111	91	93	77	79	67	69	1100

Above choice of Pitches will prove most satisfactory, although, as stated in text, the absolute magnitude of average pitch may be varied within reasonable limits.

MULTIPLE-CIRCUIT, SINGLE WINDINGS, FOR DRUM ARMATURES.

No. of Conductors	4 POLES		6 POLES		8 POLES		10 POLES		12 POLES		14 POLES		16 POLES		No. of Conductors
	F	B	F	B	F	B	F	B	F	B	F	B	F	B	
1102	275	277	183	185	137	139	109	111	91	93	77	79	67	69	1102
1104	275	277	183	185	137	139	109	111	91	93	77	79	67	69	1104
1106	275	277	183	185	137	139	109	111	91	93	77	79	69	71	1106
1108	275	277	183	185	137	139	109	111	91	93	79	81	69	71	1108
1110	277	279	183	185	137	139	109	111	91	93	79	81	69	71	1110
1112	277	279	185	187	137	139	111	113	91	93	79	81	69	71	1112
1114	277	279	185	187	139	141	111	113	91	93	79	81	69	71	1114
1116	277	279	185	187	139	141	111	113	91	93	79	81	69	71	1116
1118	279	281	185	187	139	141	111	113	93	95	79	81	69	71	1118
1120	279	281	185	187	139	141	111	113	93	95	79	81	69	71	1120
1122	279	281	185	187	139	141	111	113	93	95	79	81	69	71	1122
1124	279	281	187	189	139	141	111	113	93	95	79	81	69	71	1124
1126	281	283	187	189	139	141	111	113	93	95	79	81	69	71	1126
1128	281	283	187	189	139	141	111	113	93	95	79	81	69	71	1128
1130	281	283	187	189	141	143	111	113	93	95	79	81	69	71	1130
1132	281	283	187	189	141	143	113	115	93	95	79	81	69	71	1132
1134	283	285	187	189	141	143	113	115	93	95	79	81	69	71	1134
1136	283	285	189	191	141	143	113	115	93	95	81	83	69	71	1136
1138	283	285	189	191	141	143	113	115	93	95	81	83	71	73	1138
1140	283	285	189	191	141	143	113	115	93	95	81	83	71	73	1140
1142	285	287	189	191	141	143	113	115	95	97	81	83	71	73	1142
1144	285	287	189	191	141	143	113	115	95	97	81	83	71	73	1144
1146	285	287	189	191	143	145	113	115	95	97	81	83	71	73	1146
1148	285	287	191	193	143	145	113	115	95	97	81	83	71	73	1148
1150	287	289	191	193	143	145	113	115	95	97	81	83	71	73	1150
1152	287	289	191	193	143	145	115	117	95	97	81	83	71	73	1152
1154	287	289	191	193	143	145	115	117	95	97	81	83	71	73	1154
1156	287	289	191	193	143	145	115	117	95	97	81	83	71	73	1156
1158	289	291	191	193	143	145	115	117	95	97	81	83	71	73	1158
1160	289	291	193	195	143	145	115	117	95	97	81	83	71	73	1160
1162	289	291	193	195	145	147	115	117	95	97	81	83	71	73	1162
1164	289	291	193	195	145	147	115	117	95	97	83	85	71	73	1164
1166	291	293	193	195	145	147	115	117	97	99	83	85	71	73	1166
1168	291	293	193	195	145	147	115	117	97	99	83	85	71	73	1168
1170	291	293	195	197	145	147	115	117	97	99	83	85	73	75	1170
1172	291	293	195	197	145	147	115	117	97	99	83	85	73	75	1172
1174	293	295	195	197	145	147	117	119	97	99	83	85	73	75	1174
1176	293	295	195	197	145	147	117	119	97	99	83	85	73	75	1176
1178	293	295	195	197	147	149	117	119	97	99	83	85	73	75	1178
1180	293	295	195	197	147	149	117	119	97	99	83	85	73	75	1180
1182	295	297	195	197	147	149	117	119	97	99	83	85	73	75	1182
1184	295	297	197	199	147	149	117	119	97	99	83	85	73	75	1184
1186	295	297	197	199	147	149	117	119	97	99	83	85	73	75	1186
1188	295	297	197	199	147	149	117	119	97	99	83	85	73	75	1188
1190	297	299	197	199	147	149	117	119	99	101	83	85	73	75	1190
1192	297	299	197	199	147	149	119	121	99	101	85	87	73	75	1192
1194	297	299	197	199	149	151	119	121	99	101	85	87	73	75	1194
1196	297	299	199	201	149	151	119	121	99	101	85	87	73	75	1196
1198	299	301	199	201	149	151	119	121	99	101	85	87	73	75	1198
1200	299	301	199	201	149	151	119	121	99	101	85	87	73	75	1200

Above choice of Pitches will prove most satisfactory, although as stated in text, the absolute magnitude of average pitch may be varied within reasonable limits.

MULTIPLE-CIRCUIT, SINGLE WINDINGS, FOR DRUM ARMATURES.

No. OF CONDUCTORS	FRONT AND BACK PITCHES													No. OF CONDUCTORS
	4 POLES		6 POLES		8 POLES		10 POLES		12 POLES		14 POLES		16 POLES		
	F	B	F	B	F	B	F	B	F	B	F	B	F	B	
1202	299	301	199	201	149	151	119	121	99	101	85	87	75	77	1202
1204	299	301	199	201	149	151	119	121	99	101	85	87	75	77	1204
1206	301	303	199	201	149	151	119	121	99	101	85	87	75	77	1206
1208	301	303	201	203	149	151	119	121	99	101	85	87	75	77	1208
1210	301	303	201	203	151	153	119	121	99	101	85	87	75	77	1210
1212	301	303	201	203	151	153	121	123	99	101	85	87	75	77	1212
1214	303	305	201	203	151	153	121	123	101	103	85	87	75	77	1214
1216	303	305	201	203	151	153	121	123	101	103	85	87	75	77	1216
1218	303	305	201	203	151	153	121	123	101	103	85	87	75	77	1218
1220	303	305	203	205	151	153	121	123	101	103	87	89	75	77	1220
1222	305	307	203	205	151	153	121	123	101	103	87	89	75	77	1222
1224	305	307	203	205	151	153	121	123	101	103	87	89	75	77	1224
1226	305	307	203	205	153	155	121	123	101	103	87	89	75	77	1226
1228	305	307	203	205	153	155	121	123	101	103	87	89	75	77	1228
1230	307	309	203	205	153	155	121	123	101	103	87	89	75	77	1230
1232	307	309	205	207	153	155	123	125	101	103	87	89	75	77	1232
1234	307	309	205	207	153	155	123	125	101	103	87	89	77	79	1234
1236	307	309	205	207	153	155	123	125	101	103	87	89	77	79	1236
1238	309	311	205	207	153	155	123	125	103	105	87	89	77	79	1238
1240	309	311	205	207	153	155	123	125	103	105	87	89	77	79	1240
1242	309	311	205	207	155	157	123	125	103	105	87	89	77	79	1242
1244	309	311	207	209	155	157	123	125	103	105	87	89	77	79	1244
1246	311	313	207	209	155	157	123	125	103	105	87	89	77	79	1246
1248	311	313	207	209	155	157	123	125	103	105	89	91	77	79	1248
1250	311	313	207	209	155	157	123	125	103	105	89	91	77	79	1250
1252	311	313	207	209	155	157	127	129	103	105	89	91	77	79	1252
1254	313	315	207	209	155	157	125	127	103	105	89	91	77	79	1254
1256	313	315	209	211	155	157	125	127	103	105	89	91	77	79	1256
1258	313	315	209	211	157	159	125	127	103	105	89	91	77	79	1258
1260	313	315	209	211	157	159	125	127	103	105	89	91	77	79	1260
1262	315	317	209	211	157	159	125	127	105	107	89	91	77	79	1262
1264	315	317	209	211	157	159	125	127	105	107	89	91	77	79	1264
1266	315	317	209	211	157	159	125	127	105	107	89	91	79	81	1266
1268	315	317	211	213	157	159	125	127	105	107	89	91	79	81	1268
1270	317	319	211	213	157	159	125	127	105	107	89	91	79	81	1270
1272	317	319	211	213	157	159	127	129	105	107	89	91	79	81	1272
1274	317	319	211	213	159	161	127	129	105	107	89	91	79	81	1274
1276	317	319	211	213	159	161	127	129	105	107	91	93	79	81	1276
1278	319	321	211	213	159	161	127	129	105	107	91	93	79	81	1278
1280	319	321	213	215	159	161	127	129	105	107	91	93	79	81	1280
1282	319	321	213	215	159	161	127	129	105	107	91	93	79	81	1282
1284	319	321	213	215	159	161	127	129	105	107	91	93	79	81	1284
1286	321	323	213	215	159	161	127	129	107	109	91	93	79	81	1286
1288	321	323	213	215	159	161	127	129	107	109	91	93	79	81	1288
1290	321	323	213	215	161	163	127	129	107	109	91	93	79	81	1290
1292	321	323	215	217	161	163	129	131	107	109	91	93	79	81	1292
1294	323	325	215	217	161	163	129	131	107	109	91	93	79	81	1294
1296	323	325	215	217	161	163	129	131	107	109	91	93	79	81	1296
1298	323	325	215	217	161	163	129	131	107	109	91	93	81	83	1298
1300	323	325	215	217	161	163	129	131	107	109	91	93	81	83	1300

Above choice of Pitches will prove most satisfactory, although, as stated in text, the absolute magnitude of average pitch may be varied within reasonable limits.

MULTIPLE-CIRCUIT, SINGLE WINDINGS, FOR DRUM ARMATURES.

No. of Conductors	4 POLES		6 POLES		8 POLES		10 POLES		12 POLES		14 POLES		16 POLES		No. of Conductors
	F	B	F	B	F	B	F	B	F	B	F	B	F	B	
1302	325	327	215	217	161	163	129	131	107	109	91	93	81	83	1302
1304	325	327	217	219	161	163	129	131	107	109	93	95	81	83	1304
1306	325	327	217	219	163	165	129	131	107	109	93	95	81	83	1306
1308	325	327	217	219	163	165	129	131	107	109	93	95	81	83	1308
1310	327	329	217	219	163	165	129	131	109	111	93	95	81	83	1310
1312	327	329	217	219	163	165	131	133	109	111	93	95	81	83	1312
1314	327	329	217	219	163	165	131	133	109	111	93	95	81	83	1314
1316	327	329	219	221	163	165	131	133	109	111	93	95	81	83	1316
1318	329	331	219	221	163	165	131	133	109	111	93	95	81	83	1318
1320	329	331	219	221	163	165	131	133	109	111	93	95	81	83	1320
1322	329	331	219	221	165	167	131	133	109	111	93	95	81	83	1322
1324	329	331	219	221	165	167	131	133	109	111	93	95	81	83	1324
1326	331	333	219	221	165	167	131	133	109	111	93	95	81	83	1326
1328	331	333	221	223	165	167	131	133	109	111	93	95	81	83	1328
1330	331	333	221	223	165	167	131	133	109	111	93	95	83	85	1330
1332	331	333	221	223	165	167	133	135	109	111	95	97	83	85	1332
1334	333	335	221	223	165	167	133	135	111	113	95	97	83	85	1334
1336	333	335	221	223	165	167	133	135	111	113	95	97	83	85	1336
1338	333	335	221	223	167	169	133	135	111	113	95	97	83	85	1338
1340	333	335	223	225	167	169	133	135	111	113	95	97	83	85	1340
1342	335	337	223	225	167	169	133	135	111	113	95	97	83	85	1342
1344	335	337	223	225	167	169	133	135	111	113	95	97	83	85	1344
1346	335	337	223	225	167	169	133	135	111	113	95	97	83	85	1346
1348	335	337	223	225	167	169	133	135	111	113	95	97	83	85	1348
1350	337	339	223	225	167	169	133	135	111	113	95	97	83	85	1350
1352	337	339	225	227	167	169	135	137	111	113	95	97	83	85	1352
1354	337	339	225	227	169	171	135	137	111	113	95	97	83	85	1354
1356	337	339	225	227	169	171	135	137	111	113	95	97	83	85	1356
1358	339	341	225	227	169	171	135	137	113	115	95	97	83	85	1358
1360	339	341	225	227	169	171	135	137	113	115	97	99	83	85	1360
1362	339	341	225	227	169	171	135	137	113	115	97	99	85	87	1362
1364	339	341	227	229	169	171	135	137	113	115	97	99	85	87	1364
1366	341	343	227	229	169	171	135	137	113	115	97	99	85	87	1366
1368	341	343	227	229	169	171	135	137	113	115	97	99	85	87	1368
1370	341	343	227	229	171	173	135	137	113	115	97	99	85	87	1370
1372	341	343	227	229	171	173	137	139	113	115	97	99	85	87	1372
1374	343	345	227	229	171	173	137	139	113	115	97	99	85	87	1374
1376	343	345	229	231	171	173	137	139	113	115	97	99	85	87	1376
1378	343	345	229	231	171	173	137	139	113	115	97	99	85	87	1378
1380	343	345	229	231	171	173	137	139	113	115	97	99	85	87	1380
1382	345	347	229	231	171	173	137	139	115	117	97	99	85	87	1382
1384	345	347	229	231	171	173	137	139	115	117	97	99	85	87	1384
1386	345	347	229	231	173	175	137	139	115	117	97	99	85	87	1386
1388	345	347	231	233	173	175	137	139	115	117	99	101	85	87	1388
1390	347	349	231	233	173	175	137	139	115	117	99	101	85	87	1390
1392	347	349	231	233	173	175	139	141	115	117	99	101	85	87	1392
1394	347	349	231	233	173	175	139	141	115	117	99	101	87	89	1394
1396	347	349	231	233	173	175	139	141	115	117	99	101	87	89	1396
1398	349	351	231	233	173	175	139	141	115	117	99	101	87	89	1398
1400	349	351	233	235	173	175	139	141	115	117	99	101	87	89	1400

Above choice of Pitches will prove most satisfactory, although, as stated in text, the absolute magnitude of average pitch may be varied within reasonable limits.

MULTIPLE-CIRCUIT, SINGLE WINDINGS, FOR DRUM ARMATURES.

No OF CONDUCTORS	FRONT AND BACK PITCHES													No OF CONDUCTORS
	4 POLES		6 POLES		8 POLES		10 POLES		12 POLES		14 POLES		16 POLES		
	F	B	F	B	F	B	F	B	F	B	F	B	F	B	
1402	349	351	233	235	175	177	139	141	115	117	99	101	87	89	1402
1404	349	351	233	235	175	177	139	141	115	117	99	101	87	89	1404
1406	351	353	233	235	175	177	139	141	117	119	99	101	87	89	1406
1408	351	353	233	235	175	177	139	141	117	119	99	101	87	89	1408
1410	351	353	233	235	175	177	139	141	117	119	99	101	87	89	1410
1412	351	353	235	237	175	177	141	143	117	119	99	101	87	89	1412
1414	353	355	235	237	175	177	141	143	117	119	99	101	87	89	1414
1416	353	355	235	237	175	177	141	143	117	119	101	103	87	89	1416
1418	353	355	235	237	177	179	141	143	117	119	101	103	87	89	1418
1420	353	355	235	237	177	179	141	143	117	119	101	103	87	89	1420
1422	355	357	235	237	177	179	141	143	117	119	101	103	87	89	1422
1424	355	357	237	239	177	179	141	143	117	119	101	103	87	89	1424
1426	355	357	237	239	177	179	141	143	117	119	101	103	89	91	1426
1428	355	357	237	239	177	179	141	143	117	119	101	103	89	91	1428
1430	357	359	237	239	177	179	141	143	119	121	101	103	89	91	1430
1432	357	359	237	239	177	179	143	145	119	121	101	103	89	91	1432
1434	357	359	237	239	179	181	143	145	119	121	101	103	89	91	1434
1436	357	359	239	241	179	181	143	145	119	121	101	103	89	91	1436
1438	359	361	239	241	179	181	143	145	119	121	101	103	89	91	1438
1440	359	361	239	241	179	181	143	145	119	121	101	103	89	91	1440
1442	359	361	239	241	179	181	143	145	119	121	101	103	89	91	1442
1444	359	361	239	241	179	181	143	145	119	121	103	105	89	91	1444
1446	361	363	239	241	179	181	143	145	119	121	103	105	89	91	1446
1448	361	363	241	243	179	181	143	145	119	121	103	105	89	91	1448
1450	361	363	241	243	181	183	143	145	119	121	103	105	89	91	1450
1452	361	363	241	243	181	183	145	147	119	121	103	105	89	91	1452
1454	363	365	241	243	181	183	145	147	121	123	103	105	89	91	1454
1456	363	365	241	243	181	183	145	147	121	123	103	105	89	91	1456
1458	363	365	241	243	181	183	145	147	121	123	103	105	91	93	1458
1460	363	365	243	245	181	183	145	147	121	123	103	105	91	93	1460
1462	365	367	243	245	181	183	145	147	121	123	103	105	91	93	1462
1464	365	367	243	245	181	183	145	147	121	123	103	105	91	93	1464
1466	365	367	243	245	183	185	145	147	121	123	103	105	91	93	1466
1468	365	367	243	245	183	185	145	147	121	123	103	105	91	93	1468
1470	367	369	243	245	183	185	145	147	121	123	103	105	91	93	1470
1472	367	369	245	247	183	185	145	147	121	123	105	107	91	93	1472
1474	367	369	245	247	183	185	147	149	121	123	105	107	91	93	1474
1476	367	369	245	247	183	185	147	149	121	123	105	107	91	93	1476
1478	369	371	245	247	183	185	147	149	123	125	105	107	91	93	1478
1480	369	371	245	247	185	187	147	149	123	125	105	107	91	93	1480
1482	369	371	245	247	185	187	147	149	123	125	105	107	91	93	1482
1484	369	371	247	249	185	187	147	149	123	125	105	107	91	93	1484
1486	371	373	247	249	185	187	147	149	123	125	105	107	91	93	1486
1488	371	373	247	249	185	187	147	149	123	125	105	107	91	93	1488
1490	371	373	247	249	185	187	147	149	123	125	105	107	93	95	1490
1492	371	373	247	249	185	187	147	149	123	125	105	107	93	95	1492
1494	373	375	247	249	185	187	149	151	123	125	105	107	93	95	1494
1496	373	375	249	251	185	187	149	151	123	125	105	107	93	95	1496
1498	373	375	249	251	187	189	149	151	123	125	105	107	93	95	1498
1500	373	375	249	251	187	189	149	151	123	125	107	109	93	95	1500

Above choice of Pitches will prove most satisfactory, although, as stated in text, the absolute magnitude of average pitch may be varied within reasonable limits.

MULTIPLE-CIRCUIT, SINGLE WINDINGS, FOR DRUM ARMATURES.

No. OF CONDUCTORS	FRONT AND BACK PITCHES													No. OF CONDUCTORS
	4 POLES		6 POLES		8 POLES		10 POLES		12 POLES		14 POLES		16 POLES	
	F	B	F	B	F	B	F	B	F	B	F	B	F	B
1502	375	377	249	251	187	189	149	151	125	127	107	109	93	95	1502
1504	373	377	249	251	187	189	149	151	125	127	107	109	93	95	1504
1506	375	377	249	251	187	189	149	151	125	127	107	109	93	95	1506
1508	375	377	251	253	187	189	149	151	125	127	107	109	93	95	1508
1510	377	379	251	253	187	189	149	151	125	127	107	109	93	95	1510
1512	377	379	251	253	187	189	151	153	125	127	107	109	93	95	1512
1514	377	379	251	253	189	191	151	153	125	127	107	109	93	95	1514
1516	377	379	251	253	189	191	151	153	125	127	107	109	93	95	1516
1518	379	381	251	253	189	191	151	153	125	127	107	109	93	95	1518
1520	379	381	253	255	189	191	151	153	125	127	107	109	93	95	1520
1522	379	381	253	255	189	191	151	153	125	127	107	109	95	97	1522
1524	379	381	253	255	189	191	151	153	125	127	107	109	95	97	1524
1526	381	383	253	255	189	191	151	153	127	129	107	109	95	97	1526
1528	381	383	253	255	189	191	151	153	127	129	109	111	95	97	1528
1530	381	383	253	255	191	193	151	153	127	129	109	111	95	97	1530
1532	381	383	255	257	191	193	153	155	127	129	109	111	95	97	1532
1534	383	385	255	257	191	193	153	155	127	129	109	111	95	97	1534
1536	383	385	255	257	191	193	153	155	127	129	109	111	95	97	1536
1538	383	385	255	257	191	193	153	155	127	129	109	111	95	97	1538
1540	383	385	255	257	191	193	153	155	127	129	109	111	95	97	1540
1542	385	387	255	257	191	193	153	155	127	129	109	111	95	97	1542
1544	385	387	257	259	191	193	153	155	127	129	109	111	95	97	1544
1546	385	387	257	259	193	195	153	155	127	129	109	111	95	97	1546
1548	385	387	257	259	193	195	153	155	127	129	109	111	95	97	1548
1550	387	389	257	259	193	195	153	155	129	131	109	111	95	97	1550
1552	387	389	257	259	193	195	155	157	129	131	109	111	95	97	1552
1554	387	389	257	259	193	195	155	157	129	131	109	111	97	99	1554
1556	387	389	259	261	193	195	155	157	129	131	111	113	97	99	1556
1558	389	391	259	261	193	195	155	157	129	131	111	113	97	99	1558
1560	389	391	259	261	193	195	155	157	129	131	111	113	97	99	1560
1562	389	391	259	261	195	197	155	157	129	131	111	113	97	99	1562
1564	389	391	259	261	195	197	155	157	129	131	111	113	97	99	1564
1566	391	393	259	261	195	197	155	157	129	131	111	113	97	99	1566
1568	391	393	261	263	195	197	155	157	129	131	111	113	97	99	1568
1570	391	393	261	263	195	197	155	157	129	131	111	113	97	99	1570
1572	391	393	261	263	195	197	157	159	129	131	111	113	97	99	1572
1574	393	395	261	263	195	197	157	159	131	133	111	113	97	99	1574
1576	393	395	261	263	195	197	157	159	131	133	111	113	97	99	1576
1578	393	395	261	263	197	199	157	159	131	133	111	113	97	99	1578
1580	393	395	263	265	197	199	157	159	131	133	111	113	97	99	1580
1582	395	397	263	265	197	199	157	159	131	133	111	113	97	99	1582
1584	395	397	263	265	197	199	157	159	131	133	113	115	97	99	1584
1586	395	397	263	265	197	199	157	159	131	133	113	115	99	101	1586
1588	395	397	263	265	197	199	157	159	131	133	113	115	99	101	1588
1590	397	399	263	265	197	199	157	159	131	133	113	115	99	101	1590
1592	397	399	265	267	197	199	159	161	131	133	113	115	99	101	1592
1594	397	399	265	267	199	201	159	161	131	133	113	115	99	101	1594
1596	397	399	265	267	199	201	159	161	131	133	113	115	99	101	1596
1598	399	401	265	267	199	201	159	161	133	135	113	115	99	101	1598
1600	399	401	265	267	199	201	159	161	133	135	113	115	99	101	1600

Above choice of Pitches will prove most satisfactory, although, as stated in text, the absolute magnitude of average pitch may be varied within reasonable limits

WINDING TABLES FOR MULTIPLE–CIRCUIT, DOUBLE WINDINGS FOR DRUM ARMATURES.

MULTIPLE-CIRCUIT, DOUBLE WINDINGS, FOR DRUM ARMATURES.

RE-ENTRANCY	No. OF CONDUCTORS	4 POLES		6 POLES		8 POLES		10 POLES		12 POLES		14 POLES		16 POLES		No. OF CONDUCTORS
		F	B	F	B	F	B	F	B	F	B	F	B	F	B	
①	202	49	53	31	35	23	27	19	23	15	19	13	17	11	15	202
o o	204	49	53	31	35	23	27	19	23	15	19	13	17	11	15	204
①	206	49	53	33	37	23	27	19	23	15	19	13	17	11	15	206
o o	208	49	53	33	37	23	27	19	23	15	19	13	17	11	15	208
①	210	51	55	33	37	25	29	19	23	15	19	13	17	11	15	210
o o	212	51	55	33	37	25	29	19	23	15	19	13	17	11	15	212
①	214	51	55	33	37	25	29	19	23	15	19	13	17	11	15	214
o o	216	51	55	33	37	25	29	19	23	15	19	13	17	11	15	216
①	218	53	57	35	39	25	29	19	23	17	21	13	17	11	15	218
o o	220	53	57	35	39	25	29	19	23	17	21	13	17	11	15	220
①	222	53	57	35	39	25	29	21	25	17	21	13	17	11	15	222
o o	224	53	57	35	39	25	29	21	25	17	21	13	17	11	15	224
①	226	55	59	35	39	27	31	21	25	17	21	15	19	13	17	226
o o	228	55	59	35	39	27	31	21	25	17	21	15	19	13	17	228
①	230	55	59	37	41	27	31	21	25	17	21	15	19	13	17	230
o o	232	55	59	37	41	27	31	21	25	17	21	15	19	13	17	232
①	234	57	61	37	41	27	31	21	25	17	21	15	19	13	17	234
o o	236	57	61	37	41	27	31	21	25	17	21	15	19	13	17	236
①	238	57	61	37	41	27	31	21	25	17	21	15	19	13	17	238
o o	240	57	61	37	41	27	31	21	25	17	21	15	19	13	17	240
①	242	59	63	39	43	29	33	23	27	19	23	15	19	13	17	242
o o	244	59	63	39	43	29	33	23	27	19	23	15	19	13	17	244
①	246	59	63	39	43	29	33	23	27	19	23	15	19	13	17	246
o o	248	59	63	39	43	29	33	23	27	19	23	15	19	13	17	248
①	250	61	65	39	43	29	33	23	27	19	23	15	19	13	17	250
o o	252	61	65	39	43	29	33	23	27	19	23	15	19	13	17	252
①	254	61	65	41	45	29	33	23	27	19	23	17	21	13	17	254
o o	256	61	65	41	45	29	33	23	27	19	23	17	21	13	17	256
①	258	63	67	41	45	31	35	23	27	19	23	17	21	15	19	258
o o	260	63	67	41	45	31	35	23	27	19	23	17	21	15	19	260
①	262	63	67	41	45	31	35	25	29	19	23	17	21	15	19	262
o o	264	63	67	41	45	31	35	25	29	19	23	17	21	15	19	264
①	266	65	69	43	47	31	35	25	29	21	25	17	21	15	19	266
o o	268	65	69	43	47	31	35	25	29	21	25	17	21	15	19	268
①	270	65	69	43	47	31	35	25	29	21	25	17	21	15	19	270
o o	272	65	69	43	47	31	35	25	29	21	25	17	21	15	19	272
①	274	67	71	43	47	33	37	25	29	21	25	17	21	15	19	274
o o	276	67	71	43	47	33	37	25	29	21	25	17	21	15	19	276
①	278	67	71	45	49	33	37	25	29	21	25	17	21	15	19	278
o o	280	67	71	45	49	33	37	25	29	21	25	17	21	15	19	280
①	282	69	73	45	49	33	37	27	31	21	25	19	23	15	19	282
o o	284	69	73	45	49	33	37	27	31	21	25	19	23	15	19	284
①	286	69	73	45	49	33	37	27	31	21	25	19	23	15	19	286
o o	288	69	73	45	49	33	37	27	31	21	25	19	23	15	19	288
①	290	71	75	47	51	35	39	27	31	23	27	19	23	17	21	290
o o	292	71	75	47	51	35	39	27	31	23	27	19	23	17	21	292
①	294	71	75	47	51	35	39	27	31	23	27	19	23	17	21	294
o o	296	71	75	47	51	35	39	27	31	23	27	19	23	17	21	296
①	298	73	77	47	51	35	39	27	31	23	27	19	23	17	21	298
o o	300	73	77	47	51	35	39	27	31	23	27	19	23	17	21	300

Above choice of Pitches will prove most satisfactory, although, as stated in text, the absolute magnitude of average pitch may be varied within reasonable limits.

MULTIPLE-CIRCUIT, DOUBLE WINDING, FOR DRUM ARMATURES.

| RE-ENTRANCY | No. OF CONDUCTORS | FRONT AND BACK PITCHES | | | | | | | | | | | | | No. OF CONDUCTORS |
| | | 4 POLES | | 6 POLES | | 8 POLES | | 10 POLES | | 12 POLES | | 14 POLES | | 16 POLES | | |
		F	B	F	B	F	B	F	B	F	B	F	B	F	B	
⊕	302	73	77	49	53	35	39	29	33	23	27	19	23	17	21	302
o o	304	73	77	49	53	35	39	29	33	23	27	19	23	17	21	304
⊕	306	75	79	49	53	37	41	29	33	23	27	19	23	17	21	306
o o	308	75	79	49	53	37	41	29	33	23	27	19	23	17	21	308
⊕	310	75	79	49	53	37	41	29	33	23	27	21	25	17	21	310
o o	312	75	79	49	53	37	41	29	33	23	27	21	25	17	21	312
⊕	314	77	81	51	55	37	41	29	33	25	29	21	25	17	21	314
o o	316	77	81	51	55	37	41	29	33	25	29	21	25	17	21	316
⊕	318	77	81	51	55	37	41	29	33	25	29	21	25	17	21	318
o o	320	77	81	51	55	37	41	29	33	25	29	21	25	17	21	320
⊕	322	79	83	51	55	39	43	31	35	25	29	21	25	19	23	322
o o	324	79	83	51	55	39	43	31	35	25	29	21	25	19	23	324
⊕	326	79	83	53	57	39	43	31	35	25	29	21	25	19	23	326
o o	328	79	83	53	57	39	43	31	35	25	29	21	25	19	23	328
⊕	330	81	85	53	57	39	43	31	35	25	29	21	25	19	23	330
o o	332	81	85	53	57	39	43	31	35	25	29	21	25	19	23	332
⊕	334	81	85	53	57	39	43	31	35	25	29	21	25	19	23	334
o o	336	81	85	53	57	39	43	31	35	25	29	21	25	19	23	336
⊕	338	83	87	55	59	41	45	31	35	27	31	23	27	19	23	338
o o	340	83	87	55	59	41	45	31	35	27	31	23	27	19	23	340
⊕	342	83	87	55	59	41	45	33	37	27	31	23	27	19	23	342
o o	344	83	87	55	59	41	45	33	37	27	31	23	27	19	23	344
⊕	346	85	89	55	59	41	45	33	37	27	31	23	27	19	23	346
o o	348	85	89	55	59	41	45	33	37	27	31	23	27	19	23	348
⊕	350	85	89	57	61	41	45	35	37	27	31	23	27	19	23	350
o o	352	85	89	57	61	41	45	33	37	27	31	23	27	19	23	352
⊕	354	87	91	57	61	43	47	33	37	27	31	23	27	21	25	354
o o	356	87	91	57	61	43	47	33	37	27	31	23	27	21	25	356
⊕	358	87	91	57	61	43	47	33	37	27	31	23	27	21	25	358
o o	360	87	91	57	61	43	47	33	37	27	31	23	27	21	25	360
⊕	362	89	93	59	63	43	47	35	39	29	33	23	27	21	25	362
o o	364	89	93	59	63	43	47	35	39	29	33	23	27	21	25	364
⊕	366	89	93	59	63	43	47	35	39	29	33	25	29	21	25	366
o o	368	89	93	59	63	43	47	35	39	29	33	25	29	21	25	368
⊕	370	91	95	59	63	45	49	35	39	29	33	25	29	21	25	370
o o	372	91	95	59	63	45	49	35	39	29	33	25	29	21	25	372
⊕	374	91	95	61	65	45	49	35	39	29	33	25	29	21	25	374
o o	376	91	95	61	65	45	49	35	39	29	33	25	29	21	25	376
⊕	378	93	97	61	65	45	49	35	39	29	33	25	29	21	25	378
o o	380	93	97	61	65	45	49	35	39	29	33	25	29	21	25	380
⊕	382	93	97	61	65	45	49	37	41	29	33	25	29	21	25	382
o o	384	93	97	61	65	45	49	37	41	29	33	25	29	21	25	384
⊕	386	95	99	63	67	47	51	37	41	31	35	25	29	23	27	386
o o	388	95	99	63	67	47	51	37	41	31	35	25	29	23	27	388
⊕	390	95	99	63	67	47	51	37	41	31	35	25	29	23	27	390
o o	392	95	99	63	67	47	51	37	41	31	35	25	29	23	27	392
⊕	394	97	101	63	67	47	51	37	41	31	35	27	31	23	27	394
o o	396	97	101	63	67	47	51	37	41	31	35	27	31	23	27	396
⊕	398	97	101	65	69	47	51	37	41	31	35	27	31	23	27	398
o o	400	97	101	65	69	47	51	37	41	31	35	27	31	23	27	400

Above choice of Pitches will prove most satisfactory, although, as stated in text, the absolute magnitude of average pitch may be varied within reasonable limits.

MULTIPLE-CIRCUIT, DOUBLE WINDINGS, FOR DRUM ARMATURES.

Re-entrancy	No. of Conductors	FRONT AND BACK PITCHES													No. of Conductors
		4 POLES		6 POLES		8 POLES		10 POLES		12 POLES		14 POLES		16 POLES		
		F	B	F	B	F	B	F	B	F	B	F	B	F	B	
⊙	402	99	103	65	69	49	53	39	43	31	35	27	31	23	27	402
o o	404	99	103	65	69	49	53	39	43	31	35	27	31	23	27	404
⊙	406	99	103	65	69	49	53	39	43	31	35	27	31	23	27	406
o o	408	99	103	65	69	49	53	39	43	31	35	27	31	23	27	408
⊙	410	101	105	67	71	49	53	39	43	33	37	27	31	23	27	410
o o	412	101	105	67	71	49	53	39	43	33	37	27	31	23	27	412
⊙	414	101	105	67	71	49	53	39	43	33	37	27	31	23	27	414
o o	416	101	105	67	71	49	53	39	43	33	37	27	31	23	27	416
⊙	418	103	107	67	71	51	55	39	43	33	37	27	31	25	29	418
o o	420	103	107	67	71	51	55	39	43	33	37	27	31	25	29	420
⊙	422	103	107	69	73	51	55	41	45	33	37	29	33	25	29	422
o o	424	103	107	69	73	51	55	41	45	33	37	29	33	25	29	424
⊙	426	105	109	69	73	51	55	41	45	33	37	29	33	25	29	426
o o	428	105	109	69	73	51	55	41	45	33	37	29	33	25	29	428
⊙	430	105	109	69	73	51	55	41	45	33	37	29	33	25	29	430
o o	432	105	109	69	73	51	55	41	45	33	37	29	33	25	29	432
⊙	434	107	111	71	75	53	57	41	45	35	39	29	33	25	29	434
o o	436	107	111	71	75	53	57	41	45	35	39	29	33	25	29	436
⊙	438	107	111	71	75	53	57	41	45	35	39	29	33	25	29	438
o o	440	107	111	71	75	53	57	41	45	35	39	29	33	25	29	440
⊙	442	109	113	71	75	53	57	43	47	35	39	29	33	25	29	442
o o	444	109	113	71	75	53	57	43	47	35	39	29	33	25	29	444
⊙	446	109	113	73	77	53	57	43	47	35	39	29	33	25	29	446
o o	448	109	113	73	77	53	57	43	47	35	39	29	33	25	29	448
⊙	450	111	115	73	77	55	59	43	47	35	39	31	35	27	31	450
o o	452	111	115	73	77	55	59	43	47	35	39	31	35	27	31	452
⊙	454	111	115	73	77	55	59	43	47	35	39	31	35	27	31	454
o o	456	111	115	73	77	55	59	43	47	35	39	31	35	27	31	456
⊙	458	113	117	75	79	55	59	43	47	37	41	31	35	27	31	458
o o	460	113	117	75	79	55	59	43	47	37	41	31	35	27	31	460
⊙	462	113	117	75	79	55	59	45	49	37	41	31	35	27	31	462
o o	464	113	117	75	79	55	59	45	49	37	41	31	35	27	31	464
⊙	466	115	119	75	79	57	61	45	49	37	41	31	35	27	31	466
o o	468	115	119	75	79	57	61	45	49	37	41	31	35	27	31	468
⊙	470	115	119	77	81	57	61	45	49	37	41	31	35	27	31	470
o o	472	115	119	77	81	57	61	45	49	37	41	31	35	27	31	472
⊙	474	117	121	77	81	57	61	45	49	37	41	31	35	27	31	474
o o	476	117	121	77	81	57	61	45	49	37	41	31	35	27	31	476
⊙	478	117	121	77	81	57	61	45	49	37	41	33	37	27	31	478
o o	480	117	121	77	81	57	61	45	49	37	41	33	37	27	31	480
⊙	482	119	123	79	83	59	63	47	51	39	43	33	37	29	33	482
o o	484	119	123	79	83	59	63	47	51	39	43	33	37	29	33	484
⊙	486	119	123	79	83	59	63	47	51	39	43	33	37	29	33	486
o o	488	119	123	79	83	59	63	47	51	39	43	33	37	29	33	488
⊙	490	121	125	79	83	59	63	47	51	39	43	33	37	29	33	490
o o	492	121	125	79	83	59	63	47	51	39	43	33	37	29	33	492
⊙	494	121	125	81	85	59	63	47	51	39	43	33	37	29	33	494
o o	496	121	125	81	85	59	63	47	51	39	43	33	37	29	33	496
⊙	498	123	127	81	85	61	65	47	51	39	43	33	37	29	33	498
o o	500	123	127	81	85	61	65	47	51	39	43	33	37	29	33	500

Above choice of Pitches will prove most satisfactory, although, as stated in text, the absolute magnitude of average pitch may be varied within reasonable limits.

MULTIPLE-CIRCUIT, DOUBLE WINDINGS, FOR DRUM ARMATURES.

RE-ENTRANCY	No. OF CONDUCTORS	4 POLES		6 POLES		8 POLES		10 POLES		12 POLES		14 POLES		16 POLES		No. OF CONDUCTORS
		F	B	F	B	F	B	F	B	F	B	F	B	F	B	
②	502	123	127	81	85	61	65	49	53	39	43	33	37	29	33	502
o o	504	123	127	81	85	61	65	49	53	39	43	33	37	29	33	504
③	506	125	129	83	87	61	65	49	53	41	45	35	39	29	33	506
o o	508	125	129	83	87	61	65	49	53	41	45	35	39	29	33	508
②	510	125	129	83	87	61	65	49	53	41	45	35	39	29	33	510
o o	512	125	129	83	87	61	65	49	53	41	45	35	39	29	39	512
②	514	127	131	83	87	63	67	49	53	41	45	35	39	31	35	514
o o	516	127	131	83	87	63	67	49	53	41	45	35	39	31	35	516
②	518	127	131	85	89	63	67	49	53	41	45	35	30	31	35	518
o o	520	127	131	85	89	63	67	49	53	41	45	35	39	31	35	520
③	522	129	133	85	89	63	67	51	55	41	45	35	39	31	35	522
o o	524	129	133	85	89	63	67	51	55	41	45	35	39	31	35	524
②	526	129	133	85	89	63	67	51	55	41	45	35	39	31	35	526
o o	528	129	133	85	89	63	67	51	55	41	45	35	39	31	35	528
②	530	131	135	87	91	65	69	51	55	43	47	35	39	31	35	530
o o	532	131	135	87	91	65	69	51	55	43	47	35	39	31	35	532
③	534	131	135	87	91	65	69	51	55	43	47	37	41	31	35	534
o o	536	131	135	87	91	65	69	51	55	43	47	37	41	31	35	536
②	538	133	137	87	91	65	69	51	55	43	47	37	41	31	35	538
o o	540	133	137	87	91	65	69	51	55	43	47	37	41	31	35	540
②	542	133	137	89	93	65	69	53	57	43	47	37	41	31	35	542
o o	544	133	137	89	93	66	69	53	57	43	47	37	41	31	35	544
②	546	135	139	89	93	67	71	53	57	43	47	37	41	33	37	546
o o	548	135	139	89	93	67	71	53	57	43	47	37	41	33	37	548
③	550	135	139	89	93	67	71	53	57	43	47	37	41	33	37	550
o o	552	135	139	89	93	67	71	53	57	43	47	37	41	33	37	552
②	554	137	141	91	95	67	71	53	57	45	49	37	41	33	37	554
o o	556	137	141	91	95	67	71	53	57	45	49	37	41	33	37	556
②	558	137	141	91	95	67	71	53	57	45	49	37	41	33	37	558
o o	560	137	141	91	95	67	71	53	57	45	49	37	41	33	37	560
②	562	139	143	91	95	69	73	55	59	45	49	39	43	33	37	562
o o	564	139	143	91	95	69	73	55	59	45	49	39	43	33	37	564
②	566	139	143	93	97	69	73	55	59	45	49	39	43	33	37	566
o o	568	139	143	93	97	69	73	55	59	45	49	39	43	33	37	568
②	570	141	145	93	97	69	73	55	59	45	49	39	43	33	37	570
o o	572	141	145	93	97	69	73	55	59	45	49	39	43	33	37	572
②	574	141	145	93	97	69	73	55	59	45	49	39	43	33	37	574
o o	576	141	145	93	97	69	73	55	59	45	49	39	43	33	37	576
②	578	143	147	95	99	71	75	55	59	47	51	39	43	35	39	578
o o	580	143	147	95	99	71	75	55	59	47	51	39	43	35	39	580
②	582	143	147	95	99	71	75	57	61	47	51	39	43	35	39	582
o o	584	143	147	95	99	71	75	57	61	47	51	39	43	35	39	584
②	586	145	149	95	99	71	75	57	61	47	51	39	43	35	39	586
o o	588	145	149	95	99	71	75	57	61	47	51	39	43	35	39	588
②	590	145	149	97	101	71	75	57	61	47	51	41	45	35	39	590
o o	592	145	149	97	101	71	75	57	61	47	51	41	45	35	39	592
②	594	147	151	97	101	73	77	57	61	47	51	41	45	35	39	594
o o	596	147	151	97	101	73	77	57	61	47	51	41	45	35	39	596
②	598	147	151	97	101	73	77	57	61	47	51	41	45	35	39	598
o o	600	147	151	97	101	73	77	57	61	47	51	41	45	35	39	600

Above choice of Pitches will prove most satisfactory, although, as stated in text, the absolute
magnitude of average pitch may be varied within reasonable limits.

MULTIPLE-CIRCUIT, DOUBLE WINDINGS, FOR DRUM ARMATURES.

REENTRANCY	No. OF CONDUCTORS	4 POLES		6 POLES		8 POLES		10 POLES		12 POLES		14 POLES		16 POLES		No. OF CONDUCTORS
		F	B	F	B	F	B	F	B	F	B	F	B	F	B	
	602	149	153	99	103	73	77	59	63	49	53	41	45	35	39	602
o o	604	149	153	99	103	73	77	59	63	49	53	41	45	35	39	604
	606	149	153	99	103	73	77	59	63	49	53	41	45	35	39	606
o o	608	149	153	99	103	73	77	59	63	49	53	41	45	35	39	608
	610	151	155	99	103	75	79	59	63	49	53	41	45	37	41	610
o o	612	151	155	99	103	75	79	59	63	49	53	41	45	37	41	612
	614	151	155	101	105	75	79	59	63	49	53	41	45	37	41	614
o o	616	151	155	101	105	75	79	59	63	49	53	41	45	37	41	616
	618	153	157	101	105	75	79	59	63	49	53	43	47	37	41	618
o o	620	153	157	101	105	75	79	59	63	49	53	43	47	37	41	620
	622	153	157	101	105	75	79	61	65	49	53	43	47	37	41	622
o o	624	153	157	101	105	75	79	61	65	49	53	43	47	37	41	624
	626	155	159	103	107	77	81	61	65	51	55	43	47	37	41	626
o o	628	155	159	103	107	77	81	61	65	51	55	43	47	37	41	628
	630	155	159	103	107	77	81	61	65	51	55	43	47	37	41	630
o o	632	155	159	103	107	77	81	61	65	51	55	43	47	37	41	632
	634	157	161	103	107	77	81	61	65	51	55	43	47	37	41	634
o o	636	157	161	103	107	77	81	61	65	51	55	43	47	37	41	636
	638	157	161	105	109	77	81	61	65	51	55	43	47	37	41	638
o o	640	157	161	105	109	77	81	61	65	51	55	43	47	37	41	640
	642	159	163	105	109	79	83	63	67	51	55	43	47	39	43	642
o o	644	159	163	105	109	79	83	63	67	51	55	43	47	39	43	644
	646	159	163	105	109	79	83	63	67	51	55	45	49	39	43	646
o o	648	159	163	105	109	79	83	63	67	51	55	45	49	39	43	648
	650	161	165	107	111	79	83	63	67	53	57	45	49	39	43	650
o o	652	161	165	107	111	79	83	63	67	53	57	45	49	39	43	652
	654	161	165	107	111	79	83	63	67	53	57	45	49	39	43	654
o o	656	161	165	107	111	79	83	63	67	53	57	45	49	39	43	656
	658	163	167	107	111	81	85	63	67	53	57	45	49	39	43	658
o o	660	163	167	107	111	81	85	63	67	53	57	45	49	39	43	660
	662	163	167	109	113	81	85	65	69	53	57	45	49	39	43	662
o o	664	163	167	109	113	81	85	65	69	53	57	45	49	39	43	664
	666	165	169	109	113	81	85	65	69	53	57	45	49	39	43	666
o o	668	165	169	109	113	81	85	65	69	53	57	45	49	39	43	668
	670	165	169	109	113	81	85	65	69	53	57	45	49	39	43	670
o o	672	165	169	109	113	81	85	65	69	53	57	45	49	39	43	672
	674	167	171	111	115	83	87	65	69	55	59	47	51	41	45	674
o o	676	167	171	111	115	83	87	65	69	55	59	47	51	41	45	676
	678	167	171	111	115	83	87	65	69	55	59	47	51	41	45	678
o o	680	167	171	111	115	83	87	65	69	55	59	47	51	41	45	680
	682	169	173	111	115	83	87	67	71	55	59	47	51	41	45	682
o o	684	169	173	111	115	83	87	67	71	55	59	47	51	41	45	684
	686	169	173	113	117	83	87	67	71	55	59	47	51	41	45	686
o o	688	169	173	113	117	83	87	67	71	55	59	47	51	41	45	688
	690	171	175	113	117	85	89	67	71	55	59	47	51	41	45	690
o o	692	171	175	113	117	85	89	67	71	55	59	47	51	41	45	692
	694	171	175	113	117	85	89	67	71	55	59	47	51	41	45	694
o o	696	171	175	113	117	85	89	67	71	55	59	47	51	41	45	696
	698	173	177	115	119	85	89	67	71	57	61	47	51	41	45	698
o o	700	173	177	115	119	85	89	67	71	57	61	47	51	41	45	700

Above choice of Pitches will prove most satisfactory, although, as stated in text, the absolute magnitude of average pitch may be varied within reasonable limits.

MULTIPLE-CIRCUIT, DOUBLE WINDINGS, FOR DRUM ARMATURES.

RE-ENTRANCY	No. OF CONDUCTORS	FRONT AND BACK PITCHES													No. OF CONDUCTORS
		4 POLES		6 POLES		8 POLES		10 POLES		12 POLES		14 POLES		16 POLES		
		F	B	F	B	F	B	F	B	F	B	F	B	F	B	
⊗	702	173	177	115	119	85	89	69	73	57	61	49	53	41	45	702
oo	704	173	177	115	119	85	80	69	73	57	61	49	53	41	45	704
⊗	706	175	179	115	119	87	91	69	73	57	61	49	53	43	47	706
oo	708	175	179	115	119	87	91	69	73	57	61	49	53	43	47	708
⊗	710	175	179	117	121	87	91	69	73	57	61	49	53	43	47	710
oo	712	175	179	117	121	87	91	69	73	57	61	49	53	43	47	712
⊗	714	177	181	117	121	87	91	69	73	57	61	49	53	43	47	714
oo	716	177	181	117	121	87	91	69	73	57	61	49	53	43	47	716
⊗	718	177	181	117	121	87	91	69	73	57	61	49	53	43	47	718
oo	720	177	181	117	121	87	91	69	73	57	61	49	53	43	47	720
⊗	722	179	183	119	123	89	93	71	75	59	63	49	53	43	47	722
oo	724	179	183	119	123	89	93	71	75	59	63	49	53	43	47	724
⊗	726	179	183	119	123	89	93	71	75	59	63	49	53	43	47	726
oo	728	179	183	119	123	89	93	71	75	59	63	49	53	43	47	728
⊗	730	181	185	119	123	89	93	71	75	59	63	51	55	43	47	730
oo	732	181	185	119	123	89	93	71	75	59	63	51	55	43	47	732
⊗	734	181	185	121	125	89	93	71	75	59	63	51	55	43	47	734
oo	736	181	185	121	125	89	93	71	75	59	63	51	55	43	47	736
⊗	738	183	187	121	125	91	95	71	75	59	63	51	55	45	49	738
oo	740	183	187	121	125	91	95	71	75	59	63	51	55	45	49	740
⊗	742	183	187	121	125	91	95	73	77	59	63	51	55	45	49	742
oo	744	183	187	121	125	91	95	73	77	59	63	51	55	45	49	744
⊗	746	185	189	123	127	91	95	73	77	61	65	51	55	45	49	746
oo	748	185	189	123	127	91	95	73	77	61	65	51	55	45	49	748
⊗	750	185	189	123	127	91	95	73	77	61	65	51	55	45	49	750
oo	752	185	189	123	127	91	95	73	77	61	65	51	55	45	49	752
⊗	754	187	191	123	127	93	97	73	77	61	65	51	55	45	49	754
oo	756	187	191	123	127	93	97	73	77	61	65	51	55	45	49	756
⊗	758	187	191	125	129	93	97	73	77	61	65	53	57	45	49	758
oo	760	187	191	125	129	93	97	73	77	61	65	53	57	45	49	760
⊗	762	189	193	125	129	93	97	75	79	61	65	53	57	45	49	762
oo	764	189	193	125	129	93	97	75	79	61	65	53	57	45	49	764
⊗	766	189	193	125	129	93	97	75	79	61	65	53	57	45	49	766
oo	768	189	193	125	129	93	97	75	79	61	65	53	57	45	49	768
⊗	770	191	195	127	131	95	99	75	79	63	67	53	57	47	51	770
oo	772	191	195	127	131	95	99	75	79	63	67	53	57	47	51	772
⊗	774	191	195	127	131	95	99	75	79	63	67	53	57	47	51	774
oo	776	191	195	127	131	95	99	75	79	63	67	53	57	47	51	776
⊗	778	193	197	127	131	95	99	75	79	63	67	53	57	47	51	778
oo	780	193	197	127	131	95	99	75	79	63	67	53	57	47	51	780
⊗	782	193	197	129	133	95	99	77	81	63	67	53	57	47	51	782
oo	784	193	197	129	133	95	99	77	81	63	67	53	57	47	51	784
⊗	786	195	199	129	133	97	101	77	81	63	67	55	59	47	51	786
oo	788	195	199	129	133	97	101	77	81	63	67	55	59	47	51	788
⊗	790	195	199	129	133	97	101	77	81	63	67	55	59	47	51	790
oo	792	195	199	129	133	97	101	77	81	63	67	55	59	47	51	792
⊗	794	197	201	131	135	97	101	77	81	65	69	55	59	47	51	794
oo	796	197	201	131	135	97	101	77	81	65	69	55	59	47	51	796
⊗	798	197	201	131	135	97	101	77	81	65	69	55	59	47	51	798
oo	800	197	201	131	135	97	101	77	81	65	69	55	59	47	51	800

Above choice of Pitches will prove most satisfactory, although, as stated in text, the absolute magnitude of average pitch may be varied within reasonable limits.

MULTIPLE-CIRCUIT, DOUBLE WINDINGS, FOR DRUM ARMATURES.

RE-ENTRANCY	No. OF CONDUCTORS	4 POLES		6 POLES		8 POLES		10 POLES		12 POLES		14 POLES		16 POLES		No. OF CONDUCTORS
		F	B	F	B	F	B	F	B	F	B	F	B	F	B	
⊕	802	199	203	131	135	99	103	79	83	65	69	55	59	49	53	802
o o	804	199	203	131	135	99	103	79	83	65	69	55	59	49	53	804
⊕	806	199	203	133	137	99	103	79	83	65	69	55	59	49	53	806
o o	808	199	203	133	137	99	103	79	83	65	69	55	59	49	53	808
⊕	810	201	205	133	137	99	103	79	83	65	69	55	59	49	53	810
o o	812	201	205	133	137	99	103	79	83	65	69	55	59	49	53	812
⊕	814	201	205	133	137	99	103	79	83	66	69	57	61	49	53	814
o o	816	201	205	133	137	99	103	79	83	65	69	57	61	49	53	816
⊕	818	203	207	135	139	101	105	79	83	67	71	57	61	49	53	818
o o	820	203	207	135	139	101	105	79	83	67	71	57	61	49	53	820
⊕	822	203	207	135	139	101	105	81	85	67	71	57	61	49	53	822
o o	824	203	207	135	139	101	105	81	55	67	71	57	61	49	53	824
⊕	826	205	209	135	139	101	105	81	85	67	71	57	61	49	53	826
o o	828	205	209	135	139	101	105	81	85	67	71	57	61	49	53	828
⊕	830	205	209	137	141	101	105	81	85	67	71	57	61	49	53	830
o o	832	205	209	137	141	101	105	81	85	67	71	57	61	49	53	832
⊕	834	207	211	137	141	103	107	81	85	67	71	57	61	51	55	834
o o	836	207	211	137	141	103	107	81	85	67	71	57	61	51	55	836
⊕	838	207	211	137	141	103	107	81	85	67	71	57	61	51	55	838
o o	840	207	211	137	141	103	107	81	85	67	71	57	61	51	55	840
⊕	842	209	213	139	143	103	107	83	87	69	73	59	63	51	55	842
o o	844	209	213	139	143	103	107	83	87	69	73	59	63	51	55	844
⊕	846	209	213	139	143	103	107	83	87	69	73	59	63	51	55	846
o o	848	209	213	139	143	103	107	83	87	69	73	59	63	51	55	848
⊕	850	211	215	139	143	105	109	83	87	69	73	59	63	51	55	850
o o	852	211	215	139	143	105	109	83	87	69	73	59	63	51	55	852
⊕	854	211	215	141	145	105	109	83	87	69	73	59	63	51	55	854
o o	856	211	215	141	145	105	109	83	87	69	73	59	63	51	55	856
⊕	858	213	217	141	145	105	109	83	87	69	73	59	63	51	55	858
o o	860	213	217	141	145	105	109	83	87	69	73	59	63	51	55	860
⊕	862	213	217	141	145	105	109	85	89	69	73	59	63	51	55	862
o o	864	213	217	141	145	105	109	85	89	69	73	59	63	51	55	864
⊕	866	215	219	143	147	107	111	85	89	71	75	59	63	53	57	866
o o	868	215	219	143	147	107	111	85	89	71	75	59	63	53	57	868
⊕	870	215	219	143	147	107	111	85	89	71	75	61	65	53	57	870
o o	872	215	219	143	147	107	111	85	89	71	75	61	65	53	57	872
⊕	874	217	221	143	147	107	111	85	89	71	75	61	65	53	57	874
o o	876	217	221	143	147	107	111	85	89	71	75	61	65	53	57	876
⊕	878	217	221	145	149	107	111	85	89	71	75	61	65	53	57	878
o o	880	217	221	145	149	107	111	85	89	71	75	61	65	53	57	880
⊕	882	219	223	145	149	109	113	87	91	71	75	61	65	53	57	882
o o	884	219	223	145	149	109	113	87	91	71	75	61	65	53	57	884
⊕	886	219	223	145	149	109	113	87	91	71	75	61	65	53	57	886
o o	888	219	223	145	149	109	113	87	91	71	75	61	65	53	57	888
⊕	890	221	225	147	151	109	113	87	91	73	77	61	65	53	57	890
o o	892	221	225	147	151	109	113	87	91	73	77	61	65	53	57	892
⊕	894	221	225	147	151	109	113	87	91	73	77	61	65	53	57	894
o o	896	221	225	147	151	109	113	87	91	73	77	61	65	53	57	896
⊕	898	223	227	147	151	111	115	87	91	73	77	63	67	55	59	898
o o	900	223	227	147	151	111	115	87	91	73	77	63	67	55	59	900

Above choice of Pitches will prove most satisfactory, although, as stated in text, the absolute magnitude of average pitch may be varied within reasonable limits.

MULTIPLE-CIRCUIT, DOUBLE WINDINGS, FOR DRUM ARMATURES.

REENTRANCY	No. OF CONDUCTORS	FRONT AND BACK PITCHES													No. OF CONDUCTORS
		4 POLES		6 POLES		8 POLES		10 POLES		12 POLES		14 POLES		16 POLES		
		F	B	F	B	F	B	F	B	F	B	F	B	F	B	
⊕	902	223	227	140	153	111	115	89	93	73	77	63	67	55	59	902
oo	904	223	227	149	153	111	115	89	93	73	77	63	67	55	59	904
⊕	906	225	229	140	153	111	115	89	93	73	77	63	67	55	59	906
oo	908	225	229	149	153	111	115	89	93	73	77	63	67	55	59	908
⊕	910	225	229	140	153	111	115	89	93	73	77	63	67	55	59	910
oo	912	225	229	149	153	111	115	89	93	73	77	63	67	55	59	912
⊕	914	227	231	151	155	113	117	89	93	75	79	63	67	55	59	914
oo	916	227	231	151	155	113	117	89	93	75	79	63	67	55	59	916
⊕	918	227	231	151	155	113	117	89	93	75	79	63	67	55	59	918
oo	920	227	231	151	155	113	117	89	93	75	79	63	67	55	59	920
⊕	922	229	233	151	155	113	117	91	95	75	79	63	67	55	59	922
oo	924	229	233	151	155	113	117	91	95	75	79	63	67	55	59	924
⊕	926	229	233	153	157	113	117	91	95	75	79	65	69	55	59	926
oo	928	229	233	153	157	113	117	91	95	75	79	65	69	55	59	928
⊕	930	231	235	153	157	115	119	91	95	75	79	65	69	57	61	930
oo	932	231	235	153	157	115	119	91	95	75	79	65	69	57	61	932
⊕	934	231	235	153	157	115	119	91	95	75	79	65	69	57	61	934
oo	936	231	235	153	157	115	119	91	95	75	79	65	69	57	61	936
⊕	938	233	237	155	159	115	119	91	95	77	81	65	69	57	61	938
oo	940	233	237	155	159	115	119	91	95	77	81	65	69	57	61	940
⊕	942	233	237	155	159	115	119	93	97	77	81	65	69	57	61	942
oo	944	233	237	155	159	115	119	93	97	77	81	65	69	57	61	944
⊕	946	235	239	155	159	117	121	93	97	77	81	65	69	57	61	946
oo	948	235	239	155	159	117	121	93	97	77	81	65	69	57	61	948
⊕	950	235	239	157	161	117	121	93	97	77	81	65	69	57	61	950
oo	952	235	239	157	161	117	121	93	97	77	81	65	69	57	61	952
⊕	954	237	241	157	161	117	121	93	97	77	81	67	71	57	61	954
oo	956	237	241	157	161	117	121	93	97	77	81	67	71	57	61	956
⊕	958	237	241	157	161	117	121	93	97	77	81	67	71	57	61	958
oo	960	237	241	157	161	117	121	93	97	77	81	67	71	57	61	960
⊕	962	239	243	159	163	119	123	95	99	79	83	67	71	59	63	962
oo	964	239	243	159	163	119	123	95	99	79	83	67	71	59	63	964
⊕	966	239	243	159	163	119	123	95	99	79	83	67	71	59	63	966
oo	968	239	243	159	163	119	123	95	99	79	83	67	71	59	63	968
⊕	970	241	245	159	163	119	123	95	99	79	83	67	71	59	63	970
oo	972	241	245	159	163	119	123	95	99	79	83	67	71	59	63	972
⊕	974	241	245	161	165	119	123	95	99	79	83	67	71	59	63	974
oo	976	241	245	161	165	119	123	95	99	79	83	67	71	59	63	976
⊕	978	243	247	161	165	121	125	95	99	79	83	67	71	59	63	978
oo	980	243	247	161	165	121	125	95	99	79	83	67	71	59	63	980
⊕	982	243	247	161	165	121	125	97	101	79	83	69	73	59	63	982
oo	984	243	247	161	165	121	125	97	101	79	83	69	73	59	63	984
⊕	986	245	249	163	167	121	125	97	101	81	85	69	73	59	63	986
oo	988	245	249	163	167	121	125	97	101	81	85	69	73	59	63	988
⊕	990	245	249	163	167	121	125	97	101	81	85	69	73	59	63	990
oo	992	245	249	163	167	121	125	97	101	81	85	69	73	59	63	992
⊕	994	247	251	163	167	123	127	97	101	81	85	69	73	61	65	994
oo	996	247	251	163	167	123	127	97	101	81	85	69	73	61	65	996
⊕	998	247	251	165	169	123	127	97	101	81	85	69	73	61	65	998
oo	1000	247	251	165	169	123	127	97	101	81	85	69	73	61	65	1000

Above choice of Pitches will prove most satisfactory, although, as stated in text, the absolute magnitude of average pitch may be varied within reasonable limits.

MULTIPLE-CIRCUIT, DOUBLE WINDINGS, FOR DRUM ARMATURES.

RE-ENTRANCY	No. OF CONDUCTORS	FRONT AND BACK PITCHES														No. OF CONDUCTORS
		4 POLES		6 POLES		8 POLES		10 POLES		12 POLES		14 POLES		16 POLES		
		F	B	F	B	F	B	F	B	F	B	F	B	F	B	
⊙	1002	249	253	165	169	123	127	99	103	81	85	69	73	61	65	1002
o o	1004	249	253	165	169	123	127	99	103	81	85	69	73	61	65	1004
⊙	1006	249	253	165	169	123	127	99	103	81	85	69	73	61	65	1006
o o	1008	249	253	165	169	123	127	99	103	81	85	69	73	61	65	1008
⊙	1010	251	255	167	171	125	129	99	103	83	87	71	75	61	65	1010
o o	1012	251	255	167	171	125	129	99	103	83	87	71	75	61	65	1012
⊙	1014	251	255	167	171	125	129	99	103	83	87	71	75	61	65	1014
o o	1016	251	255	167	171	125	129	99	103	83	87	71	75	61	65	1016
⊙	1018	253	257	167	171	125	129	99	103	83	87	71	75	61	65	1018
o o	1020	253	257	167	171	125	129	99	103	83	87	71	75	61	65	1020
⊙	1022	253	257	169	173	125	129	101	105	83	87	71	75	61	65	1022
o o	1024	253	257	169	173	125	129	101	105	83	87	71	75	61	65	1024
⊙	1026	255	259	169	173	127	131	101	105	83	87	71	75	63	67	1026
o o	1028	255	259	169	173	127	131	101	105	83	87	71	75	63	67	1028
⊙	1030	255	259	169	173	127	131	101	105	83	87	71	75	63	67	1030
o o	1032	255	259	169	173	127	131	101	105	83	87	71	75	63	67	1032
⊙	1034	257	261	171	175	127	131	101	105	85	89	71	75	63	67	1034
o o	1036	257	261	171	175	127	131	101	105	85	89	71	75	63	67	1036
⊙	1038	257	261	171	175	127	131	101	105	85	89	73	77	63	67	1038
o o	1040	257	261	171	175	127	131	101	105	85	89	73	77	63	67	1040
⊙	1042	259	263	171	175	129	133	103	107	85	89	73	77	63	67	1042
o o	1044	259	263	171	175	129	133	103	107	85	89	73	77	63	67	1044
⊙	1046	259	263	173	177	129	133	103	107	85	89	73	77	63	67	1046
o o	1048	259	263	173	177	129	133	103	107	85	89	73	77	63	67	1048
⊙	1050	261	265	173	177	129	133	103	107	85	89	73	77	63	67	1050
o o	1052	261	265	173	177	129	133	103	107	85	89	73	77	63	67	1052
⊙	1054	261	265	173	177	129	133	103	107	85	89	73	77	63	67	1054
o o	1056	261	265	173	177	129	133	103	107	85	89	73	77	63	67	1056
⊙	1058	263	267	175	179	131	135	103	107	87	91	73	77	65	69	1058
o o	1060	263	267	175	179	131	135	103	107	87	91	73	77	65	69	1060
⊙	1062	263	267	175	179	131	135	105	109	87	91	73	77	65	69	1062
o o	1064	263	267	175	179	131	135	105	109	87	91	73	77	65	69	1064
⊙	1066	265	269	175	179	131	135	105	109	87	91	75	79	65	69	1066
o o	1068	265	269	175	179	131	135	105	109	87	91	75	79	65	69	1068
⊙	1070	265	269	177	181	131	135	105	109	87	91	75	79	65	69	1070
o o	1072	265	269	177	181	131	135	105	109	87	91	75	79	65	69	1072
⊙	1074	267	271	177	181	133	137	105	109	87	91	75	79	65	69	1074
o o	1076	267	271	177	181	133	137	105	109	87	91	75	79	65	69	1076
⊙	1078	267	271	177	181	133	137	105	109	87	91	75	79	65	69	1078
o o	1080	267	271	177	181	133	137	105	109	87	91	75	79	65	69	1080
⊙	1082	269	273	179	183	133	137	107	111	89	93	75	79	65	69	1082
o o	1084	269	273	179	183	133	137	107	111	89	93	75	79	65	69	1084
⊙	1086	269	273	179	183	133	137	107	111	89	93	75	79	65	69	1086
o o	1088	269	273	179	183	133	137	107	111	89	93	75	79	65	69	1088
⊙	1090	271	275	179	183	135	139	107	111	89	93	75	79	67	71	1090
o o	1092	271	275	179	183	135	139	107	111	89	93	75	79	67	71	1092
⊙	1094	271	275	181	185	135	139	107	111	89	93	77	81	67	71	1094
o o	1096	271	275	181	185	135	139	107	111	89	93	77	81	67	71	1096
⊙	1098	273	277	181	185	135	139	107	111	89	93	77	81	67	71	1098
o o	1100	273	277	181	185	135	139	107	111	89	93	77	81	67	71	1100

Above choice of Pitches will prove most satisfactory, although, as stated in text, the absolute magnitude of average pitch may be varied within reasonable limits.

MULTIPLE-CIRCUIT, DOUBLE WINDINGS, FOR DRUM ARMATURES.

RE-ENTRANCY	No. OF CONDUCTORS	4 POLES		6 POLES		8 POLES		10 POLES		12 POLES		14 POLES		16 POLES		No. OF CONDUCTORS
		F	B	F	B	F	B	F	B	F	B	F	B	F	B	
Ⓓ	1102	273	277	181	185	135	139	109	113	89	93	77	81	67	71	1102
o o	1104	273	277	181	185	135	139	109	113	89	93	77	81	67	71	1104
Ⓓ	1106	275	279	183	187	137	141	109	113	91	95	77	81	67	71	1106
o o	1108	275	279	183	187	137	141	109	113	91	95	77	81	67	71	1108
Ⓓ	1110	275	279	183	187	137	141	109	113	91	95	77	81	67	71	1110
o o	1112	275	279	183	187	137	141	109	113	91	95	77	81	67	71	1112
Ⓓ	1114	277	281	183	187	137	141	109	113	91	95	77	81	67	71	1114
o o	1116	277	281	183	187	137	141	109	113	91	95	77	81	67	71	1116
Ⓓ	1118	277	281	185	189	137	141	109	113	91	95	77	81	67	71	1118
o o	1120	277	281	185	180	137	141	109	113	91	95	77	81	67	71	1120
Ⓓ	1122	279	283	185	180	139	143	111	115	91	95	79	83	69	73	1122
o o	1124	279	283	185	180	139	143	111	115	91	95	79	83	69	73	1124
Ⓓ	1126	279	283	185	189	139	143	111	115	91	95	79	83	69	73	1126
o o	1128	279	283	185	189	139	143	111	115	91	95	79	83	69	73	1128
Ⓓ	1130	281	285	187	191	139	143	111	115	93	97	79	83	69	73	1130
o o	1132	281	285	187	191	139	143	111	115	93	97	79	83	69	73	1132
Ⓓ	1134	281	285	187	191	139	143	111	115	93	97	79	83	69	73	1134
o o	1136	281	285	187	191	139	143	111	115	93	97	79	83	69	73	1136
Ⓓ	1138	283	287	187	191	141	145	111	115	93	97	79	83	69	73	1138
o o	1140	283	287	187	191	141	145	111	115	93	97	79	83	69	73	1140
Ⓓ	1142	283	287	189	193	141	145	113	117	93	97	79	83	69	73	1142
o o	1144	283	287	189	193	141	145	113	117	93	97	79	83	69	73	1144
Ⓓ	1146	285	289	189	193	141	145	113	117	93	97	79	83	69	73	1146
o o	1148	285	289	189	193	141	145	113	117	93	97	79	83	69	73	1148
Ⓓ	1150	285	289	189	193	141	145	113	117	93	97	81	85	69	73	1150
o o	1152	285	289	189	193	141	145	113	117	93	97	81	85	69	73	1152
Ⓓ	1154	287	291	191	195	143	147	113	117	95	99	81	85	71	75	1154
o o	1156	287	291	191	195	143	147	113	117	95	99	81	85	71	75	1156
Ⓓ	1158	287	291	191	195	143	147	113	117	95	99	81	85	71	75	1158
o o	1160	287	291	191	195	143	147	113	117	95	99	81	85	71	75	1160
Ⓓ	1162	289	293	191	195	143	147	115	119	95	99	81	85	71	75	1162
o o	1164	289	293	191	195	143	147	115	119	95	99	81	85	71	75	1164
Ⓓ	1166	289	293	193	197	143	147	115	119	95	99	81	85	71	75	1166
o o	1168	289	293	193	197	143	147	115	119	95	99	81	85	71	75	1168
Ⓓ	1170	291	295	193	197	145	149	115	119	95	99	81	85	71	75	1170
o o	1172	291	295	193	197	145	149	115	119	95	99	81	85	71	75	1172
Ⓓ	1174	291	295	193	197	145	149	115	119	95	99	81	85	71	75	1174
o o	1176	291	295	193	197	145	149	115	119	95	99	81	85	71	75	1176
Ⓓ	1178	293	297	195	199	145	149	115	119	97	101	83	87	71	75	1178
o o	1180	293	297	195	199	145	149	115	119	97	101	83	87	71	75	1180
Ⓓ	1182	293	297	195	199	145	149	117	121	97	101	83	87	71	75	1182
o o	1184	293	297	195	199	145	149	117	121	97	101	83	87	71	75	1184
Ⓓ	1186	295	299	195	199	147	151	117	121	97	101	83	87	73	77	1186
o o	1188	295	299	195	199	147	151	117	121	97	101	83	87	73	77	1188
Ⓓ	1190	295	299	197	201	147	151	117	121	97	101	83	87	73	77	1190
o o	1192	295	299	197	201	147	151	117	121	97	101	83	87	73	77	1192
Ⓓ	1194	297	301	197	201	147	151	117	121	97	101	83	87	73	77	1194
o o	1196	297	301	197	201	147	151	117	121	97	101	83	87	73	77	1196
Ⓓ	1198	297	301	197	201	147	151	117	121	97	101	83	87	73	77	1198
o o	1200	207	301	197	201	147	151	117	121	97	101	83	87	73	77	1200

Above choice of Pitches will prove most satisfactory, although, as stated in text, the absolute magnitude of average pitch may be varied within reasonable limits.

MULTIPLE-CIRCUIT, DOUBLE WINDINGS, FOR DRUM ARMATURES.

RE-ENTRANCY	No. OF CONDUCTORS	FRONT AND BACK PITCHES													No. OF CONDUCTORS
		4 POLES		6 POLES		8 POLES		10 POLES		12 POLES		14 POLES		16 POLES		
		F	B	F	B	F	B	F	B	F	B	F	B	F	B	
⊙	1202	299	303	199	203	149	153	119	123	99	103	83	87	73	77	1202
o o	1204	299	303	199	203	149	153	119	123	99	103	83	87	73	77	1204
⊙	1206	299	303	199	203	149	153	119	123	99	103	85	89	73	77	1206
o o	1208	299	303	199	203	149	153	119	123	99	103	85	89	73	77	1208
⊙	1210	301	305	199	203	149	153	119	123	99	103	85	89	73	77	1210
o o	1212	301	305	199	203	149	153	119	123	99	103	85	89	73	77	1212
⊙	1214	301	305	201	205	149	153	119	123	99	103	85	89	73	77	1214
o o	1216	301	305	201	205	149	153	119	123	99	103	85	89	73	77	1216
⊙	1218	303	307	201	205	151	155	119	123	99	103	85	89	75	79	1218
o o	1220	303	307	201	205	151	155	119	123	99	103	85	89	75	79	1220
⊙	1222	303	307	201	205	151	155	121	125	99	103	85	89	75	79	1222
o o	1224	303	307	201	205	151	155	121	125	99	103	85	89	75	79	1224
⊙	1226	305	309	203	207	151	155	121	125	101	105	85	89	75	79	1226
o o	1228	305	309	203	207	151	155	121	125	101	105	85	89	75	79	1228
⊙	1230	305	309	203	207	151	155	121	125	101	105	85	89	75	79	1230
o o	1232	305	309	203	207	151	155	121	125	101	105	85	89	75	79	1232
⊙	1234	307	311	203	207	153	157	121	125	101	105	87	91	75	79	1234
o o	1236	307	311	203	207	153	157	121	125	101	105	87	91	75	79	1236
⊙	1238	307	311	205	209	153	157	121	125	101	105	87	91	75	79	1238
o o	1240	307	311	205	209	153	157	121	125	101	105	87	91	75	79	1240
⊙	1242	309	313	205	209	153	157	123	127	101	105	87	91	75	79	1242
o o	1244	309	313	205	209	153	157	123	127	101	105	87	91	75	79	1244
⊙	1246	309	313	205	209	153	157	123	127	101	105	87	91	75	79	1246
o o	1248	309	313	205	209	153	157	123	127	101	105	87	91	75	79	1248
⊙	1250	311	315	207	211	155	159	123	127	103	107	87	91	77	81	1250
o o	1252	311	315	207	211	155	159	123	127	103	107	87	91	77	81	1252
⊙	1254	311	315	207	211	155	159	123	127	103	107	87	91	77	81	1254
o o	1256	311	315	207	211	155	159	123	127	103	107	87	91	77	81	1256
⊙	1258	313	317	207	211	155	159	123	127	103	107	87	91	77	81	1258
o o	1260	313	317	207	211	155	159	123	127	103	107	87	91	77	81	1260
⊙	1262	313	317	209	213	155	159	125	129	103	107	89	93	77	81	1262
o o	1264	313	317	209	213	155	159	125	129	103	107	89	93	77	81	1264
⊙	1266	315	319	209	213	157	161	125	129	103	107	89	93	77	81	1266
o o	1268	315	319	209	213	157	161	125	129	103	107	89	93	77	81	1268
⊙	1270	315	319	209	213	157	161	125	129	103	107	89	93	77	81	1270
o o	1272	315	319	209	213	157	161	125	129	103	107	89	93	77	81	1272
⊙	1274	317	321	211	215	157	161	125	129	105	109	89	93	77	81	1274
o o	1276	317	321	211	215	157	161	125	129	105	109	89	93	77	81	1276
⊙	1278	317	321	211	215	157	161	125	129	105	109	89	93	77	81	1278
o o	1280	317	321	211	215	157	161	125	129	105	109	89	93	77	81	1280
⊙	1282	319	323	211	215	159	163	127	131	105	109	89	93	79	83	1282
o o	1284	319	323	211	215	159	163	127	131	105	109	89	93	79	83	1284
⊙	1286	319	323	213	217	159	163	127	131	105	109	89	93	79	83	1286
o o	1288	319	323	213	217	159	163	127	131	105	109	89	93	79	83	1288
⊙	1290	321	325	213	217	159	163	127	131	105	109	91	95	79	83	1290
o o	1292	321	325	213	217	159	163	127	131	105	109	91	95	79	83	1292
⊙	1294	321	325	213	217	159	163	127	131	105	109	91	95	79	83	1294
o o	1296	321	325	213	217	159	163	127	131	105	109	91	95	79	83	1296
⊙	1298	323	327	215	219	161	165	127	131	107	111	91	95	79	83	1298
o o	1300	323	327	215	219	161	165	127	131	107	111	91	95	79	83	1300

Above choice of Pitches will prove most satisfactory, although, as stated in text, the absolute magnitude of average pitch may be varied within reasonable limits.

MULTIPLE-CIRCUIT, DOUBLE WINDINGS, FOR DRUM ARMATURES.

RE-ENTRANCY	No. OF CONDUCTORS	FRONT AND BACK PITCHES													No. OF CONDUCTORS
		4 POLES		6 POLES		8 POLES		10 POLES		12 POLES		14 POLES		16 POLES		
		F	B	F	B	F	B	F	B	F	B	F	B	F	B	
②	1302	323	327	215	219	161	165	129	133	107	111	91	95	79	83	1302
oo	1304	323	327	215	219	161	165	129	133	107	111	91	95	79	83	1304
②	1306	325	329	215	219	161	165	129	133	107	111	91	95	79	83	1306
oo	1308	325	329	215	219	161	165	129	133	107	111	91	95	79	83	1308
②	1310	325	329	217	221	161	165	129	133	107	111	91	95	79	83	1310
oo	1312	325	329	217	221	161	165	129	133	107	111	91	95	79	83	1312
②	1314	327	331	217	221	163	167	129	133	107	111	91	95	81	85	1314
oo	1316	327	331	217	221	163	167	129	133	107	111	91	95	81	85	1316
②	1318	327	331	217	221	163	167	129	133	107	111	93	97	81	85	1318
oo	1320	327	331	217	221	163	167	129	133	107	111	93	97	81	85	1320
②	1322	329	333	219	223	163	167	131	135	109	113	93	97	81	85	1322
oo	1324	329	333	219	223	163	167	131	135	109	113	93	97	81	85	1324
②	1326	329	333	219	223	165	167	131	135	109	113	95	97	81	85	1326
oo	1328	329	333	219	223	163	167	131	135	109	113	93	97	81	85	1328
②	1330	331	335	219	223	165	169	131	135	109	113	93	97	81	85	1330
oo	1332	331	335	219	223	165	169	131	135	109	113	93	97	81	85	1332
②	1334	331	335	221	225	165	169	131	135	109	113	93	97	81	85	1334
oo	1336	331	335	221	225	165	169	131	135	109	113	93	97	81	85	1336
②	1338	333	337	221	225	165	169	131	135	109	113	93	97	81	85	1338
oo	1340	333	337	221	225	165	169	131	135	109	113	93	97	81	85	1340
②	1342	333	337	221	225	165	169	133	137	109	113	93	97	81	85	1342
oo	1344	333	337	221	225	165	169	133	137	109	113	93	97	81	85	1344
②	1346	335	339	223	227	167	171	133	137	111	115	95	99	83	87	1346
oo	1348	335	339	223	227	167	171	133	137	111	115	95	99	83	87	1348
②	1350	335	339	223	227	167	171	133	137	111	115	95	99	83	87	1350
oo	1352	335	339	223	227	167	171	133	137	111	115	95	99	83	87	1352
②	1354	337	341	223	227	167	171	133	137	111	115	95	99	83	87	1354
oo	1356	337	341	223	227	167	171	133	137	111	115	95	99	83	87	1356
②	1358	337	341	225	229	167	171	133	137	111	115	95	99	83	87	1358
oo	1360	337	341	225	229	167	171	133	137	111	115	95	99	83	87	1360
②	1362	339	343	225	229	169	173	135	139	111	115	95	99	83	87	1362
oo	1364	339	343	225	229	169	173	135	139	111	115	95	99	83	87	1364
②	1366	339	343	225	229	169	173	135	139	111	115	95	99	83	87	1366
oo	1368	339	343	225	229	169	173	135	139	111	115	95	99	83	87	1368
②	1370	341	345	227	231	169	173	135	139	113	117	95	99	83	87	1370
oo	1372	341	345	227	231	169	173	135	139	113	117	95	99	83	87	1372
②	1374	341	345	227	231	169	173	135	139	113	117	97	101	83	87	1374
oo	1376	341	345	227	231	169	173	135	139	113	117	97	101	83	87	1376
②	1378	343	347	227	231	171	175	135	139	113	117	97	101	85	89	1378
oo	1380	343	347	227	231	171	175	135	139	113	117	97	101	85	89	1380
②	1382	343	347	229	233	171	175	137	141	113	117	97	101	85	89	1382
oo	1384	343	347	229	233	171	175	137	141	113	117	97	101	85	89	1384
②	1386	345	349	229	233	171	175	137	141	113	117	97	101	85	89	1386
oo	1388	345	349	229	233	171	175	137	141	113	117	97	101	85	89	1388
②	1390	345	349	229	233	171	175	137	141	113	117	97	101	85	89	1390
oo	1392	345	349	229	233	171	175	137	141	113	117	97	101	85	89	1392
②	1394	347	351	231	235	173	177	137	141	115	119	97	101	85	89	1394
oo	1396	347	351	231	235	173	177	137	141	115	119	97	101	85	89	1396
②	1398	347	351	231	235	173	177	137	141	115	119	97	101	85	89	1398
oo	1400	347	351	231	235	173	177	137	141	115	119	97	101	85	89	1400

Above choice of Pitches will prove most satisfactory, although, as stated in text, the absolute
magnitude of average pitch may be varied within reasonable limits.

MULTIPLE-CIRCUIT, DOUBLE WINDINGS, FOR DRUM ARMATURES.

RE-ENTRANCY	No. OF CONDUCTORS	FRONT AND BACK PITCHES													No. OF CONDUCTORS
		4 POLES		6 POLES		8 POLES		10 POLES		12 POLES		14 POLES		16 POLES		
		F	B	F	B	F	B	F	B	F	B	F	B	F	B	
⊗	1402	349	353	231	235	173	177	139	143	115	119	99	103	85	89	1402
oo	1404	349	353	231	235	173	177	139	143	115	119	99	103	85	89	1404
⊗	1406	349	353	233	237	173	177	139	143	115	119	99	103	85	89	1406
oo	1408	349	353	233	237	173	177	139	143	115	119	95	103	85	89	1408
⊗	1410	351	355	233	237	175	179	139	143	115	119	99	103	87	91	1410
oo	1412	351	355	233	237	175	179	139	143	115	119	99	103	87	91	1412
⊗	1414	351	355	233	237	175	179	139	143	115	119	99	103	87	91	1414
oo	1416	351	355	233	237	175	179	139	143	115	119	99	103	87	91	1416
⊗	1418	353	357	235	239	175	179	139	143	117	121	99	103	87	91	1418
oo	1420	353	357	235	239	175	179	139	143	117	121	99	103	87	91	1420
⊗	1422	353	357	235	239	175	179	141	145	117	121	99	103	87	91	1422
oo	1424	353	357	235	239	175	179	141	145	117	121	99	103	87	91	1424
⊗	1426	355	359	235	239	177	181	141	145	117	121	99	103	87	91	1426
oo	1428	355	359	235	239	177	181	141	145	117	121	99	103	87	91	1428
⊗	1430	355	359	237	241	177	181	141	145	117	121	101	105	87	91	1430
oo	1432	355	359	237	241	177	181	141	145	117	121	101	105	87	91	1432
⊗	1434	357	361	237	241	177	181	141	145	117	121	101	105	87	91	1434
oo	1436	357	361	237	241	177	181	141	145	117	121	101	105	87	91	1436
⊗	1438	357	361	237	241	177	181	141	145	117	121	101	105	87	91	1438
oo	1440	357	361	237	241	177	181	141	145	117	121	101	105	87	91	1440
⊗	1442	359	363	239	243	179	183	143	147	119	123	101	105	89	93	1442
oo	1444	359	363	239	243	179	183	143	147	119	123	101	105	89	93	1444
⊗	1446	359	363	239	243	179	183	143	147	119	123	101	105	89	93	1446
oo	1448	359	363	239	243	179	183	143	147	119	123	101	105	89	93	1448
⊗	1450	361	365	239	243	179	183	143	147	119	123	101	105	89	93	1450
oo	1452	361	365	239	243	179	183	143	147	119	123	101	105	89	93	1452
⊗	1454	361	365	241	245	179	183	143	147	119	123	101	105	89	93	1454
oo	1456	361	365	241	245	179	183	143	147	119	123	101	105	89	93	1456
⊗	1458	363	367	241	245	181	185	143	147	119	123	103	107	89	93	1458
oo	1460	363	367	241	245	181	185	143	147	119	123	103	107	89	93	1460
⊗	1462	363	367	241	245	181	185	145	149	119	123	103	107	89	93	1462
oo	1464	363	367	241	245	181	185	145	149	119	123	103	107	89	93	1464
⊗	1466	365	369	243	247	181	185	145	149	121	125	103	107	89	93	1466
oo	1468	365	369	243	247	181	185	145	149	121	125	103	107	89	93	1468
⊗	1470	365	369	243	247	181	185	145	149	121	125	103	107	89	93	1470
oo	1472	365	369	243	247	181	185	145	149	121	125	103	107	89	93	1472
⊗	1474	367	371	243	247	183	187	145	149	121	125	103	107	91	95	1474
oo	1476	367	371	243	247	183	187	145	149	121	125	103	107	91	95	1476
⊗	1478	367	371	245	249	183	187	145	149	121	125	103	107	91	95	1478
oo	1480	367	371	245	249	183	187	145	149	121	125	103	107	91	95	1480
⊗	1482	369	373	245	249	183	187	147	151	121	125	103	107	91	95	1482
oo	1484	369	373	245	249	183	187	147	151	121	125	103	107	91	95	1484
⊗	1486	369	373	245	249	183	187	147	151	121	125	105	109	91	95	1486
oo	1488	369	373	245	249	183	187	147	151	121	125	105	109	91	95	1488
⊗	1490	371	375	247	251	185	189	147	151	123	127	105	109	91	95	1490
oo	1492	371	375	247	251	185	189	147	151	123	127	105	109	91	95	1492
⊗	1494	371	375	247	251	185	189	147	151	123	127	105	109	91	95	1494
oo	1496	371	375	247	251	185	189	147	151	123	127	105	109	91	95	1496
⊗	1498	373	377	247	251	185	189	147	151	123	127	105	109	91	95	1498
oo	1500	373	377	247	251	185	189	147	151	123	127	105	109	91	95	1500

Above choice of Pitches will prove most satisfactory, although, as stated in text, the absolute magnitude of average pitch may be varied within reasonable limits.

MULTIPLE-CIRCUIT, DOUBLE WINDINGS, FOR DRUM ARMATURES.

FRONT AND BACK PITCHES

REENTRANCY	No. OF CONDUCTORS	4 POLES F	4 POLES B	6 POLES F	6 POLES B	8 POLES F	8 POLES B	10 POLES F	10 POLES B	12 POLES F	12 POLES B	14 POLES F	14 POLES B	16 POLES F	16 POLES B	No. OF CONDUCTORS
⊗	1502	373	377	249	253	185	189	149	153	123	127	105	109	91	95	1502
oo	1504	373	377	249	253	185	189	149	153	123	127	105	109	91	95	1504
⊗	1506	375	379	249	253	187	191	149	153	123	127	105	109	93	97	1506
oo	1508	375	379	249	253	187	191	149	153	123	127	105	109	93	97	1508
⊗	1510	375	379	249	253	187	191	149	153	123	127	105	109	93	97	1510
oo	1512	375	379	249	253	187	191	149	153	123	127	105	109	93	97	1512
⊗	1514	377	381	251	255	187	191	149	153	125	129	107	111	93	97	1514
oo	1516	377	381	251	255	187	191	149	153	125	129	107	111	93	97	1516
⊗	1518	377	381	251	255	187	191	149	153	125	129	107	111	93	97	1518
oo	1520	377	381	251	255	187	191	149	153	125	129	107	111	93	97	1520
⊗	1522	379	383	251	255	189	193	151	155	125	129	107	111	93	97	1522
oo	1524	379	383	251	255	189	193	151	155	125	129	107	111	93	97	1524
⊗	1526	379	383	253	257	189	193	151	155	125	129	107	111	93	97	1526
oo	1528	379	383	253	257	189	193	151	155	125	129	107	111	93	97	1528
⊗	1530	381	385	253	257	189	193	151	155	125	129	107	111	93	97	1530
oo	1532	381	385	253	257	189	193	151	155	125	129	107	111	93	97	1532
⊗	1534	381	385	253	257	189	193	151	155	125	129	107	111	93	97	1534
oo	1536	381	385	253	257	189	193	151	155	125	129	107	111	93	97	1536
⊗	1538	383	387	255	259	191	195	151	155	127	131	107	111	95	99	1538
oo	1540	383	387	255	259	191	195	151	155	127	131	107	111	95	99	1540
⊗	1542	383	387	255	259	191	195	153	157	127	131	109	113	95	99	1542
oo	1544	383	387	255	259	191	195	153	157	127	131	109	113	95	99	1544
⊗	1546	385	389	255	259	191	195	153	157	127	131	109	113	95	99	1546
oo	1548	385	389	255	259	191	195	153	157	127	131	109	113	95	99	1548
⊗	1550	385	389	257	261	191	195	153	157	127	131	109	113	95	99	1550
oo	1552	385	389	257	261	191	195	153	157	127	131	109	113	95	99	1552
⊗	1554	387	391	257	261	193	197	153	157	127	131	109	113	95	99	1554
oo	1556	387	391	257	261	193	197	153	157	127	131	109	113	95	99	1556
⊗	1558	387	391	257	261	193	197	153	157	127	131	109	113	95	99	1558
oo	1560	387	391	257	261	193	197	153	157	127	131	109	113	95	99	1560
⊗	1562	389	393	259	263	193	197	155	159	129	133	109	113	95	99	1562
oo	1564	389	393	259	263	193	197	155	159	129	133	109	113	95	99	1564
⊗	1566	389	393	259	263	193	197	155	159	129	133	109	113	95	99	1566
oo	1568	389	393	259	263	193	197	155	159	129	133	109	113	95	99	1568
⊗	1570	391	395	259	263	195	199	155	159	129	133	111	115	97	101	1570
oo	1572	391	395	259	263	195	199	155	159	129	133	111	115	97	101	1572
⊗	1574	391	395	261	265	195	199	155	159	129	133	111	115	97	101	1574
oo	1576	391	395	261	265	195	199	155	159	129	133	111	115	97	101	1576
⊗	1578	393	397	261	265	195	199	155	159	129	133	111	115	97	101	1578
oo	1580	393	397	261	265	195	199	155	159	129	133	111	115	97	101	1580
⊗	1582	393	397	261	265	195	199	157	161	129	133	111	115	97	101	1582
oo	1584	393	397	261	265	195	199	157	161	129	133	111	115	97	101	1584
⊗	1586	395	399	263	267	197	201	157	161	131	135	111	115	97	101	1586
oo	1588	395	399	263	267	197	201	157	161	131	135	111	115	97	101	1588
⊗	1590	395	399	263	267	197	201	157	161	131	135	111	115	97	101	1590
oo	1592	395	399	263	267	197	201	157	161	131	135	111	115	97	101	1592
⊗	1594	397	401	263	267	197	201	157	161	131	135	111	115	97	101	1594
oo	1596	397	401	263	267	197	201	157	161	131	135	111	115	97	101	1596
⊗	1598	397	401	265	269	197	201	157	161	131	135	113	117	97	101	1598
oo	1600	397	401	265	269	197	201	157	161	131	135	113	117	97	101	1600

Above choice of Pitches will prove most satisfactory, although, as stated in text, the absolute magnitude of average pitch may be varied within reasonable limits.

WINDING TABLES FOR MULTIPLE-CIRCUIT, TRIPLE WINDINGS
FOR DRUM ARMATURES.

MULTIPLE-CIRCUIT, TRIPLE WINDINGS, FOR DRUM ARMATURES.

| RE-ENTRANCY | No. OF CONDUCTORS | FRONT AND BACK PITCHES | | | | | | | | | | | | | No. OF CONDUCTORS |
| | | 4 POLES | | 6 POLES | | 8 POLES | | 10 POLES | | 12 POLES | | 14 POLES | | 16 POLES | | |
		F	B	F	B	F	B	F	B	F	B	F	B	F	B	
ⓓ	202	47	53	31	37	23	29	17	23	13	19	11	17	9	15	202
ooo	204	47	53	31	37	23	29	17	23	13	19	11	17	9	15	204
ⓓ	206	49	55	31	37	23	29	17	23	15	21	11	17	9	15	206
ⓓ	208	49	55	31	37	23	29	17	23	15	21	11	17	9	15	208
ooo	210	49	55	31	37	23	29	17	23	15	21	11	17	11	17	210
ⓓ	212	49	55	33	39	23	29	19	25	15	21	13	19	11	17	212
ⓓ	214	51	57	33	39	23	29	19	25	15	21	13	19	11	17	214
ooo	216	51	57	33	39	23	29	19	25	15	21	13	19	11	17	216
ⓓ	218	51	57	33	39	25	31	19	25	15	21	13	19	11	17	218
ⓓ	220	51	57	33	39	25	31	19	25	15	21	13	19	11	17	220
ooo	222	53	59	33	39	25	31	19	25	15	21	13	19	11	17	222
ⓓ	224	53	59	35	41	25	31	19	25	15	21	13	19	11	17	224
ⓓ	226	53	59	35	41	25	31	19	25	15	21	13	19	11	17	226
ooo	228	53	59	35	41	25	31	19	25	15	21	13	19	11	17	228
ⓓ	230	55	61	35	41	25	31	19	25	17	23	13	19	11	17	230
ⓓ	232	55	61	35	41	25	31	21	27	17	23	13	19	11	17	232
ooo	234	55	61	35	41	27	33	21	27	17	23	13	19	11	17	234
ⓓ	236	55	61	37	43	27	33	21	27	17	23	13	19	11	17	236
ⓓ	238	57	63	37	43	27	33	21	27	17	23	13	19	11	17	238
ooo	240	57	63	37	43	27	33	21	27	17	23	15	21	11	17	240
ⓓ	242	57	63	37	43	27	33	21	27	17	23	15	21	13	19	242
ⓓ	244	57	63	37	43	27	33	21	27	17	23	15	21	13	19	244
ooo	246	59	65	37	43	27	33	21	27	17	23	15	21	13	19	246
ⓓ	248	59	65	39	45	27	33	21	27	17	23	15	21	13	19	248
ⓓ	250	59	65	39	45	29	35	21	27	17	23	15	21	13	19	250
ooo	252	59	65	39	45	29	35	23	29	17	23	15	21	13	19	252
ⓓ	254	61	67	39	45	29	35	23	29	19	25	15	21	13	19	254
ⓓ	256	61	67	39	45	29	35	23	29	19	25	15	21	13	19	256
ooo	258	61	67	39	45	29	35	23	29	19	25	15	21	13	19	258
ⓓ	260	61	67	41	47	29	35	23	29	19	25	15	21	13	19	260
ⓓ	262	63	69	41	47	29	35	23	29	19	25	15	21	13	19	262
ooo	264	63	69	41	47	29	35	23	29	19	25	15	21	13	19	264
ⓓ	266	63	69	41	47	31	37	23	29	19	25	15	21	13	19	266
ⓓ	268	63	69	41	47	31	37	23	29	19	25	17	23	13	19	268
ooo	270	65	71	41	47	31	37	23	29	19	25	17	23	13	19	270
ⓓ	272	65	71	43	49	31	37	25	31	19	25	17	23	13	19	272
ⓓ	274	65	71	43	49	31	37	25	31	19	25	17	23	15	21	274
ooo	276	65	71	43	49	31	37	25	31	19	25	17	23	15	21	276
ⓓ	278	67	73	43	49	31	37	25	31	21	27	17	23	15	21	278
ⓓ	280	67	73	43	49	31	37	25	31	21	27	17	23	15	21	280
ooo	282	67	73	43	49	33	39	25	31	21	27	17	23	15	21	282
ⓓ	284	67	73	45	51	33	39	25	31	21	27	17	23	15	21	284
ⓓ	286	69	75	45	51	33	39	25	31	21	27	17	23	15	21	286
ooo	288	69	75	45	51	33	39	25	31	21	27	17	23	15	21	288
ⓓ	290	69	75	45	51	33	39	25	31	21	27	17	23	15	21	290
ⓓ	292	69	75	45	51	33	39	27	33	21	27	17	23	15	21	292
ooo	294	71	77	45	51	33	39	27	33	21	27	17	23	15	21	294
ⓓ	296	71	77	47	53	33	39	27	33	21	27	19	25	15	21	296
ⓓ	298	71	77	47	53	35	41	27	33	21	27	19	25	15	21	298
ooo	300	71	77	47	53	35	41	27	33	21	27	19	25	15	21	300

Above choice of Pitches will prove most satisfactory, although, as stated in text, the absolute magnitude of average pitch may be varied within reasonable limits.

MULTIPLE-CIRCUIT, TRIPLE WINDINGS, FOR DRUM ARMATURES.

RE-ENTRANCY	No. OF CONDUCTORS	FRONT AND BACK PITCHES													No. OF CONDUCTORS
		4 POLES		6 POLES		8 POLES		10 POLES		12 POLES		14 POLES		16 POLES	
		F	B	F	B	F	B	F	B	F	B	F	B	F	B	
	302	73	79	47	53	35	41	27	33	23	29	19	25	15	21	302
	304	73	79	47	53	35	41	27	33	23	29	19	25	15	21	304
	306	73	79	47	53	35	41	27	33	23	29	19	25	17	23	306
	308	73	79	49	55	35	41	27	33	23	29	19	25	17	23	308
	310	75	81	49	55	35	41	27	33	23	29	19	25	17	23	310
	312	75	81	49	55	35	41	29	35	23	29	19	25	17	23	312
	314	75	81	49	55	37	43	29	35	23	29	19	25	17	23	314
	316	75	81	49	55	37	43	29	35	23	29	19	25	17	23	316
	318	77	83	49	55	37	43	29	35	23	29	19	25	17	23	318
	320	77	83	51	57	37	43	29	35	23	29	19	25	17	23	320
	322	77	83	51	57	37	43	29	35	23	29	19	25	17	23	322
	324	77	83	51	57	37	43	29	35	23	29	21	27	17	23	324
	326	79	85	51	57	37	43	29	35	25	31	21	27	17	23	326
	328	79	85	51	57	37	43	29	35	25	31	21	27	17	23	328
	330	79	85	51	57	39	45	29	35	25	31	21	27	17	23	330
	332	79	85	53	59	39	45	31	37	25	31	21	27	17	23	332
	334	81	87	53	59	39	45	31	37	25	31	21	27	17	23	334
	336	81	87	53	59	39	45	31	37	25	31	21	27	17	23	336
	338	81	87	53	59	39	45	31	37	25	31	21	27	19	25	338
	340	81	87	53	59	39	45	31	37	25	31	21	27	19	25	340
	342	83	89	53	59	39	45	31	37	25	31	21	27	19	25	342
	344	83	89	55	61	39	45	31	37	25	31	21	27	19	25	344
	346	83	89	55	61	41	47	31	37	25	31	21	27	19	25	346
	348	83	89	55	61	41	47	31	37	25	31	21	27	19	25	348
	350	85	91	55	61	41	47	31	37	27	33	21	27	19	25	350
	352	85	91	55	61	41	47	33	39	27	33	23	29	19	25	352
	354	85	91	55	61	41	47	33	39	27	33	23	29	19	25	354
	356	85	91	57	63	41	47	33	39	27	33	23	29	19	25	356
	358	87	93	57	63	41	47	33	39	27	33	23	29	19	25	358
	360	87	93	57	63	41	47	33	39	27	33	23	29	19	25	360
	362	87	93	57	63	43	49	33	39	27	33	23	29	19	25	362
	364	87	93	57	63	43	49	33	39	27	33	23	29	19	25	364
	366	89	95	57	63	43	49	33	39	27	33	23	29	19	25	366
	368	89	95	59	65	43	49	33	39	27	33	23	29	19	25	368
	370	89	95	59	65	43	49	33	39	27	33	23	29	21	27	370
	372	89	95	59	65	43	49	35	41	27	33	23	29	21	27	372
	374	91	97	59	65	43	49	35	41	29	35	23	29	21	27	374
	376	91	97	59	65	43	49	35	41	29	35	23	29	21	27	376
	378	91	97	59	65	45	51	35	41	29	35	23	29	21	27	378
	380	91	97	61	67	45	51	35	41	29	35	25	31	21	27	380
	382	93	99	61	67	45	51	35	41	29	35	25	31	21	27	382
	384	93	99	61	67	45	51	35	41	29	35	25	31	21	27	384
	386	93	99	61	67	45	51	35	41	29	35	25	31	21	27	386
	388	93	99	61	67	45	51	35	41	29	35	25	31	21	27	388
	390	95	101	61	67	45	51	35	41	29	35	25	31	21	27	390
	392	95	101	63	69	45	51	37	43	29	35	25	31	21	27	392
	394	95	101	63	69	47	53	37	43	29	35	25	31	21	27	394
	396	95	101	63	69	47	53	37	43	29	35	25	31	21	27	396
	398	97	103	63	69	47	53	37	43	31	37	25	31	21	27	398
	400	97	103	63	69	47	53	37	43	31	37	25	31	21	27	400

Above choice of Pitches will prove most satisfactory, although, as stated in text, the absolute magnitude of average pitch may be varied within reasonable limits.

MULTIPLE-CIRCUIT TRIPLE WINDINGS, FOR DRUM ARMATURES.

REENTRANCY	No. OF CONDUCTORS	FRONT AND BACK PITCHES													No. OF CONDUCTORS
		4 POLES		6 POLES		8 POLES		10 POLES		12 POLES		14 POLES		16 POLES		
		F	B	F	B	F	B	F	B	F	B	F	B	F	B	
ooo	402	97	103	63	69	47	53	37	43	31	37	25	31	23	29	402
②	404	97	103	65	71	47	53	37	43	31	37	25	31	23	29	404
②	406	99	105	65	71	47	53	37	43	31	37	25	31	23	29	406
ooo	408	99	105	65	71	47	53	37	43	31	37	27	33	23	29	408
②	410	99	105	65	71	49	55	37	43	31	37	27	33	23	29	410
②	412	99	105	65	71	49	55	39	45	31	37	27	33	23	29	412
ooo	414	101	107	65	71	49	55	39	45	31	37	27	33	25	29	414
②	416	101	107	67	73	49	55	39	45	31	37	27	33	23	29	416
②	418	101	107	67	73	49	55	39	45	31	37	27	33	23	29	418
ooo	420	101	107	67	73	49	55	39	45	31	37	27	33	23	29	420
②	422	103	109	67	73	49	55	39	45	33	39	27	33	23	29	422
②	424	103	109	67	73	49	55	39	45	33	39	27	33	23	29	424
ooo	426	103	109	67	73	51	57	39	45	33	39	27	33	23	29	426
②	428	103	109	69	75	51	57	39	45	33	39	27	33	23	29	428
②	430	105	111	69	75	51	57	39	45	33	39	27	33	23	29	430
ooo	432	105	111	69	75	51	57	41	47	33	39	27	33	23	29	432
②	434	105	111	69	75	51	57	41	47	33	39	27	33	25	31	434
②	436	105	111	69	75	51	57	41	47	33	39	29	35	25	31	436
ooo	438	107	113	69	75	51	57	41	47	33	39	29	35	25	31	438
②	440	107	113	71	77	51	57	41	47	33	39	29	35	25	31	440
②	442	107	113	71	77	53	59	41	47	33	39	29	35	25	31	442
ooo	444	107	113	71	77	53	59	41	47	33	39	29	35	25	31	444
②	446	109	115	71	77	53	59	41	47	35	41	29	35	25	31	446
②	448	109	115	71	77	53	59	41	47	35	41	29	35	25	31	448
ooo	450	109	115	71	77	53	59	41	47	35	41	29	35	25	31	450
②	452	109	115	73	79	53	59	43	49	35	41	29	35	25	31	452
②	454	111	117	73	79	53	59	43	49	35	41	29	35	25	31	454
ooo	456	111	117	73	79	53	59	43	49	35	41	29	35	25	31	456
②	458	111	117	73	79	55	61	43	49	35	41	29	35	25	31	458
②	460	111	117	73	79	55	61	43	49	35	41	29	35	25	31	460
ooo	462	113	119	73	79	55	61	43	49	35	41	29	35	25	31	462
②	464	113	119	75	81	55	61	43	49	35	41	31	37	25	31	464
②	466	113	119	75	81	55	61	43	49	35	41	31	37	27	33	466
ooo	468	113	119	75	81	55	61	43	49	35	41	31	37	27	33	468
②	470	115	121	75	81	55	61	43	49	37	43	31	37	27	33	470
②	472	115	121	75	81	55	61	45	51	37	43	31	37	27	33	472
ooo	474	115	121	75	81	57	63	45	51	37	43	31	37	27	33	474
②	476	115	121	77	83	57	63	45	51	37	43	31	37	27	33	476
②	478	117	123	77	83	57	63	45	51	37	43	31	37	27	33	478
ooo	480	117	123	77	83	57	63	45	51	37	43	31	37	27	33	480
②	482	117	123	77	83	57	63	45	51	37	43	31	37	27	33	482
②	484	117	123	77	83	57	63	45	51	37	43	31	37	27	33	484
ooo	486	119	125	77	83	57	63	45	51	37	43	31	37	27	33	486
②	488	119	125	79	85	57	63	45	51	37	43	31	37	27	33	488
②	490	119	125	79	85	59	65	45	51	37	43	31	37	27	33	490
ooo	492	119	125	79	85	59	65	47	53	37	43	33	39	27	33	492
②	494	121	127	79	85	59	65	47	53	39	45	33	39	27	33	494
②	496	121	127	79	85	59	65	47	53	39	45	33	39	27	33	496
ooo	498	121	127	79	85	59	65	47	53	39	45	33	39	29	35	498
②	500	121	127	81	87	59	65	47	53	39	45	33	39	29	35	500

Above choice of Pitches will prove most satisfactory, although, as stated in text, the absolute magnitude of average pitch may be varied within reasonable limits.

MULTIPLE-CIRCUIT, TRIPLE WINDINGS, FOR DRUM ARMATURES.

RE-ENTRANCY	No. OF CONDUCTORS	FRONT AND BACK PITCHES													No. OF CONDUCTORS
		4 POLES		6 POLES		8 POLES		10 POLES		12 POLES		14 POLES		16 POLES		
		F	B	F	B	F	B	F	B	F	B	F	B	F	B	
⊚	502	123	129	81	87	59	65	47	53	39	45	33	39	29	35	502
000	504	123	129	81	87	59	65	47	53	39	45	33	39	29	35	504
⊚	506	123	129	81	87	61	67	47	53	39	45	33	39	29	35	506
⊚	508	123	129	81	87	61	67	47	53	39	45	33	39	29	35	508
000	510	125	131	81	87	61	67	47	53	39	45	33	39	29	35	510
⊚	512	125	131	83	89	61	67	49	55	39	45	33	39	29	35	512
⊚	514	125	131	83	89	61	67	49	55	39	45	33	39	29	35	514
000	516	125	131	83	89	61	67	49	55	39	45	33	39	29	35	516
⊚	518	127	133	83	89	61	67	49	55	41	47	33	39	29	35	518
⊚	520	127	133	83	89	61	67	49	55	41	47	35	41	29	35	520
000	522	127	133	83	89	63	69	49	55	41	47	35	41	29	35	522
⊚	524	127	133	85	91	63	69	49	55	41	47	35	41	29	35	524
⊚	526	129	135	85	91	63	69	49	55	41	47	35	41	29	35	526
000	528	129	135	85	91	63	69	49	55	41	47	35	41	29	35	528
⊚	530	129	135	85	91	63	69	49	55	41	47	35	41	31	37	530
⊚	532	129	135	85	91	63	69	51	57	41	47	35	41	31	37	532
000	534	131	137	85	91	63	69	51	57	41	47	35	41	31	37	534
⊚	536	131	137	87	93	63	69	51	57	41	47	35	41	31	37	536
⊚	538	131	137	87	93	65	71	51	57	41	47	35	41	31	37	538
000	540	131	137	87	93	65	71	51	57	41	47	35	41	31	37	540
⊚	542	133	139	87	93	65	71	51	57	43	49	35	41	31	37	542
⊚	544	133	139	87	93	65	71	51	57	43	49	35	41	31	37	544
000	546	133	139	87	93	65	71	51	57	43	49	35	41	31	37	546
⊚	548	133	139	89	95	65	71	51	57	43	49	37	43	31	37	548
⊚	550	135	141	89	95	65	71	51	57	43	49	37	43	31	37	550
000	552	135	141	89	95	65	71	53	59	43	49	37	43	31	37	552
⊚	554	135	141	89	95	67	73	53	59	43	49	37	43	31	37	554
⊚	556	135	141	89	95	67	73	53	59	43	49	37	43	31	37	556
000	558	137	143	89	95	67	73	53	59	43	49	37	43	31	37	558
⊚	560	137	143	91	97	67	73	53	59	43	49	37	43	31	37	560
⊚	562	137	143	91	97	67	73	53	59	43	49	37	43	33	39	562
000	564	137	143	91	97	67	73	53	59	43	49	37	43	33	39	564
⊚	566	139	145	91	97	67	73	53	59	45	51	37	43	33	39	566
⊚	568	139	145	91	97	67	73	53	59	45	51	37	43	33	39	568
000	570	139	145	91	97	69	75	53	59	45	51	37	43	33	39	570
⊚	572	139	145	93	99	69	75	55	61	45	51	37	43	33	39	572
⊚	574	141	147	93	99	69	75	55	61	45	51	37	43	33	39	574
000	576	141	147	93	99	69	75	55	61	45	51	39	45	33	39	576
⊚	578	141	147	93	99	69	75	55	61	45	51	39	45	33	39	578
⊚	580	141	147	93	99	69	75	55	61	45	51	39	45	33	39	580
000	582	143	149	93	99	69	75	55	61	45	51	39	45	33	39	582
⊚	584	143	149	95	101	69	75	55	61	45	51	39	45	33	39	584
⊚	586	143	149	95	101	71	77	55	61	45	51	39	45	33	39	586
000	588	143	149	95	101	71	77	55	61	45	51	39	45	33	39	588
⊚	590	145	151	95	101	71	77	55	61	47	53	39	45	33	39	590
⊚	592	145	151	95	101	71	77	57	63	47	53	39	45	33	39	592
000	594	145	151	95	101	71	77	57	63	47	53	39	45	35	41	594
⊚	596	145	151	97	103	71	77	57	63	47	53	39	45	35	41	596
⊚	598	147	153	97	103	71	77	57	63	47	53	39	45	35	41	598
000	600	147	153	97	103	71	77	57	63	47	53	39	45	35	41	600

Above choice of Pitches will prove most satisfactory, although, as stated in text, the absolute
magnitude of average pitch may be varied within reasonable limits.

MULTIPLE-CIRCUIT, TRIPLE WINDINGS, FOR DRUM ARMATURES.

RE-ENTRANCY	No. OF CONDUCTORS	FRONT AND BACK PITCHES													No. OF CONDUCTORS
		4 POLES		6 POLES		8 POLES		10 POLES		12 POLES		14 POLES		16 POLES		
		F	B	F	B	F	B	F	B	F	B	F	B	F	B	
	602	147	153	97	103	73	79	57	63	47	53	39	45	35	41	602
	604	147	153	97	103	73	79	57	63	47	53	41	47	35	41	604
	606	149	155	97	103	73	79	57	63	47	53	41	47	35	41	606
	608	149	155	99	105	73	79	57	63	47	53	41	47	35	41	608
	610	149	155	99	105	73	79	57	63	47	53	41	47	35	41	610
	612	149	155	99	105	73	79	59	65	47	53	41	47	35	41	612
	614	151	157	99	105	73	79	59	65	49	55	41	47	35	41	614
	616	151	157	99	105	73	79	59	65	49	55	41	47	35	41	616
	618	151	157	99	105	75	81	59	65	49	55	41	47	35	41	618
	620	151	157	101	107	75	81	59	65	49	55	41	47	35	41	620
	622	153	159	101	107	75	81	59	65	49	55	41	47	35	41	622
	624	153	159	101	107	75	81	59	65	49	55	41	47	35	41	624
	626	153	159	101	107	75	81	59	65	49	55	41	47	37	43	626
	628	153	159	101	107	75	81	59	65	49	55	41	47	37	43	628
	630	155	161	101	107	75	81	59	65	49	55	41	47	37	43	630
	632	155	161	103	109	75	81	61	67	49	55	43	49	37	43	632
	634	155	161	103	109	77	83	61	67	49	55	43	49	37	43	634
	636	155	161	103	109	77	83	61	67	49	55	43	49	37	43	636
	638	157	163	103	109	77	83	61	67	51	57	43	49	37	43	638
	640	157	163	103	109	77	83	61	67	51	57	43	49	37	43	640
	642	157	163	103	109	77	83	61	67	51	57	43	49	37	43	642
	644	157	163	105	111	77	83	61	67	51	57	43	49	37	43	644
	646	159	165	105	111	77	83	61	67	51	57	43	49	37	43	646
	648	159	165	105	111	77	83	61	67	51	57	43	49	37	43	648
	650	159	165	105	111	79	85	61	67	51	57	43	49	37	43	650
	652	159	165	105	111	79	85	63	69	51	57	43	49	37	43	652
	654	161	167	105	111	79	85	63	69	51	57	43	49	37	43	654
	656	161	167	107	113	79	85	63	69	51	57	43	49	37	43	656
	658	161	167	107	113	79	85	63	69	51	57	43	49	39	45	658
	660	161	167	107	113	79	85	63	69	51	57	43	49	39	45	660
	662	163	169	107	113	79	85	63	69	53	59	45	51	39	45	662
	664	163	169	107	113	79	85	63	69	53	59	45	51	39	45	664
	666	163	169	107	113	81	87	63	69	53	59	45	51	39	45	666
	668	163	169	109	115	81	87	63	69	53	59	45	51	39	45	668
	670	165	171	109	115	81	87	63	69	53	59	45	51	39	45	670
	672	165	171	109	115	81	87	65	71	53	59	45	51	39	45	672
	674	165	171	109	115	81	87	65	71	53	59	45	51	39	45	674
	676	165	171	109	115	81	87	65	71	53	59	45	51	39	45	676
	678	167	173	109	115	81	87	65	71	53	59	45	51	39	45	678
	680	167	173	111	117	81	87	65	71	53	59	45	51	39	45	680
	682	167	173	111	117	83	89	65	71	53	59	45	51	39	45	682
	684	167	173	111	117	83	89	65	71	53	59	45	51	39	45	684
	686	169	175	111	117	83	89	65	71	55	61	45	51	39	45	686
	688	169	175	111	117	83	89	65	71	55	61	47	53	39	45	688
	690	169	175	111	117	83	89	65	71	55	61	47	53	41	47	690
	692	169	175	113	119	83	89	67	73	55	61	47	53	41	47	692
	694	171	177	113	119	83	89	67	73	55	61	47	53	41	47	694
	696	171	177	113	119	83	89	67	73	55	61	47	53	41	47	696
	698	171	177	113	119	85	91	67	73	55	61	47	53	41	47	698
	700	171	177	113	119	86	91	67	73	55	61	47	53	41	47	700

Above choice of Pitches will prove most satisfactory, although, as stated in text, the absolute magnitude of average pitch may be varied within reasonable limits.

MULTIPLE-CIRCUIT, TRIPLE WINDINGS, FOR DRUM ARMATURES.

RE-ENTRANCY	No. OF CONDUCTORS	FRONT AND BACK PITCHES													No. OF CONDUCTORS
		4 POLES		6 POLES		8 POLES		10 POLES		12 POLES		14 POLES		16 POLES		
		F	B	F	B	F	B	F	B	F	B	F	B	F	B	
000	702	178	179	113	119	85	91	67	73	55	61	47	53	41	47	702
⊕⊕	704	173	179	115	121	85	91	67	73	55	61	47	53	41	47	704
⊕⊕	706	173	179	115	121	85	91	67	73	55	61	47	53	41	47	706
000	708	173	179	115	121	85	91	67	73	55	61	47	53	41	47	708
⊕⊕	710	175	181	115	121	85	91	67	73	57	63	47	53	41	47	710
⊕⊕	712	175	181	115	121	85	91	69	75	57	63	47	53	41	47	712
000	714	175	181	115	121	87	93	69	75	57	63	47	53	41	47	714
⊕⊕	716	175	181	117	123	87	93	69	75	57	63	49	55	41	47	716
⊕⊕	718	177	183	117	123	87	93	69	75	57	63	49	55	41	47	718
000	720	177	183	117	123	87	93	69	75	57	63	49	55	41	47	720
⊕⊕	722	177	183	117	123	87	93	69	75	57	63	49	55	43	49	722
⊕⊕	724	177	183	117	123	87	93	69	75	57	63	49	55	43	49	724
000	726	179	185	117	123	87	93	69	75	57	63	49	55	43	49	726
⊕⊕	728	179	185	119	125	87	93	69	75	57	63	49	55	43	49	728
⊕⊕	730	179	185	119	125	89	95	69	75	57	63	49	55	43	49	730
000	732	179	185	119	125	89	95	71	77	57	63	49	55	43	49	732
⊕⊕	734	181	187	119	125	89	95	71	77	59	65	49	55	43	49	734
⊕⊕	736	181	187	119	125	89	95	71	77	59	65	49	55	43	49	736
000	738	181	187	119	125	89	95	71	77	59	65	49	55	43	49	738
⊕⊕	740	181	187	121	127	89	95	71	77	59	65	49	55	43	49	740
⊕⊕	742	183	189	121	127	89	95	71	77	59	65	49	55	43	49	742
000	744	183	189	121	127	89	95	71	77	59	65	51	57	43	49	744
⊕⊕	746	183	189	121	127	91	97	71	77	59	65	51	57	43	49	746
⊕⊕	748	183	189	121	127	91	97	71	77	59	65	51	57	43	49	748
000	750	185	191	121	127	91	97	71	77	59	65	51	57	43	49	750
⊕⊕	752	185	191	123	129	91	97	73	79	59	65	51	57	43	49	752
⊕⊕	754	185	191	123	129	91	97	73	79	59	65	51	57	45	51	754
000	756	185	191	123	129	91	97	73	79	59	65	51	57	45	51	756
⊕⊕	758	187	193	123	129	91	97	73	79	61	67	51	57	45	51	758
⊕⊕	760	187	193	123	129	91	97	73	79	61	67	51	57	45	51	760
000	762	187	193	123	129	93	99	73	79	61	67	51	57	45	51	762
⊕⊕	764	187	193	125	131	93	99	73	79	61	67	51	57	45	51	764
⊕⊕	766	189	195	125	131	93	99	73	79	61	67	51	57	45	51	766
000	768	189	195	125	131	93	99	73	79	61	67	51	57	45	51	768
⊕⊕	770	189	195	125	131	93	99	73	79	61	67	51	57	45	51	770
⊕⊕	772	189	195	125	131	93	99	75	81	61	67	53	59	45	51	772
000	774	191	197	125	131	93	99	75	81	61	67	53	59	45	51	774
⊕⊕	776	191	197	127	133	93	99	75	81	61	67	53	59	45	51	776
⊕⊕	778	191	197	127	133	95	101	75	81	61	67	53	59	45	51	778
000	780	191	197	127	133	95	101	75	81	61	67	53	59	45	51	780
⊕⊕	782	193	199	127	133	95	101	75	81	63	69	53	59	45	51	782
⊕⊕	784	193	199	127	133	95	101	75	81	63	69	53	59	45	51	784
000	786	193	199	127	133	95	101	75	81	63	69	53	59	47	53	786
⊕⊕	788	193	199	129	135	95	101	75	81	63	69	53	59	47	53	788
⊕⊕	790	195	201	129	135	95	101	75	81	63	69	53	59	47	53	790
000	792	195	201	129	135	95	101	77	83	63	69	53	59	47	53	792
⊕⊕	794	195	201	129	135	97	103	77	83	63	69	53	59	47	53	794
⊕⊕	796	195	201	129	135	97	103	77	83	63	69	53	59	47	53	796
000	798	197	203	129	135	97	103	77	83	63	69	53	59	47	53	798
⊕⊕	800	197	203	131	137	97	103	77	83	63	69	55	61	47	53	800

Above choice of Pitches will prove most satisfactory, although, as stated in text, the absolute magnitude of average pitch may be varied within reasonable limits.

MULTIPLE-CIRCUIT, TRIPLE WINDINGS, FOR DRUM ARMATURES.

RE-ENTRANCY	No. OF CONDUCTORS	4 POLES		6 POLES		8 POLES		10 POLES		12 POLES		14 POLES		16 POLES		No. OF CONDUCTORS
		F	B	F	B	F	B	F	B	F	B	F	B	F	B	
②	802	197	203	131	137	97	103	77	83	63	69	55	61	47	53	802
○○○	804	197	203	131	137	97	103	77	83	63	69	55	61	47	53	804
②	806	199	205	131	137	97	103	77	83	65	71	55	61	47	53	806
②	808	199	205	131	137	97	103	77	83	65	71	55	61	47	53	808
○○○	810	199	205	131	137	99	105	77	83	65	71	55	61	47	53	810
②	812	199	205	133	139	99	105	79	85	65	71	55	61	47	53	812
②	814	201	207	133	139	99	105	79	85	65	71	55	61	47	53	814
○○○	816	201	207	133	139	99	105	79	85	65	71	55	61	47	53	816
②	818	201	207	133	139	99	105	79	85	65	71	55	61	49	55	818
②	820	201	207	133	139	99	105	79	85	65	71	55	61	49	55	820
○○○	822	203	209	133	139	99	105	79	85	65	71	55	61	49	55	822
②	824	203	209	135	141	99	105	79	85	65	71	55	61	49	55	824
②	826	203	209	135	141	101	107	79	85	65	71	55	61	49	55	826
○○○	828	203	209	135	141	101	107	79	85	65	71	57	63	49	55	828
②	830	205	211	135	141	101	107	79	85	67	73	57	63	49	55	830
②	832	205	211	135	141	101	107	81	87	67	73	57	63	49	55	832
○○○	834	205	211	135	141	101	107	81	87	67	73	57	63	49	55	834
②	836	205	211	137	143	101	107	81	87	67	73	57	63	49	55	836
②	838	207	213	137	143	101	107	81	87	67	73	57	63	49	55	838
○○○	840	207	213	137	143	101	107	81	87	67	73	57	63	49	55	840
②	842	207	213	137	143	103	109	81	87	67	73	57	63	49	55	842
②	844	207	213	137	143	103	109	81	87	67	73	57	63	49	55	844
○○○	846	209	215	137	143	103	109	81	87	67	73	57	63	49	55	846
②	848	209	215	139	145	103	109	81	87	67	73	57	63	49	55	848
②	850	209	215	139	145	103	109	81	87	67	73	57	63	51	57	850
○○○	852	209	215	139	145	103	109	83	89	67	73	57	63	51	57	852
②	854	211	217	139	145	103	109	83	89	69	75	57	63	51	57	854
②	856	211	217	139	145	103	109	83	89	69	75	59	65	51	57	856
○○○	858	211	217	139	145	105	111	83	89	69	75	59	65	51	57	858
②	860	211	217	141	147	105	111	83	89	69	75	59	65	51	57	860
②	862	213	219	141	147	105	111	83	89	69	75	59	65	51	57	862
○○○	864	213	219	141	147	105	111	83	89	69	75	59	65	51	57	864
②	866	213	219	141	147	105	111	83	89	69	75	59	65	51	57	866
②	868	213	219	141	147	105	111	83	89	69	75	59	65	51	57	868
○○○	870	215	221	141	147	105	111	83	89	69	75	59	65	51	57	870
②	872	215	221	143	149	105	111	85	91	69	75	59	65	51	57	872
②	874	215	221	143	149	107	113	85	91	69	75	59	65	51	57	874
○○○	876	215	221	143	149	107	113	85	91	69	75	59	65	51	57	876
②	878	217	223	143	149	107	113	85	91	71	77	59	65	51	57	878
②	880	217	223	143	149	107	113	85	91	71	77	59	65	51	57	880
○○○	882	217	223	143	149	107	113	85	91	71	77	59	65	53	59	882
②	884	217	223	145	151	107	113	85	91	71	77	61	67	53	59	884
②	886	219	225	145	151	107	113	85	91	71	77	61	67	53	59	886
○○○	888	219	225	145	151	107	113	85	91	71	77	61	67	53	59	888
②	890	219	225	145	151	109	115	85	91	71	77	61	67	53	59	890
②	892	219	225	145	151	109	115	87	93	71	77	61	67	53	59	892
○○○	894	221	227	145	151	109	115	87	93	71	77	61	67	53	59	894
②	896	221	227	147	153	109	115	87	93	71	77	61	67	53	59	896
②	898	221	227	147	153	109	115	87	93	71	77	61	67	53	59	898
○○○	900	221	227	147	153	109	115	87	93	71	77	61	67	53	59	900

Above choice of pitches will prove most satisfactory, although, as stated in text, the absolute magnitude of average pitch may be varied within reasonable limits.

MULTIPLE-CIRCUIT, TRIPLE WINDINGS, FOR DRUM ARMATURES.

No. OF CONDUCTORS	4 POLES F	B	6 POLES F	B	8 POLES F	B	10 POLES F	B	12 POLES F	B	14 POLES F	B	16 POLES F	B	No. OF CONDUCTORS
902	223	229	147	153	109	115	87	93	73	79	61	67	53	59	902
904	223	229	147	153	109	115	87	93	73	79	61	67	53	59	904
906	223	229	147	153	111	117	87	93	73	79	61	67	53	59	906
908	223	229	149	155	111	117	87	93	73	79	61	67	53	59	908
910	225	231	149	155	111	117	87	93	73	79	61	67	53	59	910
912	225	231	149	155	111	117	89	95	73	79	63	69	53	59	912
914	225	231	149	155	111	117	89	95	73	79	63	69	55	61	914
916	225	231	149	155	111	117	89	95	73	79	63	69	55	61	916
918	227	233	149	155	111	117	89	95	73	79	63	69	55	61	918
920	227	233	151	157	111	117	89	95	73	79	63	69	55	61	920
922	227	233	151	157	113	119	89	95	73	79	63	69	55	61	922
924	227	233	151	157	113	119	89	95	73	79	63	69	55	61	924
926	229	235	151	157	113	119	89	95	75	81	63	69	55	61	926
928	229	235	151	157	113	119	89	95	75	81	63	69	55	61	928
930	229	235	151	157	113	119	89	95	75	81	63	69	55	61	930
932	229	235	153	159	113	119	91	97	75	81	63	69	55	61	932
934	231	237	153	159	113	119	91	97	75	81	63	69	55	61	934
936	231	237	153	159	113	119	91	97	75	81	63	69	55	61	936
938	231	237	153	159	115	121	91	97	75	81	63	69	55	61	938
940	231	237	153	159	115	121	91	97	75	81	65	71	55	61	940
942	233	239	153	159	115	121	91	97	75	81	65	71	55	61	942
944	233	239	155	161	115	121	91	97	75	81	65	71	55	61	944
946	233	239	155	161	115	121	91	97	75	81	65	71	57	63	946
948	233	239	155	161	115	121	91	97	75	81	65	71	57	63	948
950	235	241	155	161	115	121	91	97	77	83	65	71	57	63	950
952	235	241	155	161	115	121	93	99	77	83	65	71	57	63	952
954	235	241	155	161	117	123	93	99	77	83	65	71	57	63	954
956	235	241	157	163	117	123	93	99	77	83	65	71	57	63	956
958	237	243	157	163	117	123	93	99	77	83	65	71	57	63	958
960	237	243	157	163	117	123	93	99	77	83	65	71	57	63	960
962	237	243	157	163	117	123	93	99	77	83	65	71	57	63	962
964	237	243	157	163	117	123	93	99	77	83	65	71	57	63	964
966	239	245	157	163	117	123	93	99	77	83	65	71	57	63	966
968	239	245	159	165	117	123	93	99	77	83	67	73	57	63	968
970	239	245	159	165	119	125	93	99	77	83	67	73	57	63	970
972	239	245	159	165	119	125	95	101	77	83	67	73	57	63	972
974	241	247	159	165	119	125	95	101	79	85	67	73	57	63	974
976	241	247	159	165	119	125	95	101	79	85	67	73	57	63	976
978	241	247	159	165	119	125	95	101	79	85	67	73	59	65	978
980	241	247	161	167	119	125	95	101	79	85	67	73	59	65	980
982	243	249	161	167	119	125	95	101	79	85	67	73	59	65	982
984	243	249	161	167	119	125	95	101	79	85	67	73	59	65	984
986	243	249	161	167	121	127	95	101	79	85	67	73	59	65	986
988	243	249	161	167	121	127	95	101	79	85	67	73	59	65	988
990	245	251	161	167	121	127	95	101	79	85	67	73	59	65	990
992	245	251	163	169	121	127	97	103	79	85	67	73	59	65	992
994	245	251	163	169	121	127	97	103	79	85	67	73	59	65	994
996	245	251	163	169	121	127	97	103	79	85	69	75	59	65	996
998	247	253	163	169	121	127	97	103	81	87	69	75	59	65	998
1000	247	253	163	169	121	127	97	103	81	87	69	75	59	65	1000

Above choice of Pitches will prove most satisfactory, although, as stated in text, the absolute magnitude of average pitch may be varied within reasonable limits.

MULTIPLE-CIRCUIT, TRIPLE WINDINGS, FOR DRUM ARMATURES.

| REENTRANCY | No. OF CONDUCTORS | FRONT AND BACK PITCHES | | | | | | | | | | | | | No. OF CONDUCTORS |
| | | 4 POLES | | 6 POLES | | 8 POLES | | 10 POLES | | 12 POLES | | 14 POLES | | 16 POLES | | |
		F	B	F	B	F	B	F	B	F	B	F	B	F	B	
	1002	247	253	163	169	123	129	97	103	81	87	69	75	59	65	1002
	1004	247	253	165	171	123	129	97	103	81	87	69	75	59	65	1004
	1006	249	255	165	171	123	129	97	103	81	87	69	75	59	65	1006
	1008	249	255	165	171	123	129	97	103	81	87	69	75	59	65	1008
	1010	249	255	165	171	123	129	97	103	81	87	69	75	61	67	1010
	1012	249	255	165	171	123	129	99	105	81	87	69	75	61	67	1012
	1014	251	257	165	171	123	129	99	105	81	87	69	75	61	67	1014
	1016	251	257	167	173	123	129	99	105	81	87	69	75	61	67	1016
	1018	251	257	167	173	125	131	99	105	81	87	69	75	61	67	1018
	1020	251	257	167	173	125	131	99	105	81	87	69	75	61	67	1020
	1022	253	259	167	173	125	131	99	105	83	89	69	75	61	67	1022
	1024	253	259	167	173	125	131	99	105	83	89	71	77	61	67	1024
	1026	253	259	167	173	125	131	99	105	83	89	71	77	61	67	1026
	1028	253	259	169	175	125	131	99	105	83	89	71	77	61	67	1028
	1030	255	261	169	175	125	131	99	105	83	89	71	77	61	67	1030
	1032	255	261	169	175	125	131	101	107	83	89	71	77	61	67	1032
	1034	255	261	169	175	127	133	101	107	83	89	71	77	61	67	1034
	1036	255	261	169	175	127	133	101	107	83	89	71	77	61	67	1036
	1038	257	263	169	175	127	133	101	107	83	89	71	77	61	67	1038
	1040	257	263	171	177	127	133	101	107	83	89	71	77	61	67	1040
	1042	257	263	171	177	127	133	101	107	83	89	71	77	63	69	1042
	1044	257	263	171	177	127	133	101	107	83	89	71	77	63	69	1044
	1046	259	265	171	177	127	133	101	107	85	91	71	77	63	69	1046
	1048	259	265	171	177	127	133	101	107	85	91	71	77	63	69	1048
	1050	259	265	171	177	129	135	101	107	85	91	71	77	63	69	1050
	1052	259	265	173	179	129	135	103	109	85	91	73	79	63	69	1052
	1054	261	267	173	179	129	135	103	109	85	91	73	79	63	69	1054
	1056	261	267	173	179	129	135	103	109	85	91	73	79	63	69	1056
	1058	261	267	173	179	129	135	103	109	85	91	73	79	63	69	1058
	1060	261	267	173	179	129	135	103	109	85	91	73	79	63	69	1060
	1062	263	269	173	179	129	135	103	109	85	91	73	79	63	69	1062
	1064	263	269	175	181	129	135	103	109	85	91	73	79	63	69	1064
	1066	263	269	175	181	131	137	103	109	85	91	73	79	63	69	1066
	1068	263	269	175	181	131	137	103	109	85	91	73	79	63	69	1068
	1070	265	271	175	181	131	137	103	109	87	93	73	79	63	69	1070
	1072	265	271	175	181	131	137	105	111	87	93	73	79	63	69	1072
	1074	265	271	175	181	131	137	105	111	87	93	73	79	65	71	1074
	1076	265	271	177	183	131	137	105	111	87	93	73	79	65	71	1076
	1078	267	273	177	183	131	137	105	111	87	93	73	79	65	71	1078
	1080	267	273	177	183	131	137	105	111	87	93	75	81	65	71	1080
	1082	267	273	177	183	133	139	105	111	87	93	75	81	65	71	1082
	1084	267	273	177	183	133	139	105	111	87	93	75	81	65	71	1084
	1086	269	275	177	183	133	139	105	111	87	93	75	81	65	71	1086
	1088	269	275	179	185	133	139	105	111	87	93	75	81	65	71	1088
	1090	269	275	179	185	133	139	105	111	87	93	75	81	65	71	1090
	1092	269	275	179	185	133	139	107	113	87	93	75	81	65	71	1092
	1094	271	277	179	185	133	139	107	113	89	95	75	81	65	71	1094
	1096	271	277	179	185	133	139	107	113	89	95	75	81	65	71	1096
	1098	271	277	179	185	135	141	107	113	89	95	75	81	65	71	1098
	1100	271	277	181	187	135	141	107	113	89	95	75	81	65	71	1100

Above choice of Pitches will prove most satisfactory, although, as stated in text, the absolute magnitude of average pitch may be varied within reasonable limits.

MULTIPLE-CIRCUIT, TRIPLE WINDINGS, FOR DRUM ARMATURES.

RE-ENTRANCY	No. OF CONDUCTORS	FRONT AND BACK PITCHES													No. OF CONDUCTORS
		4 POLES		6 POLES		8 POLES		10 POLES		12 POLES		14 POLES		16 POLES	
		F	B	F	B	F	B	F	B	F	B	F	B	F	B	
	1102	273	279	181	187	135	141	107	113	89	95	75	81	65	71	1102
	1104	273	279	181	187	135	141	107	113	89	95	75	81	65	71	1104
	1106	273	279	181	187	135	141	107	113	89	95	75	81	67	73	1106
	1108	273	279	181	187	135	141	107	113	89	95	77	83	67	73	1108
	1110	275	281	181	187	135	141	107	113	89	95	77	83	67	73	1110
	1112	275	281	183	189	135	141	109	115	89	95	77	83	67	73	1112
	1114	275	281	183	189	137	143	109	115	89	95	77	83	67	73	1114
	1116	275	281	183	189	137	143	109	115	89	95	77	83	67	73	1116
	1118	277	283	183	189	137	143	109	115	91	97	77	83	67	73	1118
	1120	277	283	183	189	137	143	109	115	91	97	77	83	67	73	1120
	1122	277	283	183	189	137	143	109	115	91	97	77	83	67	73	1122
	1124	277	283	185	191	137	143	109	115	91	97	77	83	67	73	1124
	1126	279	285	185	191	137	143	109	115	91	97	77	83	67	73	1126
	1128	279	285	185	191	137	143	109	115	91	97	77	83	67	73	1128
	1130	279	285	185	191	139	145	109	115	91	97	77	83	67	73	1130
	1132	279	285	185	191	139	145	111	117	91	97	77	83	67	73	1132
	1134	281	287	185	191	139	145	111	117	91	97	77	83	67	73	1134
	1136	281	287	187	193	139	145	111	117	91	97	79	85	67	73	1136
	1138	281	287	187	193	139	145	111	117	91	97	79	85	69	75	1138
	1140	281	287	187	193	139	145	111	117	91	97	79	85	69	75	1140
	1142	283	289	187	193	139	145	111	117	93	99	79	85	69	75	1142
	1144	283	289	187	193	139	145	111	117	93	99	79	85	69	75	1144
	1146	283	289	187	193	141	147	111	117	93	99	79	85	69	75	1146
	1148	283	289	189	195	141	147	111	117	93	99	79	85	69	75	1148
	1150	285	291	189	195	141	147	111	117	93	99	79	85	69	75	1150
	1152	285	291	189	195	141	147	113	119	93	99	79	85	69	75	1152
	1154	285	291	189	195	141	147	113	119	93	99	79	85	69	75	1154
	1156	285	291	189	195	141	147	113	119	93	99	79	85	69	75	1156
	1158	287	293	189	195	141	147	113	119	93	99	79	85	69	75	1158
	1160	287	293	191	197	141	147	113	119	93	99	79	85	69	75	1160
	1162	287	293	191	197	143	149	113	119	93	99	79	85	69	75	1162
	1164	287	293	191	197	143	149	113	119	93	99	81	87	69	75	1164
	1166	289	295	191	197	143	149	113	119	95	101	81	87	69	75	1166
	1168	289	295	191	197	143	149	113	119	95	101	81	87	69	75	1168
	1170	289	295	191	197	143	149	113	119	95	101	81	87	71	77	1170
	1172	289	295	193	199	143	149	115	121	95	101	81	87	71	77	1172
	1174	291	297	193	199	143	149	115	121	95	101	81	87	71	77	1174
	1176	291	297	193	199	143	149	115	121	95	101	81	87	71	77	1176
	1178	291	297	193	199	145	151	115	121	95	101	81	87	71	77	1178
	1180	291	297	193	199	145	151	115	121	95	101	81	87	71	77	1180
	1182	293	299	193	199	145	151	115	121	95	101	81	87	71	77	1182
	1184	293	299	195	201	145	151	115	121	95	101	81	87	71	77	1184
	1186	293	299	195	201	145	151	115	121	95	101	81	87	71	77	1186
	1188	293	299	195	201	145	151	115	121	95	101	81	87	71	77	1188
	1190	295	301	195	201	145	151	115	121	97	103	81	87	71	77	1190
	1192	295	301	195	201	145	151	117	123	97	103	83	89	71	77	1192
	1194	295	301	195	201	147	153	117	123	97	103	83	89	71	77	1194
	1196	295	301	197	203	147	153	117	123	97	103	83	89	71	77	1196
	1198	297	303	197	203	147	153	117	123	97	103	83	89	71	77	1198
	1200	297	303	197	203	147	153	117	123	97	103	83	89	71	77	1200

Above choice of Pitches will prove most satisfactory, although, as stated in text, the absolute
magnitude of average pitch may be varied within reasonable limits.

MULTIPLE-CIRCUIT, TRIPLE WINDINGS, FOR DRUM ARMATURES.

RE-ENTRANCY	No. OF CONDUCTORS	4 POLES F	4 POLES B	6 POLES F	6 POLES B	8 POLES F	8 POLES B	10 POLES F	10 POLES B	12 POLES F	12 POLES B	14 POLES F	14 POLES B	16 POLES F	16 POLES B	No. OF CONDUCTORS
②	1202	297	303	197	203	147	153	117	123	97	103	83	89	73	79	1202
②	1204	297	303	197	203	147	153	117	123	97	103	83	89	73	79	1204
ooo	1206	299	305	197	203	147	153	117	123	97	103	83	89	73	79	1206
②	1208	299	305	199	205	147	153	117	123	97	103	83	89	73	79	1208
②	1210	299	305	199	205	149	155	117	123	97	103	83	89	73	79	1210
ooo	1212	299	305	199	205	149	155	119	125	97	103	83	89	73	79	1212
②	1214	301	307	199	205	149	155	119	125	99	105	83	89	73	79	1214
②	1216	301	307	199	205	149	155	119	125	99	105	83	89	73	79	1216
ooo	1218	301	307	199	205	149	155	119	125	99	105	83	89	73	79	1218
②	1220	301	307	201	207	149	155	119	125	99	105	85	91	73	79	1220
②	1222	303	309	201	207	149	155	119	125	99	105	85	91	73	79	1222
ooo	1224	303	309	201	207	149	155	119	125	99	105	85	91	73	79	1224
②	1226	303	309	201	207	151	157	119	125	99	105	85	91	73	79	1226
②	1228	303	309	201	207	151	157	119	125	99	105	85	91	73	79	1228
ooo	1230	305	311	201	207	151	157	119	125	99	105	85	91	73	79	1230
②	1232	305	311	203	209	151	157	121	127	99	105	85	91	73	79	1232
②	1234	305	311	203	209	151	157	121	127	99	105	85	91	75	81	1234
ooo	1236	305	311	203	209	151	157	121	127	99	105	85	91	75	81	1236
②	1238	307	313	203	209	151	157	121	127	101	107	85	91	75	81	1238
②	1240	307	313	203	209	151	157	121	127	101	107	85	91	75	81	1240
ooo	1242	307	313	203	209	153	159	121	127	101	107	85	91	75	81	1242
②	1244	307	313	205	211	153	159	121	127	101	107	85	91	75	81	1244
②	1246	309	315	205	211	153	159	121	127	101	107	85	91	75	81	1246
ooo	1248	309	315	205	211	153	159	121	127	101	107	87	93	75	81	1248
②	1250	309	315	205	211	153	159	121	127	101	107	87	93	75	81	1250
②	1252	309	315	205	211	153	159	123	129	101	107	87	93	75	81	1252
ooo	1254	311	317	205	211	153	159	123	129	101	107	87	93	75	81	1254
②	1256	311	317	207	213	153	159	123	129	101	107	87	93	75	81	1256
②	1258	311	317	207	213	155	161	123	129	101	107	87	93	75	81	1258
ooo	1260	311	317	207	213	155	161	123	129	101	107	87	93	75	81	1260
②	1262	313	319	207	213	155	161	123	129	103	109	87	93	75	81	1262
②	1264	313	319	207	213	155	161	123	129	103	109	87	93	75	81	1264
ooo	1266	313	319	207	213	155	161	123	129	103	109	87	93	77	83	1266
②	1268	313	319	209	215	155	161	123	129	103	109	87	93	77	83	1268
②	1270	315	321	209	215	155	161	123	129	103	109	87	93	77	83	1270
ooo	1272	315	321	209	215	155	161	123	129	103	109	87	93	77	83	1272
②	1274	315	321	209	215	157	163	125	131	103	109	87	93	77	83	1274
②	1276	315	321	209	215	157	163	125	131	103	109	89	95	77	83	1276
ooo	1278	317	323	209	215	157	163	125	131	103	109	89	95	77	83	1278
②	1280	317	323	211	217	157	163	125	131	103	109	89	95	77	83	1280
②	1282	317	323	211	217	157	163	125	131	103	109	89	95	77	83	1282
ooo	1284	317	323	211	217	157	163	125	131	103	109	89	95	77	83	1284
②	1286	319	325	211	217	157	163	125	131	105	111	89	95	77	83	1286
②	1288	319	325	211	217	157	163	125	131	105	111	89	95	77	83	1288
ooo	1290	319	325	211	217	159	165	125	131	105	111	89	95	77	83	1290
②	1292	319	325	213	219	159	165	127	133	105	111	89	95	77	83	1292
②	1294	321	327	213	219	159	165	127	133	105	111	89	95	77	83	1294
ooo	1296	321	327	213	219	159	165	127	133	105	111	89	95	77	83	1296
②	1298	321	327	213	219	159	165	127	133	105	111	89	95	79	85	1298
②	1300	321	327	213	219	159	165	127	133	105	111	89	95	79	85	1300

Above choice of Pitches will prove most satisfactory, although, as stated in text, the absolute magnitude of average pitch may be varied within reasonable limits.

MULTIPLE-CIRCUIT, TRIPLE WINDINGS, FOR DRUM ARMATURES.

REENTRANCY	No. OF CONDUCTORS	4 POLES F	4 POLES B	6 POLES F	6 POLES B	8 POLES F	8 POLES B	10 POLES F	10 POLES B	12 POLES F	12 POLES B	14 POLES F	14 POLES B	16 POLES F	16 POLES B	No. OF CONDUCTORS
ooo	1302	323	329	213	219	159	165	127	133	105	111	89	95	79	85	1302
⊕⊕	1304	323	329	215	221	159	165	127	133	105	111	91	97	79	85	1304
⊕⊕	1306	323	329	215	221	161	167	127	133	105	111	91	97	79	85	1306
ooo	1308	323	329	215	221	161	167	127	133	105	111	91	97	79	85	1308
⊕⊕	1310	325	331	215	221	161	167	127	133	107	113	91	97	79	85	1310
⊕⊕	1312	325	331	215	221	161	167	129	135	107	113	91	97	79	85	1312
ooo	1314	325	331	215	221	161	167	129	135	107	113	91	97	79	85	1314
⊕⊕	1316	325	331	217	223	161	167	129	135	107	113	91	97	79	85	1316
⊕⊕	1318	327	333	217	223	161	167	129	135	107	113	91	97	79	85	1318
ooo	1320	327	333	217	223	161	167	129	135	107	113	91	97	79	85	1320
⊕⊕	1322	327	333	217	223	163	169	129	135	107	113	91	97	79	85	1322
⊕⊕	1324	327	333	217	223	163	169	129	135	107	113	91	97	79	85	1324
ooo	1326	329	335	217	223	163	169	129	135	107	113	91	97	79	85	1326
⊕⊕	1328	329	335	219	225	163	169	129	135	107	113	91	97	79	85	1328
⊕⊕	1330	329	335	219	225	163	169	129	135	107	113	91	97	81	87	1330
ooo	1332	329	335	219	225	163	169	131	137	107	113	93	99	81	87	1332
⊕⊕	1334	331	337	219	225	163	169	131	137	109	115	93	99	81	87	1334
⊕⊕	1336	331	337	219	225	163	169	131	137	109	115	93	99	81	87	1336
ooo	1338	331	337	219	225	165	171	131	137	109	115	93	99	81	87	1338
⊕⊕	1340	331	337	221	227	165	171	131	137	109	115	93	99	81	87	1340
⊕⊕	1342	333	339	221	227	165	171	131	137	109	115	93	99	81	87	1342
ooo	1344	333	339	221	227	165	171	131	137	109	115	93	99	81	87	1344
⊕⊕	1346	333	339	221	227	165	171	131	137	109	115	93	99	81	87	1346
⊕⊕	1348	333	339	221	227	165	171	131	137	109	115	93	99	81	87	1348
ooo	1350	335	341	221	227	165	171	131	137	109	115	93	99	81	87	1350
⊕⊕	1352	335	341	223	229	165	171	133	139	109	115	93	99	81	87	1352
⊕⊕	1354	335	341	223	229	167	173	133	139	109	115	93	99	81	87	1354
ooo	1356	335	341	223	229	167	173	133	139	109	115	93	99	81	87	1356
⊕⊕	1358	337	343	223	229	167	173	133	139	111	117	93	99	81	87	1358
⊕⊕	1360	337	343	223	229	167	173	133	139	111	117	95	101	81	87	1360
ooo	1362	337	343	223	229	167	173	133	139	111	117	95	101	83	89	1362
⊕⊕	1364	337	343	223	231	167	173	133	139	111	117	95	101	83	89	1364
⊕⊕	1366	339	345	225	231	167	173	135	139	111	117	95	101	83	89	1366
ooo	1368	339	345	225	231	167	173	133	139	111	117	95	101	83	89	1368
⊕⊕	1370	339	345	225	231	169	175	133	139	111	117	95	101	83	89	1370
⊕⊕	1372	339	345	225	231	169	175	135	141	111	117	95	101	83	89	1372
ooo	1374	341	347	225	231	169	175	135	141	111	117	95	101	83	89	1374
⊕⊕	1376	341	347	227	233	169	175	135	141	111	117	95	101	83	89	1376
⊕⊕	1378	341	347	227	233	169	175	135	141	111	117	95	101	83	89	1378
ooo	1380	341	347	227	233	169	175	135	141	111	117	95	101	83	89	1380
⊕⊕	1382	343	349	227	233	169	175	135	141	113	119	95	101	83	89	1382
⊕⊕	1384	343	349	227	233	169	175	135	141	113	119	95	101	83	89	1384
ooo	1386	343	349	227	233	171	177	135	141	113	119	95	101	83	89	1386
⊕⊕	1388	343	349	229	235	171	177	135	141	113	119	97	103	83	89	1388
⊕⊕	1390	345	351	229	235	171	177	135	141	113	119	97	103	83	89	1390
ooo	1392	345	351	229	235	171	177	137	143	113	119	97	103	83	89	1392
⊕⊕	1394	345	351	229	235	171	177	137	143	113	119	97	103	85	91	1394
⊕⊕	1396	345	351	229	235	171	177	137	143	113	119	97	103	85	91	1396
ooo	1398	347	353	229	235	171	177	137	143	113	119	97	103	85	91	1398
⊕⊕	1400	347	353	231	237	171	177	137	143	113	119	97	103	85	91	1400

Above choice of Pitches will prove most satisfactory, although, as stated in text, the absolute magnitude of average pitch may be varied within reasonable limits.

MULTIPLE-CIRCUIT, TRIPLE WINDINGS, FOR DRUM ARMATURES.

RE-ENTRANCY	No. OF CONDUCTORS	4 POLES F	4 POLES B	6 POLES F	6 POLES B	8 POLES F	8 POLES B	10 POLES F	10 POLES B	12 POLES F	12 POLES B	14 POLES F	14 POLES B	16 POLES F	16 POLES B	No. OF CONDUCTORS
	1402	347	353	231	237	173	179	137	143	113	119	97	103	85	91	1402
	1404	347	353	231	237	173	179	137	143	113	119	97	103	85	91	1404
	1406	349	355	231	237	173	179	137	143	115	121	97	103	85	91	1406
	1408	349	355	231	237	173	179	137	143	115	121	97	103	85	91	1408
	1410	349	355	231	237	173	179	137	143	115	121	97	103	85	91	1410
	1412	349	355	233	239	173	179	139	145	115	121	97	103	85	91	1412
	1414	351	357	233	239	173	179	139	145	115	121	97	103	85	91	1414
	1416	351	357	233	239	173	179	139	145	115	121	99	105	85	91	1416
	1418	351	357	233	239	175	181	139	145	115	121	99	105	85	91	1418
	1420	351	357	233	239	175	181	139	145	115	121	99	105	85	91	1420
	1422	353	359	233	239	175	181	139	145	115	121	99	105	85	91	1422
	1424	353	359	235	241	175	181	139	145	115	121	99	105	85	91	1424
	1426	353	359	235	241	175	181	139	145	115	121	99	105	87	93	1426
	1428	353	359	235	241	175	181	139	145	115	121	99	105	87	93	1428
	1430	355	361	235	241	175	181	139	145	117	123	99	105	87	93	1430
	1432	355	361	235	241	175	181	141	147	117	123	99	105	87	93	1432
	1434	355	361	235	241	177	183	141	147	117	123	99	105	87	93	1434
	1436	355	361	237	243	177	183	141	147	117	123	99	105	87	93	1436
	1438	357	363	237	243	177	183	141	147	117	123	99	105	87	93	1438
	1440	357	363	237	243	177	183	141	147	117	123	99	105	87	93	1440
	1442	357	363	237	243	177	183	141	147	117	123	99	105	87	93	1442
	1444	357	363	237	243	177	183	141	147	117	123	99	105	87	93	1444
	1446	359	365	237	243	177	183	141	147	117	123	101	107	87	93	1446
	1448	359	365	239	245	177	183	141	147	117	123	101	107	87	93	1448
	1450	359	365	239	245	179	185	141	147	117	123	101	107	87	93	1450
	1452	359	365	239	245	179	185	143	149	117	123	101	107	87	93	1452
	1454	361	367	239	245	179	185	143	149	119	125	101	107	87	93	1454
	1456	361	367	239	245	179	185	143	149	119	125	101	107	87	93	1456
	1458	361	367	239	245	179	185	143	149	119	125	101	107	89	95	1458
	1460	361	367	241	247	179	185	143	149	119	125	101	107	89	95	1460
	1462	363	369	241	247	179	185	143	149	119	125	101	107	89	95	1462
	1464	363	369	241	247	179	185	143	149	119	125	101	107	89	95	1464
	1466	363	369	241	247	181	187	143	149	119	125	101	107	89	95	1466
	1468	363	369	241	247	181	187	143	149	119	125	101	107	89	95	1468
	1470	365	371	241	247	181	187	143	149	119	125	101	107	89	95	1470
	1472	365	371	243	249	181	187	145	151	119	125	103	109	89	95	1472
	1474	365	371	243	249	181	187	145	151	119	125	103	109	89	95	1474
	1476	365	371	243	249	181	187	145	151	119	125	103	109	89	95	1476
	1478	367	373	243	249	181	187	145	151	121	127	103	109	89	95	1478
	1480	367	373	243	249	181	187	145	151	121	127	103	109	89	95	1480
	1482	367	373	243	249	183	189	145	151	121	127	103	109	89	95	1482
	1484	367	373	245	251	183	189	145	151	121	127	103	109	89	95	1484
	1486	369	375	245	251	183	189	145	151	121	127	103	109	89	95	1486
	1488	369	375	245	251	183	189	145	151	121	127	103	109	89	95	1488
	1490	369	375	245	251	183	189	145	151	121	127	103	109	91	97	1490
	1492	369	375	245	251	183	189	147	153	121	127	103	109	91	97	1492
	1494	371	377	245	251	183	189	147	153	121	127	103	109	91	97	1494
	1496	371	377	247	253	183	189	147	153	121	127	103	109	91	97	1496
	1498	371	377	247	253	185	191	147	153	121	127	103	109	91	97	1498
	1500	371	377	247	253	185	191	147	153	121	127	105	111	91	97	1500

Above choice of Pitches will prove most satisfactory, although, as stated in text, the absolute magnitude of average pitch may be varied within reasonable limits.

MULTIPLE-CIRCUIT, TRIPLE WINDINGS, FOR DRUM ARMATURES.

RE-ENTRANCY	No. OF CONDUCTORS	FRONT AND BACK PITCHES													No. OF CONDUCTORS
		4 POLES		6 POLES		8 POLES		10 POLES		12 POLES		14 POLES		16 POLES		
		F	B	F	B	F	B	F	B	F	B	F	B	F	B	
③③	1502	373	370	247	253	185	191	147	153	123	129	105	111	91	97	1502
③③	1504	373	370	247	253	185	191	147	153	123	129	105	111	91	97	1504
ooo	1506	373	379	247	253	185	191	147	153	123	129	105	111	91	97	1506
③③	1508	373	379	249	255	185	191	147	153	123	129	105	111	91	97	1508
③③	1510	375	381	249	255	185	191	147	153	123	129	105	111	91	97	1510
ooo	1512	375	381	249	255	185	191	149	155	123	129	105	111	91	97	1512
③③	1514	375	381	249	255	187	193	149	155	123	129	105	111	91	97	1514
③③	1516	375	381	249	255	187	193	149	155	123	129	105	111	91	97	1516
ooo	1518	377	383	249	255	187	193	149	155	123	129	105	111	91	97	1518
③③	1520	377	383	251	257	187	193	149	155	123	129	105	111	91	97	1520
③③	1522	377	383	251	257	187	193	149	155	123	129	105	111	93	99	1522
ooo	1524	377	383	251	257	187	193	149	155	123	129	105	111	93	99	1524
③③	1526	379	385	251	257	187	193	149	155	125	131	105	111	93	99	1526
③③	1528	379	385	251	257	187	193	149	155	125	131	107	113	93	99	1528
ooo	1530	379	385	251	257	189	195	149	155	125	131	107	113	93	99	1530
③③	1532	379	385	253	259	189	195	151	157	125	131	107	113	93	99	1532
③③	1534	381	387	253	259	189	195	151	157	125	131	107	113	93	99	1534
ooo	1536	381	387	253	259	189	195	151	157	125	131	107	113	93	99	1536
③③	1538	381	387	253	259	189	195	151	157	125	131	107	113	93	99	1538
③③	1540	381	387	253	259	189	195	151	157	125	131	107	113	93	99	1540
ooo	1542	383	389	253	259	189	195	151	157	125	131	107	113	93	99	1542
③③	1544	383	389	255	261	189	195	151	157	125	131	107	113	93	99	1544
③③	1546	383	389	255	261	191	197	151	157	125	131	107	113	93	99	1546
ooo	1548	383	389	255	261	191	197	151	157	125	131	107	113	93	99	1548
③③	1550	385	391	255	261	191	197	151	157	127	133	107	113	93	99	1550
③③	1552	385	391	255	261	191	197	153	159	127	133	107	113	93	99	1552
ooo	1554	385	391	255	261	191	197	153	159	127	133	107	113	95	101	1554
③③	1556	385	391	257	263	191	197	153	159	127	133	109	115	95	101	1556
③③	1558	387	393	257	263	191	197	153	159	127	133	109	115	95	101	1558
ooo	1560	387	393	257	263	191	197	153	159	127	133	109	115	95	101	1560
③③	1562	387	393	257	263	193	199	153	159	127	133	109	115	95	101	1562
③③	1564	387	393	257	263	193	199	153	159	127	133	109	115	95	101	1564
ooo	1566	389	395	257	263	193	199	153	159	127	133	109	115	95	101	1566
③③	1568	389	395	259	265	193	199	153	159	127	133	109	115	95	101	1568
③③	1570	389	395	259	265	193	199	153	159	127	133	109	115	95	101	1570
ooo	1572	389	395	259	265	193	199	155	161	127	133	109	115	95	101	1572
③③	1574	391	397	259	265	193	199	155	161	129	135	109	115	95	101	1574
③③	1576	391	397	259	265	193	199	155	161	129	135	109	115	95	101	1576
ooo	1578	391	397	259	265	195	201	155	161	129	135	109	115	95	101	1578
③③	1580	391	397	261	267	195	201	155	161	129	135	109	115	95	101	1580
③③	1582	393	399	261	267	195	201	155	161	129	135	109	115	95	101	1582
ooo	1584	393	399	261	267	195	201	155	161	129	135	111	117	95	101	1584
③③	1586	393	399	261	267	195	201	155	161	129	135	111	117	97	103	1586
③③	1588	393	399	261	267	195	201	155	161	129	135	111	117	97	103	1588
ooo	1590	395	401	261	267	195	201	155	161	129	135	111	117	97	103	1590
③③	1592	395	401	263	269	195	201	157	163	129	135	111	117	97	103	1592
③③	1594	395	401	263	269	197	203	157	163	129	135	111	117	97	103	1594
ooo	1596	395	401	263	269	197	203	157	163	129	135	111	117	97	103	1596
③③	1598	397	403	263	269	197	203	157	163	131	137	111	117	97	103	1598
③③	1600	397	403	263	269	197	203	157	163	131	137	111	117	97	103	1600

Above choice of Pitches will prove most satisfactory, although, as stated in text, the absolute magnitude of average pitch may be varied within reasonable limits.

LIST OF WORKS

ON

ELECTRICAL SCIENCE.

PUBLISHED AND FOR SALE BY

D. VAN NOSTRAND COMPANY,

23 Murray and 27 Warren Streets, New York.

ABBOTT, A. V. The Electrical Transmission of Energy. A Manual for the Design of Electrical Circuits. 8vo, cloth. (*In press.*)

ARNOLD, E. Armature Windings of Direct Current Dynamos. Extension and application of a general winding rule. Translated from the original German by Francis B. DeGress, M. E. (*In press*)

ATKINSON, PHILIP. Elements of Static Electricity, with full description of the Holtz and Topler Machines, and their mode of operating. Illustrated. 12mo, cloth. $1.50.

> **The Elements of Dynamic Electricity and Magnetism.** Second Edition. Illustrated. 12mo, cloth. $2.00.

> **Elements of Electric Lighting,** including Electric Generation, Measurement, Storage, and Distribution. Seventh Edition. Fully revised and new matter added. Illustrated. 8vo, cloth. $1.50.

> **The Electric Transformation of Power and Its Application by the Electric Motor,** including Electric Railway Construction. Illustrated. 12mo, cloth. $2.00.

BADT, F. B. Dynamo Tender's Handbook. 70 Illustrations. 16mo, cloth. $1.00.

> **Electric Transmission Handbook.** Illustrations and Tables. 16mo, cloth. $1.00.

> **Incandescent Wiring Handbook.** Fourth Edition. Illustrations and Tables. 12mo, cloth. $1.00.

> **Bell Hanger's Handbook.** Illustrated. 12mo, cloth. $1.00.

BIGGS, C. H. W. First Principles of Electrical Engineering. Being an attempt to provide an Elementary Book for those who are intending to enter the profession of Electrical Engineering. Second Edition. Illustrated. 12mo, cloth. $1.00.

BLAKESLEY, T. H. Papers on Alternating Currents of Electricity. For the use of Students and Engineers. Third Edition, enlarged. 12mo, cloth. $1.50.

BOTTONE, S. R. Electrical Instrument-Making for Amateurs. A Practical Handbook. Fourth Edition. Enlarged by a chapter on "The Telephone." With 48 Illustrations. 12mo, cloth. 50 cents.

> **Electric Bells, and All about Them.** A Practical Book for Practical Men. With over 100 Illustrations. 12mo, cloth. 50 cents.

> **The Dynamo: How Made and How Used.** A Book for Amateurs. Sixth Edition. 100 Illustrations. 12mo, cloth. $1.00.

> **Electro-Motors: How Made and How Used.** A Handbook for Amateurs and Practical Men. Illustrated. 12mo, cloth. 50 cents.

(1)

CLARK, D. K. Tramways: Their Construction and Working. Embracing a Comprehensive History of the System, with Accounts of the Various Modes of Traction, a Description of the Varieties of Rolling Stock, and Ample Details of Cost and Working Expenses; with Special Reference to the Tramways of the United Kingdom. Second Edition. Revised and rewritten. With over 400 Illustrations. Thick 8vo, cloth. $9.00.

CROCKER, F. B., and WHEELER, S. S. The Practical Management of Dynamos and Motors. Third Edition. Illustrated. 12mo, cloth. $1.00.

CROCKER, F. B. Electric Lighting. (In press.)

CUMMING, LINNÆUS, M.A. Electricity Treated Experimentally. For the Use of Schools and Students. Third Edition. 12mo, cloth. $1.50.

DESMOND, CHAS. Electricity for Engineers. Part I.: Constant Current. Part II.: Alternate Current. Revised Edition. Illustrated. 12mo, cloth. $2.50.

DU MONCEL, Count TH. Electro-Magnets: The Determination of the Elements of their Construction. 16mo, cloth. (No. 64 Van Nostrand's Science Series.) 50 cents.

DYNAMIC ELECTRICITY. Its Modern Use and Measurement, chiefly in its application to Electric Lighting and Telegraphy, including: 1. Some Points in Electric Lighting, by Dr. John Hopkinson. 2. On the Treatment of Electricity for Commercial Purposes, by J. N. Schoolbred. 3. Electric-Light Arithmetic, by R. E. Day, M.E. 18mo, boards. (No. 71 Van Nostrand's Science Series.) 50 cents.

EMMETT, WM. L. Alternating Current Wiring and Distribution. 16mo, cloth. Illustrated. $1.00.

EWING, J. A. Magnetic Induction in Iron and Other Metals. Second issue. Illustrated. 8vo, cloth. $4.00.

FISKE, Lieut. BRADLEY A., U.S.N. Electricity in Theory and Practice: or, The Elements of Electrical Engineering. Eighth Edition. 8vo, cloth. $2.50.

FLEMING, Prof. J. A. The Alternate-Current Transformer in Theory and Practice. Vol. I.: The Induction of Electric Currents. 500 pp. Second Edition. Illustrated. 8vo, cloth. $3.00. Vol. II.: The Utilization of Induced Currents. 594 pp. Illustrated. 8vo, cloth. $5.00.

Electric Lamps and Electric Lighting. 8vo, cloth. $3.00.

GORDON, J. E. H. School Electricity. 12mo, cloth. $2.00.

GORE, Dr. GEORGE. The Art of Electrolytic Separation of Metals (Theoretical and Practical). Illustrated. 8vo, cloth. $3.50.

GUILLEMIN, AMÉDÉE. Electricity and Magnetism. Translated, revised, and edited by Prof. Silvanus P. Thompson, 600 Illustrations and several Plates. Large 8vo, cloth. $8.00.

GUY, ARTHUR F. Electric Light and Power, giving the result of practical experience in Central-Station Work. 8vo, cloth. Illustrated. $2.50.

HASKINS, C. H. The Galvanometer and Its Uses. A Manual for Electricians and Students. Fourth Edition, revised. 12mo, morocco. $1.50.

Transformers: Their Theory, Construction, and Application Simplified. Illustrated. 12mo, cloth. $1.25.

HAWKINS, C. C., and WALLIS, F. The Dynamo: Its Theory, Design, and Manufacture. 190 Illustrations. 8vo, cloth. $3.00.

HOBBES, W. R. P. The Arithmetic of Electrical Measurements. With numerous examples, fully worked. New Edition. 12mo, cloth. 50 cents.

HOSPITALIER, E. Polyphased Alternating Currents. Illustrated. 8vo, cloth. $1.40.

HOUSTON, Prof. E. J. A Dictionary of Electrical Words, Terms, and Phrases. Third Edition. Rewritten and greatly enlarged. Large 8vo, 570 illustrations, cloth. $5.00.

INCANDESCENT ELECTRIC LIGHTING. A Practical Description of the Edison System, by H. Latimer. To which is added, The Design and Operation of Incandescent Stations, by C. J. Field; A Description of the Edison Electrolyte Meter, by A. E. Kennelly; and a Paper on the Maximum Efficiency of Incandescent Lamps, by T. W. Howells. Illustrated. 16mo, cloth. (No. 57 Van Nostrand's Science Series.) 50 cents.

INDUCTION COILS: How Made and How Used. Fifth Edition. 16mo, cloth. (No. 53 Van Nostrand's Science Series.) 50 cents.

KAPP, GISBERT, C.E. Electric Transmission of Energy and Its Transformation, Subdivision, and Distribution. A Practical Handbook. Fourth Edition, thoroughly revised. 12mo, cloth. $3.50.

Alternate-Current Machinery. 190 pp. Illustrated. (No. 96 Van Nostrand's Science Series.) 50 cents.

Dynamos, Alternators, and Transformers. Illustrated. 8vo, cloth. $4.00.

KEMPE, H. R. The Electrical Engineer's Pocket-Book: Modern Rules, Formulæ, Tables, and Data. 32mo, leather. $1.75.

A Handbook of Electrical Testing. Fifth edition. 200 illustrations. 8vo, cloth. $7.25.

KENNELLY, A. E. Theoretical Elements of Electro-Dynamic Machinery. Vol. I. Illustrated. 8vo, cloth. $1.50.

KILGOUR, M. H., and SWAN, H., and BIGGS, C. H. W. Electrical Distribution: Its Theory and Practice. Illustrated. 8vo, cloth. $4.00.

LOCKWOOD, T. D. Electricity, Magnetism, and Electro-Telegraphy. A Practical Guide and Handbook of General Information for Electrical Students, Operators, and Inspectors. Fourth Edition. Illustrated. 8vo, cloth. $2.50.

LORING, A. E. A Handbook of the Electro-Magnetic Telegraph. 16mo, cloth. (No. 39 Van Nostrand's Science Series.) 50 cents.

MARTIN, T. C., and WETZLER, J. The Electro-Motor and Its Applications. Third Edition. With an Appendix on the Development of the Electric Motor since 1888, by Dr. L. Bell. 300 Illustrations. 4to, cloth. $3.00.

MORROW, J. T., and REID, T. Arithmetic of Magnetism and Electricity. 12mo, cloth. $1.00.

MUNRO, JOHN, C.E., and JAMIESON, ANDREW, C.E. A Pocket-Book of Electrical Rules and Tables. For the use of Electricians and Engineers. Tenth Edition. Revised and enlarged. With numerous diagrams. Pocket size, leather. $2.50.

NIPHER, FRANCIS E., A.M. Theory of Magnetic Measurements. With an Appendix on the Method of Least Squares. 12mo, cloth. $1.00.

(3)

NOAD, H. M. The Student's Text-Book of Electricity. A New Edition. Carefully revised by W. H. Preece. 12mo, cloth. Illustrated. $4.00.

OHM, Dr. G. S. The Galvanic Circuit Investigated Mathematically. Berlin, 1827. Translated by William Francis. With Preface and Notes by the Editor, Thos. D. Lockwood. 12mo, cloth. (No. 102 Van Nostrand's Science Series.) 50 cents.

PALAZ, A. Treatise on Industrial Photometry. Specially applied to Electric Lighting. Translated from the French by G. W. Patterson, Jr., Assistant Professor of Physics in the University of Michigan, and M. R. Patterson, B.A. Fully Illustrated. 8vo, cloth. $4.00.

PERRY, NELSON W. Electric Railway Motors. Their Construction, Operation, and Maintenance. An Elementary Practical Handbook for those engaged in the management and operation of Electric Railway Apparatus, with Rules and Instructions for Motormen. 12mo, cloth. $1.00.

PLANTÉ, GASTON. The Storage of Electrical Energy, and Researches in the Effects created by Currents combining Quantity with High Tension. Translated from the French by Paul B. Elwell. 89 Illustrations. 8vo. $4.00.

POOLE, J. The Practical Telephone Handbook. Illustrated. 8vo, cloth. $1.00.

POPE, F. L. Modern Practice of the Electric Telegraph. A Handbook for Electricians and Operators. An entirely new work, revised and enlarged, and brought up to date throughout. Illustrations. 8vo, cloth. $1.50.

PREECE, W. H., and STUBBS, A. J. Manual of Telephony. Illustrated. 12mo, cloth. $4.50.

RECKENZAUN, A. Electric Traction. Illustrated. 8vo, cloth. $4.00.

RUSSELL, STUART A. Electric-Light Cables and the Distribution of Electricity. 107 Illustrations. 8vo, cloth. $2.25.

SALOMONS, Sir DAVID, M.A. Electric-Light Installations. A Practical Handbook. Seventh Edition, revised and enlarged. Vol. I.: Management of Accumulators. Illustrated. 12mo, cloth. $1.50. Vol. II.: Apparatus. Illustrated. 12mo, cloth. $2.25. Vol. III.: Application. Illustrated. 12mo, cloth. $1.50.

SCHELLEN, Dr. H. Magneto-Electric and Dynamo-Electric Machines. Their Construction and Practical Application to Electric Lighting and the Transmission of Power. Translated from the third German edition by N. S. Keith and Percy Neymann, Ph.D. With very large Additions and Notes relating to American Machines, by N. S. Keith. Vol. I. with 353 Illustrations. Second Edition. $5.00.

SLOANE, Prof. T. O'CONOR. Standard Electrical Dictionary. 300 Illustrations. 8vo, cloth. $3.00.

SNELL, ALBION T. Electric Motive Power. The Transmission and Distribution of Electric Power by Continuous and Alternate Currents. With a Section on the Applications of Electricity to Mining Work. Illustrated. 8vo, cloth. $4.00.

SWINBURNE, JAS., and WORDINGHAM, C. H. The Measurement of Electric Currents. Electrical Measuring Instruments. Meters for Electrical Energy. Edited, with Preface, by T. Commerford Martin. Folding Plate and numerous Illustrations. 16mo, cloth. 50 cents.

THOM, C., and JONES, W. H. **Telegraphic Connections,** embracing recent methods in Quadruplex Telegraphy. Twenty colored Plates. 8vo, cloth. $1.50.

THOMPSON, EDWARD P. **How to Make Inventions; or, Inventing as a Science and an Art.** An Inventor's Guide. Second Edition. Revised and Enlarged. Illustrated. 8vo, paper. $1.00.

THOMPSON, Prof. S. P. **Dynamo-Electric Machinery.** With an Introduction and Notes by Frank L. Pope and H. R. Butler. Fully Illustrated. (No. 66 Van Nostrand's Science Series.) 50 cents.

Recent Progress in Dynamo-Electric Machines. Being a Supplement to "Dynamo-Electric Machinery." Illustrated. 12mo, cloth. (No. 75 Van Nostrand's Science Series.) 50 cents.

The Electro-Magnet and Electro-Magnetic Mechanism. Second Edition, revised. 213 Illustrations. 8vo, cloth. $6.00.

TREVERT, E. **Practical Directions for Armature and Field-Magnet Winding.** Illustrated. 12mo, cloth. $1.50.

How to Build Dynamo-Electric Machinery. Embracing Theory Designing and the Construction of Dynamos and Motors. With Appendices on Field-Magnet and Armature Winding. Management of Dynamos and Motors, and useful Tables of Wire Gauges. Illustrated. 8vo, cloth. $2.50.

TUMLIRZ, Dr. **Potential, and its Application to the Explanation of Electrical Phenomena.** Translated by D. Robertson, M.D. 12mo, cloth. $1.25.

TUNZELMANN, G. W. de. **Electricity in Modern Life.** Illustrated. 12mo, cloth. $1.25.

URQUHART, J. W. **Dynamo Construction.** A Practical Handbook for the Use of Engineer Constructors and Electricians in Charge. Illustrated. 12mo, cloth. $3.00.

WALKER, FREDERICK. **Practical Dynamo-Building for Amateurs.** How to Wind for any Output. Illustrated. 16mo, cloth. (No. 98 Van Nostrand's Science Series.) 50 cents.

WALMSLEY, R. M. **The Electric Current.** How Produced and How Used. With 379 Illustrations. 12mo, cloth. $3.00.

WEBB, H. L. **A Practical Guide to the Testing of Insulated Wires and Cables.** Illustrated. 12mo, cloth. $1.00.

WORMELL, R. **Electricity in the Service of Man.** A Popular and Practical Treatise on the Application of Electricity in Modern Life. From the German, and edited, with copious additions, by R. Wormell, and an Introduction by Prof. J. Perry. With nearly 850 Illustrations. Royal 8vo, cloth. $6.00.

WEYMOUTH, F. MARTEN. **Drum Armatures and Commutators.** (Theory and Practice.) A complete treatise on the theory and construction of drum-winding, and of commutators for closed-coil armatures, together with a full résumé of some of the principal points involved in their design; and an exposition of armature reactions and sparking. Illustrated. 8vo, cloth. $3.00.